The Protozoan Phylum Apicomplexa

Volume I

Author

Norman D. Levine

Department of Veterinary Pathobiology
College of Veterinary Medicine
University of Illinois
Urbana, Illinois

CRC Press
Taylor & Francis Group
Boca Raton London New York

CRC Press is an imprint of the
Taylor & Francis Group, an **informa** business

Library of Congress Cataloging-in-Publication Data

Levine, Norman D.
    The protozoan phylum Apicomplexa.

    Bibliography: p.
    Includes index.
    1. Apicomplexa. 2. Apicomplexa—Classification.
3. Veterinary protozoology.   I. Title. [DNLM:
1. Protozoan Infections-veterinary.   SF 780.6 L665p]
SF780.6.L485   1988   593.1'9   87-8005
ISBN 0-8493-4652-5 (set)
ISBN 0-8493-4653-3 (v. 1)
ISBN 0-8493-4654-1 (v. 2)

A Library of Congress record exists under LC control number: 87008005

ISBN 13: 978-1-315-89700-4 (hbk)
ISBN 13: 978-1-351-07610-4 (ebk)

Visit the Taylor & Francis Web site at http://www.taylorandfrancis.com and the
CRC Press Web site at http://www.crcpress.com

THE AUTHOR

Norman Dion Levine, Ph.D. is a Professor Emeritus in the Department of Veterinary Pathobiology, in the College of Veterinary Medicine and the Illinois Agricultural Experiment Station, University of Illinois, Urbana. He is also a former director of the Center for Human Ecology (University of Illinois), a former member of the Department of Zoology (University of Illinois), and an affiliate of the Illinois Natural History Survey.

Dr. Levine received his B.S. degree in Zoology and Entomology from Iowa State University and his Ph.D. degree in Zoology (Physico-Chemical Biology) from the University of California (Berkeley).

He joined the staff of the University of Illinois and Illinois State Department of Agriculture in 1937, and moved up through the ranks to his present position. He has been President of the American Microscopical Society, Society of Protozoologists, American Society of Parasitologists, Illinois State Academy of Science, and American Society of Professional Biologists, and is an honorary member of the American Microscopical Society, Illinois State Academy of Science, Society of Protozoologists, Phi Sigma (honorary biology fraternity), and World Association for the Advancement of Veterinary Parasitology. He has received the Distinguished Achievement Citation of Iowa State University, and has been elected a Distinguished Veterinary Parasitologist by the American Association of Veterinary Parasitologists. He was Editor of the *Journal of Protozoology* from 1965 to 1971. He has been a member of the editorial boards of the *American Journal of Veterinary Research, Transactions of the American Microscopical Society, American Midland Naturalist,* and *Laboratory Animal Science.* He has been chairman of the National Institutes of Health Tropical Medicine and Parasitology Study Section, a member of the Board of Governors of the American Academy of Microbiology, a Committeeman-at-Large for Section F (Zoological Science) and Council Member of the American Association for the Advancement of Science, a member of the Governing Board of the American Board of Microbiology, a member of the National Academy of Science-National Research Council, and is a patron for the International Commission on Zoological Nomenclature of the International Trust for Zoological Nomenclature. He has served on the organizing committees of many International Congresses of Protozoology and of Parasitology, and was one of the organizers of the World Federation of Parasitologists.

Dr. Levine is the author or editor of 15 books and about 550 publications in scientific journals. His special fields of research are protozoology, parasitology, and human ecology.

TABLE OF CONTENTS

Volume I

Chapter 1
Introduction 1

Chapter 2
Class Perkinsasida 9
   Family Perkinsidae 9
      Genus *Perkinsus* 9

Chapter 3
The Gregarines: Archigregarines 11
Class Conoidasida 11
Subclass Gregarinasina 11
   Order Archigregarinorida 11
      Family Selenidioididae 11
         Genus *Selenidioides* 11
         Genus *Meroselenidium* 12
         Genus *Merogregarina* 12
      Family Exoschizonidae 12
         Genus *Exoschizon* 12

Chapter 4
The Gregarines: Blastogregarines 13
   Order Eugregarinorida 13
   Suborder Blastogregarinorina 13
      Family Siedleckiidae 13
         Genus *Siedleckia* 13

Chapter 5
The Gregarines: Aseptate Eugregarines 15
   Suborder Aseptatorina 15
      Family Selenidiidae 15
         Genus *Selenidium* 15
         Genus *Selenocystis* 17
         Genus *Ditrypanocystis* 17
         Genus *Heterospora* 17
      Family Lecudinidae 17
         Genus *Lecudina* 17
         Genus *Polyrhabdina* 19
         Genus *Ulivina* 19
         Genus *Sycia* 20
         Genus *Pontesia* 20
         Genus *Bhatiella* 20
         Genus *Viviera* 20
         Genus *Cochleomeritus* 20
         Genus *Ancora* 20
         Genus *Hentschelia* 21

Genus *Lecythion* 21
Genus *Hyperidion* 21
Genus *Zygosoma* 21
Genus *Filipodium* 21
Genus *Diplauxis* 22
Genus *Chlamydocystis* 22
Genus *Contortiocorpa* 22
Genus *Lankesteria* 22
Genus *Monocystella* 23
Genus *Ascogregarina* 24
Genus *Kofoidina* 25
Genus *Ophioidina* 25
Genus *Paraophioidina* 25
Genus *Lateroprotomeritus* 26
Genus *Extremocystis* 26
Family Urosporidae 26
Genus *Urospora* 26
Genus *Gonospora* 27
Genus *Lithocystis* 28
Genus *Ceratospora* 29
Genus *Pterospora* 29
Genus *Paragonospora* 29
Family Aikinetocystidae 29
Genus *Aikinetocystis* 29
Genus *Nellocystis* 30
Family Monocystidae 30
Subfamily Monocystinae 30
Genus *Monocystis* 30
Genus *Nematocystis* 33
Genus *Rhabdocystis* 34
Genus *Apolocystis* 34
Genus *Cephalocystis* 35
Genus *Trigonepimerus* 35
Genus *Mastocystis* 35
Subfamily Zygocystinae 36
Genus *Zygocystis* 36
Genus *Adelphocystis* 36
Genus *Pleurocystis* 36
Subfamily Rhynchocystinae 37
Genus *Rhynchocystis* 37
Genus *Dirhynchocystis* 37
Genus *Grayallia* 38
Subfamily Stomatophorinae 38
Genus *Stomatophora* 38
Genus *Craterocystis* 38
Genus *Astrocystella* 38
Genus *Albertisella* 38
Genus *Beccaricystis* 39
Genus *Parachoanocystoides* 39
Genus *Choanocystoides* 39
Genus *Zeylanocystis* 39

    Genus *Arborocystis* 39
    Genus *Chakravartiella* 39
  Subfamily Oligochaetocystinae 40
    Genus *Oligochaetocystis* 40
    Genus *Echiurocystis* 40
    Genus *Neomonocystis* 40
    Genus *Acarogregarina* 40
  Family Diplocystidae 40
    Genus *Diplocystis* 40
  Family Allantocystidae 41
    Genus *Allantocystis* 41
  Family Schaudinnellidae 41
    Genus *Schaudinnella* 41
  Family Enterocystidae 41
    Genus *Enterocystis* 41
  Family Ganymedidae 42
    Genus *Ganymedes* 42

Chapter 6
The Gregarines: Septate Eugregarines

The Gregarines: Septate Eugregarines 43
  Suborder Septatorina 43
  Superfamily Porosporicae 43
  Family Porosporidae 43
    Genus *Porospora* 43
    Genus *Nematopsis* 43
    Genus *Pachyporospora* 45
  Superfamily Gregarinicae 45
  Family Cephaloidophoridae 45
    Genus *Cephaloidophora* 45
    Genus *Caridohabitans* 47
    Genus *Rotundula* 48
  Family Cephalolobidae 48
    Genus *Cephalolobus* 48
    Genus *Callynthrochlamys* 48
  Family Uradiophoridae 49
    Genus *Uradiophora* 49
    Genus *Heliospora* 49
    Genus *Pyxinioides* 49
    Genus *Nematoides* 50
    Genus *Bifilida* 50
  Family Gregarinidae 50
    Genus *Gregarina* 50
    Genus *Erhardovina* 62
    Genus *Gymnospora* 62
    Genus *Triseptata* 63
    Genus *Gamocystis* 63
    Genus *Anisolobus* 63
    Genus *Garnhamia* 63
    Genus *Torogregarina* 64
    Genus *Faucispora* 64
    Genus *Spinispora* 64

Genus *Bolivia*... 64
Genus *Degiustia* 64
Genus *Cirrigregarina* 64
Genus *Molluskocystis* 64
Family Metameridae.. 65
Genus *Metamera*... 65
Genus *Gopaliella* 65
Genus *Deuteromera*... 65
Genus *Cognettiella*.. 65
Family Didymophyidae.. 65
Genus *Didymophyes* 65
Genus *Liposcelisus*.. 67
Genus *Quadruhyalodiscus* 67
Family Hirmocystidae 67
Genus *Hirmocystis* 67
Genus *Hyalospora* 68
Genus *Tettigonospora*.. 69
Genus *Euspora* 69
Genus *Tintinospora* 69
Genus *Arachnocystis*.. 69
Genus *Protomagalhaensia* 70
Genus *Pintospora* 70
Genus *Endomycola* 70
Genus *Retractocephalus* 70
Superfamily Stenophoricae 70
Family Stenophoridae 70
Genus *Stenophora* 71
Genus *Fonsecaia*... 74
Genus *Hyalosporina* 74
Family Leidyanidae 74
Genus *Leidyana*.. 74
Family Cnemidosporidae 75
Genus *Cnemidospora* 75
Family Monoductidae 76
Genus *Monoductus*.. 76
Genus *Stenoductus*.. 76
Genus *Phleobum*... 76
Family Sphaerocystidae 77
Genus *Sphaerocystis* 77
Genus *Schneideria* 77
Genus *Paraschneideria* 77
Genus *Neoschneideria* 77
Family Trichorhynchidae 78
Genus *Trichorhynchus* 78
Family Dactylophoridae 78
Genus *Dactylophorus* 78
Genus *Echinomera*.. 78
Genus *Grebnickiella* 79
Genus *Rhopalonia* 79
Genus *Acutispora* 79
Genus *Seticephalus* 80

Genus *Dendrorhynchus* 80
Genus *Mecistophora* 80
Family Stylocephalidae 80
Genus *Stylocephalus* 80
Genus *Stylocephaloides* 82
Genus *Cystocephalus* 82
Genus *Bulbocephalus* 82
Genus *Xiphocephalus* 83
Genus *Lophocephalus* 83
Genus *Lophocephaloides* 83
Genus *Sphaerorhynchus* 84
Genus *Oocephalus* 84
Genus *Campanacephalus* 84
Genus *Clavicephalus* 84
Genus *Cystocephaloides* 84
Genus *Orbocephalus* 85
Genus *Lepismatophila* 85
Genus *Colepismatophila* 85
Family Actinocephalidae 85
Subfamily Actinocephalinae 85
Genus *Actinocephalus* 85
Genus *Caulocephalus* 87
Genus *Cornimeritus* 87
Genus *Umbracephalus* 88
Genus *Urnaepimeritus* 88
Genus *Asterophora* 88
Genus *Pileocephalus* 88
Genus *Gemmicephalus* 89
Genus *Pilidiophora* 89
Genus *Geneiorhynchus* 89
Genus *Acanthoepimeritus* 90
Genus *Phialoides* 90
Genus *Legeria* 90
Genus *Pyxinia* 90
Genus *Discorhynchus* 91
Genus *Steinina* 91
Genus *Bothriopsides* 92
Genus *Pomania* 92
Genus *Stictospora* 92
Genus *Coleorhynchus* 93
Genus *Stylocystis* 93
Genus *Amphoroides* 93
Genus *Taeniocystis* 93
Genus *Sciadiophora* 94
Genus *Anthorhynchus* 94
Genus *Agrippina* 94
Genus *Globulocephalus* 94
Genus *Alaspora* 95
Genus *Ascocephalus* 95
Genus *Amphorocephalus* 95
Genus *Tricystis* 95

Genus *Thalicola* 95
Genus *Epicavus*.. 96
Genus *Gryllotalpia*.. 96
Genus *Chilogregarina* 96
Genus *Crucocephalus*.. 96
Genus *Harendraia* 97
Genus *Levinea* 97
Subfamily Acanthosporinae 97
Genus *Acanthospora*... 97
Genus *Grenoblia*.. 97
Genus *Corycella* 97
Genus *Ancyrophora*.. 97
Genus *Rhizionella*... 99
Genus *Cometoides* 99
Genus *Prismatospora*.. 99
Genus *Tetraedrospora* 99
Genus *Ramicephalus*... 99
Genus *Coronoepimeritus*... 100
Genus *Dinematospora* 101
Genus *Doliospora*... 101
Genus *Acanthosporidium* 101
Genus *Quadruspinospora* 101
Genus *Contospora* 102
Genus *Tetractinospora*... 102
Genus *Echinoocysta* 102
Genus *Mukundaella*.. 102
Genus *Tetrameridionospinispora*.................................... 102
Subfamily Menosporinae 102
Genus *Menospora*.. 102
Genus *Hoplorhynchus*.. 103
Genus *Odonaticola* 103
Family Brustiosporidae 103
Genus *Brustiospora* 104
Family Acutidae... 104
Genus *Acuta* 104
Genus *Apigregarina* 104
Family Monoicidae... 104
Genus *Monoica* 104
Superfamily Fusionicae.. 104
Family Fusionidae... 104
Genus *Fusiona*.. 104

Chapter 7
The Gregarines: Neogregarines...................................... 105
Order Neogregarinorida 105
Family Gigaductidae 105
Genus *Gigaductus* 105
Family Ophryoystidae 105
Genus *Ophryocystis*... 105
Family Schizocystidae 106
Genus *Schizocystis*.. 106

Genus *Machadoella*..106
Genus *Lymphotropha* 107
Family Caulleryellidae 107
Genus *Caulleryella* 107
Genus *Tipulocystis*...107
Family Syncystidae..107
Genus *Syncystis* 107
Family Lipotrophidae...107
Genus *Lipotropha* 108
Genus *Menzbieria*...108
Genus *Mattesia* 108
Genus *Lipocystis*...109
Genus *Farinocystis* 109
Incertae Sedis 109
Genus *Sawayella*...109

Chapter 8

The Coccidia: Agamococcidiorida, Protococcidiorida, and Ixorheorida 111
Subclass Coccidiasina 111
Order Agamococcidiorida...111
Family Rhytidocystidae...111
Genus *Rhytidocystis*...111
Order Ixorheorida...111
Family Ixorheidae 111
Genus *Ixorheis* 111
Order Protococcidiorida...111
Family Grellidae 111
Genus *Grellia*...112
Genus *Coelotropha* 112
Family Myriosporidae 112
Genus *Myriospora* 112
Genus *Myriosporides* 112
Genus *Mackinnonia*..112
Family Angeiocystidae 113
Genus *Angeiocystis* 113
Family Eleutheroschizonidae 113
Genus *Eleutheroschizon*..113

Chapter 9

The Coccidia: Adeleinorina...115
Order Eucoccidiorida 115
Suborder Adeleorina 115
Family Adeleidae...115
Genus *Adelea*...115
Genus *Adelina* 115
Genus *Klossia* 116
Genus *Orcheobius* 116
Genus *Chagasella*...117
Genus *Ithania*...117
Genus *Rasajeyna*..117
Genus *Ganapatiella*..117

Genus *Gibbsia* 117
Family Legerellidae 117
Genus *Legerella* 117
Family Haemogregarinidae 118
Genus *Haemogregarina* 118
Genus *Karyolysus* 128
Genus *Hepatozoon* 129
Genus *Cyrilia* 133
Family Klossiellidae 133
Genus *Klossiella* 133

Chapter 10
The Coccidia: Eimeriorina 135
Suborder Eimeriorina 135
Family Spirocystidae 135
Genus *Spirocystis* 135
Family Selenococcidiidae 135
Genus *Selenococcidium* 135
Family Dobellidae 135
Genus *Dobellia* 135
Family Aggregatidae 135
Genus *Aggregata* 135
Genus *Merocystis* 136
Genus *Pseudoklossia* 137
Genus *Grasseella* 137
Genus *Ovivora* 137
Genus *Selysina* 137
Family Caryotrophidae 137
Genus *Caryotropha* 137
Genus *Dorisiella* 138
Family Cryptosporidiidae 138
Genus *Cryptosporidium* 138
Family Pfeifferinellidae 138
Genus *Pfeifferinella* 138
Family Eimeriidae 138
Genus *Tyzzeria* 139
Genus *Alveocystis* 139
Genus *Eimeria* 139
Genus *Epieimeria* 178
Genus *Mantonella* 178
Genus *Cyclospora* 178
Genus *Caryospora* 178
Genus *Isospora* 180
Genus *Dorisa* 188
Genus *Wenyonella* 189
Genus *Octosporella* 190
Genus *Hoarella* 190
Genus *Sivatoshella* 190
Genus *Pythonella* 190
Genus *Gousseffia* 190
Genus *Skrjabinella* 190

Genus *Diaspora*..190
Family Barrouxiidae...191
  Genus *Barrouxia*..191
  Genus *Goussia*..191
  Genus *Defretinella*...193
  Genus *Crystallospora*.......................................194
Family Atoxoplasmatidae..194
  Genus *Atoxoplasma*..194
Family Lankesterellidae 195
  Genus *Lankesterella* 195
Family Dactylosomatidae..196
  Genus *Dactylosoma*..196
  Genus *Schellackia* 196
Family Calyptosporidae...197
  Genus *Calyptospora* 197

Index 199

Volume II

Chapter 11
Predator-Prey Coccidia: The Sarcocystidae...................... 1

  Family Sarcocystidae.. 1
  Subfamily Sarcocystinae.. 1
    Genus *Sarcocystis*.. 1
    Genus *Frenkelia* 8
    Genus *Arthrocystis*....................................... 8
  Subfamily Toxoplasmatinae 8
    Genus *Toxoplasma*... 8
    Genus *Besnoitia*.. 9

Chapter 12
Blood Parasites: The Malaria and Related Parasites 11

Class Aconoidasida..11
  Order Haemospororida..11
   Family Plasmodiidae 11
    Genus *Plasmodium* 11
    Genus *Nycteria* 21
    Genus *Polychromophilus*...................................22
    Genus *Dionisia* 22
    Genus *Mesnilium* 22
    Genus *Hepatocystis* 22
    Genus *Rayella* 24
    Genus *Haemoproteus* 24
    Genus *Leucocytozoon* 31
    Genus *Saurocytozoon* 34

Chapter 13
Blood Parasites: The Piroplasms 35

  Order Piroplasmorida 35

Family Anthemosomatidae 35
    Genus *Anthemosoma* 35
Family Babesiidae 35
    Genus *Babesia* 35
    Genus *Echinozoon* 41
Family Theileriidae 42
    Genus *Theileria* 42
Family Haemohormidiidae 44
    Genus *Haemohormidium* 44
    Genus *Sauroplasma* 45

Chapter 14
Parasites of Uncertain Affinities (Incertae Sedis) 47
    Genus *Cristalloidophora* 47
    Genus *Echinococcidium* 47
    Genus *Elleipsisoma* 47
    Genus *Globidiellum* 47
    Genus *Joyeuxella* 47
    Genus *Rhabdospora* 47
    Genus *Serpentoplasma* 47
    Genus *Spermatobium* 48
    Genus *Spiriopsis* 48
    Genus *Spirogregarina* 48
    Genus *Toxocystis* 48
    Genus *Trophosphaera* 48

References 49

Appendix 1
***Nomina Dubia, Nomina Nuda,* Non-Apicomplexa, Etc.** 145

Appendix 2
Superseded Generic Names 149

Index 153

Chapter 1

INTRODUCTION

Protozoa are the most abundant of all living things. They have flowed into every possible niche. They can be found in all habitats, from tropical tree leaves and oceans to Arctic snows. They are the food of fish, their skeletons form the white cliffs of Dover and other chalk deposits, and they teem in the soil. One of the habitats they have invaded is the bodies of animals. Protozoa can be found in body cavities such as the lumen of the intestine, in blood plasma, in cells such as those of the blood and other tissues, and even in the nuclei of cells. Some of them cause disease, but the better adjusted do not. After all, when an organism kills its host, it must die too.

Protozoa are more primitive than animals. Almost all are microscopic and one-celled. They are usually thought of as simple, but they are far from that — they are complex. Protozoa have their own methods of carrying out all the bodily functions necessary for existence. Of the million-plus species of living things, they are among the most ancient, yet they are among the least known.

The Protozoa form a subkingdom of the kingdom Protista. There are about 65,000 named species, about half of them fossil. In its latest classification (Levine et al., 1980), the Society of Protozoologists recognized seven phyla; two are very small and so far relatively unimportant. There are only a few species of Labyrinthomorpha; they feed on eelgrass. There are only a few Ascetospora (the former Haplospora), some of which are parasites of snails; little is known about this group. There are about 550 named species of Microspora, mostly parasites of insects. Their spores have a polar filament, a long hollow tube through which their sporoplasms emerge. Some are being studied carefully because they may be useful in the biological control of insect vectors of disease. There are probably 500 named species of Myxozoa, although apparently no one has counted them. They have a polar filament too, but it is solid. The great majority are parasites of fish, and a few are important as causes of death in fingerlings in fish hatcheries.

The largest phyla of protozoa are the Sarcomastigophora, Ciliophora, and Apicomplexa. The Sarcomastigophora include both flagellates and sarcodines. The flagellates move by means of long, whip-like flagella, the sarcodines by means of pseudopods. They have been placed together because there are some transitional protozoa which have flagella at some times and pseudopods at others. The Ciliophora move by means of eyelash-like cilia (which are actually sets of small flagella), but their most important feature is the possession of two types of nucleus, a diploid micronucleus and a polyploid macronucleus; all the other phyla have a vesicular nucleus. There are about 4300 named species of flagellates (about 2000 of them parasitic), about 41,600 of sarcodines (about 30,000 fossil and 250 parasitic), and about 7200 of ciliates (about 2200 of them parasitic).

The Apicomplexa move by means of longitudinal ridges or, as I think, by means of subpellicular microtubules. The microgametes of some of them have flagella, but no other stage does. There are almost 4600 named species, but perhaps more than ten times as many are still to be named. All are parasitic. Most do not injure their hosts, but some do. Among the Apicomplexa are the malaria parasites of man and other animals, the coccidia, hemogregarines and piroplasms of domestic and other animals, and the gregarines of invertebrates. Malaria is still the most important disease of man, coccidiosis causes heavy losses in domestic animals throughout the world while the piroplasms do the same in many countries, and gregarines may become increasingly important as biological control agents for disease vectors.

About the turn of the century, the Apicomplexa plus some other groups were called Sporozoa because it was thought that they all have spores. They do not. With the advent of the electron microscope, it was realized that most "Sporozoa" have an apical complex. Those which do not (the Microspora, Myxozoa, and Ascetospora) were shifted out, and the group became natural, not artificial.

The apical complex is present at the anterior end at some stage or other. It consists of a number of structures, some of which have been lost by some groups. (1) the first structure is one or more electron-dense **polar rings**; (2) inside the polar rings is a **conoid,** a hollow truncated cone composed of a number of spirally coiled microtubules; (3) passing through the conoid are the necks of a number of **rhoptries** (from the Greek word for club), electron-dense long-necked bags of uncertain function which perhaps may secrete enzymes used in cell penetration; (4) alongside the rhoptries in the cytoplasm are a large number of short rod-like **micronemes**, which some workers believe are attached to the rhoptries; and (5) running backward from one of the polar rings are **subpellicular micro-tubules**, whose number varies with the group and which probably have to do with both locomotion and support. In addition, there are one or more **micropores**, again visible only with the electron microscope, which are apparently used for feeding.

The Apicomplexa have a vesicular nucleus like that of the Sarcomastigophora. It may or may not contain a nucleolus. They have no cilia or flagella, except for flagellated micro-gametes in some groups. They have a Golgi apparatus and mitochondria of the usual pro-tozoan type, and they also have endoplasmic reticulum and an assortment of granules and vacuoles in the cytoplasm. Their carbohydrate granules seem to be amylopectin, at least in the few species in which their composition has been determined.

As mentioned previously, about 4600 species of Apicomplexa have been named, but perhaps more than ten times as many are still to be named. They are divided into over 300 genera and more than 60 families, but this division is deceiving. Most of these groups actually contain only one or a few species. There are over 130 monospecific genera and 13 monospecific families. There are fewer than 50 genera with 10 or more named species, and only 8 with 100 or more. These 8 genera (*Eimeria, Haemogregarina, Gregarina, Isospora, Haemoproteus, Plasmodium, Sarcocystis,* and *Babesia*) comprise more than half of the species.

Most of the important Apicomplexa fall into five main groups. Perhaps the most primitive of these are the **gregarines**, with about 1600 species, 225 genera, and 40 families. They are parasites of invertebrates. Most of them have been seen and described only once, some as long ago as the 19th century. We are becoming more interested in them because some may turn out to be valuable in biological control of mosquitoes and other disease vectors. People are becoming increasingly afraid of insecticides because of their side effects, and some disease vectors have become resistant to some insecticides, so this is a logical step.

A second group is composed of the **hemogregarines**, with about 400 species, 4 genera, and 1 family. Despite their name, they are not gregarines. They live in the blood cells of fish, reptiles, and other vertebrates. Some are known to be transmitted by leeches, mites, ticks, or mosquitoes, but we do not know how most of them get from one animal to another. In fact, relatively little is known about the hemogregarines. Few appear to be pathogenic, and they occur mostly in lower vertebrates, so there has been little incentive to study them. More work undoubtedly will be done in the future.

The third group of Apicomplexa is composed of the **coccidia**, with about 2000 species, 65 genera, and 22 families. They are important parasites of domestic animals and man. Chickens, turkeys, and other birds, cattle, sheep, goats, pigs, dogs, cats, rabbits, rats, and mice are especially affected, but coccidia occur in all other vertebrates that have been examined for them, including fish and man, and also in a few invertebrates. Each host has several species of its own; the great majority of coccidia are not transmissible even from cattle to sheep or from chickens to turkeys. They multiply inside the intestinal cells, destroy them, and produce heavy-walled cysts (oocysts) which pass the infection from one animal to another.

It took the electron microscope and some recent cross-transmission research to reveal that the ubiquitous parasites *Toxoplasma* and *Sarcocystis* are coccidia, but with rather unusual life cycles. Most other coccidia are homoxenous, having a single type of host, but *Toxoplasma, Sarcocystis*, and related genera are heteroxenous, with a predator-prey life cycle in which the asexual stages occur in a prey animal and the sexual ones in a predator. About one third of the people in the U.S. have been infected with *Toxoplasma*, and infants may be born with serious brain damage or blindness. *Sarcocystis* is common in domestic animals; we still do not know what effect it has on most of its hosts, but we know that some species harm them.

A fourth group of Apicomplexa, the **hemosporids**, is composed of blood parasites, including the malaria parasites. They comprise about 450 species and 10 genera, all in a single family. They multiply in the blood cells and are transmitted by mosquitoes or other biting flies. Malaria occurs in man, monkeys, rodents, many birds, and reptiles. It is still the most important disease of man, attacking about 150 million people a year and killing about 1.5 million. There used to be several million cases a year in the U.S., but malaria has been practically eliminated here and pushed back to the tropics and subtropics. More has been written about malaria parasites than about any other protozoa, although they were unknown until almost the turn of the century.

The fifth group of Apicomplexa, the **piroplasms**, is as important in livestock as the malaria parasites are in man. It comprises about 170 species and 6 genera in 3 families. They, too, multiply in the RBC, but they are transmitted by ticks rather than by mosquitoes or biting flies. Piroplasms cause serious diseases of cattle in India, Africa, South and Central America, Australia, and the U.S.S.R.

In addition, there are many small groups or single species of transitional or dead-end forms. One might call the latter failures of evolution, but they still exist, while the true failures do not. Given a change in conditions, perhaps some of them might start new and more successful lines.

The basic pattern of apicomplexan reproduction is alternation of sexual and asexual phases. This pattern is modified in different groups, and some lack certain phases. A zygote is formed by fusion of gametes (syngamy). It divides by multiple fission (schizogony) to form sporozoites (so-called because they were all traditionally thought to be formed within spores; actually, some are and some are not). Sporozoites are infective trophozoites. They enter host cells, grow, and become meronts. These divide asexually by multiple fission, endodyogeny, or endopolygeny, a process known as merogony, to produce a number of merozoites (so-called because the meront that gives rise to them divides into parts — Greek *mero*). There may be one or more generations of meronts and merozoites. If there is more than one, the merozoites enter new host cells, become meronts, and produce more merozoites. Finally, some merozoites become macrogamonts or macrogametes, while others become microgamonts. The macrogamonts, if formed, divide by multiple fission to form a number of macrogametes while the microgamonts divide by binary or multiple fission to form two or more microgametes. A microgamete fuses with a macrogamete to form a zygote. These produce sporozoites, and the cycle is repeated.

The Apicomplexa are haploid throughout their life cycles except for the zygote. This stage is diploid, but the first division after its formation is meiotic, so that all other stages are haploid.

There are thus three multiplications in the basic apicomplexan life cycle. These are merogony, gamogony, and sporogony. This basic life cycle is modified in different groups, as will be mentioned below.

The origin of the Apicomplexa is conjectural and never will be proven. There are no fossils. The discovery that *Perkinsus marinus*, a parasite of the American oyster (Perkins, 1976), is an apicomplexan with an incomplete conoid and with flagella on its "zoospores" makes one wonder if it could be a connecting link. However, it is not well enough known to justify more than speculation. The trypanosomatids have subpellicular microtubules, and some of them are marine. Another possibility is that the Apicomplexa arose from the dinoflagellates. Indeed, Ray (1930) found an organism that he thought was a dinoflagellate in the marine polychete *Scolelepis fulignosa* at Plymouth, England, and it was only later that Dibb (1938) showed that it was a stage of the primitive gregarine *Selenocystis foliata*.

Leaving *Perkinsus* aside, I think that the Apicomplexa originated in marine polychetes. First came the gregarines and coccidia. This origin seems most likely since polychetes have both primitive gregarines and primitive coccidia, and it is sometimes hard to decide which group a particular parasite belongs.

From the polychetes, some Apicomplexa (primarily gregarines) passed to other marine invertebrates — sipunculids, ascidians, and others. Some entered the related oligochetes — earthworms, etc. All these were associated with water, either sea or fresh. The gregarines then passed to arthropods, at first to those associated with water such as mayflies and mosquitoes, whose larvae live in it, and then to those not associated with it such as cockroaches and bees. Some beetle hosts, by the way, are associated with water while others are not.

In the meantime, another branch, the coccidia, traveled somewhat the same road, but specialized in vertebrate rather than invertebrate hosts. The plasmodiids presumably developed from them. They occur in all classes of vertebrates except (with one exception) fish; they use insect vectors so far as is known. The piroplasms, in turn, probably arose from the plasmodiids, and developed primarily in ruminants and rodents. They use arthropod vectors other than insects, so far as is known.

It would be useless at this time to attempt to discuss the evolution of these primary groups. It is perfectly possible to create schemas for each of them, but how correct they would be is questionable. Indeed, the same thing could be said, although perhaps less emphatically, of my own schema.

The classification of the phylum which is used in these volumns is basically that given by Levine et al. (1980). Their classification went only to suborders, whereas this one goes to species and gives diagnoses for all taxa above the species level.

A question that I have not addressed is that of the difference between species and strains. Are all of the species that have been named from certain hosts truly separate species, or are some of them simply strains of some other species? The ability of Mayberry et al. (1982) to infect some but not other strains of the mouse *Mus musculus* with *Eimeria separata* of the Norway rat *Rattus norvegicus* is a case in point. Another is the report of Crum, Fayer, and Prestwood (1981) that they were able to produce light infections with *Sarcocystis odocoilecanis* (normally a parasite of the white-tailed deer *Odocoideus virginianus*) in the ox *Bos taurus* and sheep *Ovis aries*. Is this evidence of evolution in progress? Future research undoubtedly will be needed to clarify (or ''murkify'') this situation. However, the species concept is man-made and species are not immutable entities. Taxonomic names are really matters of convenience regardless of what some taxonomists seem to think. Taxonomy is not a science of absolutes: changes in our concepts are inevitable, and Evolution has not ceased.

The last listing of the "Sporozoa" was by Labbé in 1899. He included 178 species and 70 genera in his order Gregarinida, and 60 species and 22 genera in his order "Coccidiida". In addition, he listed ten species of *Sarcocystis*, six of *Haemogregarina*, two each of *Piroplasma* and *Babesia*, and one each of *Plasmodium*, *Laverania*, *Haemoproteus*, and *Halteridium*. We have progressed considerably in the intervening 80 or so years. Instead of 252 species, about 4600 now have been named. They are listed in the following chapters.

The taxonomy of the Apicomplexa furnishes an excellent example of the incrementalism that is characteristic of modern science. Their discovery and naming was not the work of a few but of many. Among the gregarines, 310 persons first saw and gave names to the species;

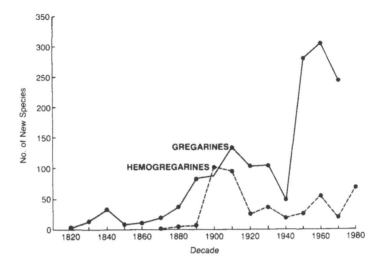

FIGURE 1. Number of new species of gregarines and hemogregarines named each decade from 1820 to 1980.

among the hemogregarines, 195 persons; among the coccidia, 614; among the hemosporins, 290; and among the piroplasms, 175. Most of these persons participated in naming very few new species and most specialized in only a single group. Some of them were interested primarily in some other group of parasites, and studied the Apicomplexa only as they came to attention, while others wrote no other scientific papers. Of the large number of authors, 161 named species in two apicomplexan groups, 45 in three, 14 in four, and 5 in all five.

The countries to which the people who named apicomplexan species belonged represent all regions of the world — North and South America, Europe, the U.S.S.R., Asia, Africa, and Australia.

In addition to giving diagnoses of the genera and higher taxa and the names of the species of the phylum Apicomplexa, I have given appendices of the *nomina nuda, nomina dubia,* non-Apicomplexa, and superseded generic names in the phylum (see Levine, 1984).

The first apicomplexan species was named in 1826, and the number named in each decade has increased progressively since then, except that there were sharp decreases as results of World War I and II. About 45% of the total were named between 1959 and 1979 (Figures 1 and 2).

It is of interest to record the progress that has been made so far in determining the numbers of apicomplexan protozoa that have been reported from various animals. The number of species of mammals has been taken from Nowak and Paradiso (1983), the number of species of birds from Bennett, Whiteway, and Woodworth-Lynas (1982), the number of species of insects from Borror and DeLong (1964), and the number of species of reptiles, amphibia, fish, and invertebrates from Parker (1982).

The approximate numbers and percentages of the species of the various host groups from which Apicomplexa have been reported are given in Table 1. It seems best to comment on the vertebrates and invertebrates separately. Gregarines have been reported only from invertebrates, while the other four apicomplexan groups are almost exclusively parasites of vertebrates. Although the total number of apicomplexan species that have been named is impressive, it is obvious that there is a long way to go before the task is complete. Gregarines have been reported from only 0.32% of the invertebrate species, and the other four groups from only 0.46 to 3.5% of the vertebrates. Among vertebrates, coccidia have been reported from 10.0% of the mammalian species and hemosporins from 12.7% of the bird species. The other percentages tail off, averaging 1.1%. Even among the birds, hemosporins have

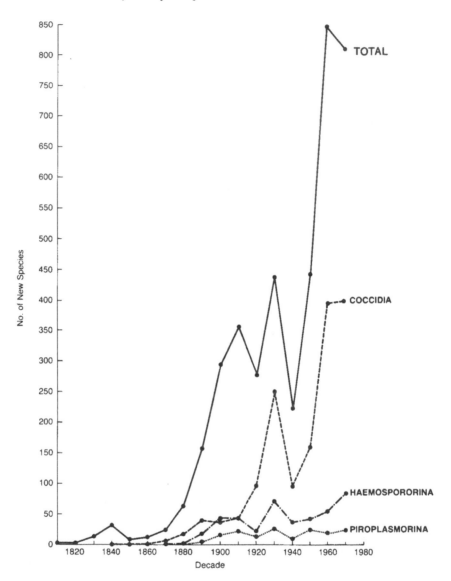

FIGURE 2.   Number of new species of coccidia, hemospororins, piroplasms, and total Apicomplexa named each decade from 1820 to 1980.

not been reported from 53 of the 150 families. Insects are by far the most numerous of invertebrate species, with about 686,000 named species; gregarines have been reported from only 0.32% of them. There are about 277,000 named species of beetles, and gregarines have been reported from about 850 (0.31%) of them. If we assume that there is one species of apicomplexan parasite for every species of host, there are still perhaps 860,000 species to be named. Even if we were to assume that there is one species of apicomplexan for every 10 host species, there are still some 86,000 apicomplexan species to go. The great majority would be gregarines, of course, but even if we limit ourselves to vertebrates, there still would be perhaps 60,000 apicomplexan species yet to be named.

I have emended names throughout the text where they have been given incorrectly. For example, the genitive case of second-declension Latin words (of which the nominative case ends in us) ends in i, not usi or ae. The situation is similar for other Latin declensions, although of course the endings are different (see Table 2, Ride et al., 1985). Further, if a

Table 1
HOSTS SO FAR REPORTED FOR NAMED SPECIES OF APICOMPLEXAN PROTOZOA

Host species from which named parasite species have been reported

Host group	Number of species	Gregarines		Hemogregarines		Coccidia		Hemospororids		Piroplasms		Other	
		Number	%	Number	%	Number	%	Number	%	Number	%	Number	%
Mammals	4,200	0	0.0	66	1.6	432	10.0	131	3.1	128	3.05	2	0.05
Birds	9,400	0	0.0	13	0.14	299	3.1	1,186	12.7	23	0.55	1	0.1
Reptiles	6,000	0	0.0	300	5.0	185	3.1	118	1.97	5	0.01	1	0.02
Amphibia	3,140	0	0.0	25	0.89	46	1.5	1	0.03	1	0.005	0	0.0
Fish	19,000	0	0.0	181	0.05	154	0.81	1	0.005	35	0.18	4	0.02
Total vertebrates	41,740	0	0.0	585	1.4	1,116	2.7	1,437	3.5	192	0.46	8	0.02
Invertebrates	828,000	3,124	0.38	2	0.0002	185	0.02	0	0.0	0	0.0	9	0.001
Grand total	869,740	3,124	0.37	587	0.07	1,301	1.5	1,437	0.17	192	0.002	17	0.002

species name is given in honor of some person, it should end in i if the person is a male, and in ae if the person is a female.

I make no pretense that the scientific names of hosts herein are necessarily correct. They are for the most part simply the names used by the various authors, and are subject to change.

So far as apicomplexan names are concerned, it is likely, for instance, that some of the species assigned at present to *Isospora* actually may belong to *Sarcocystis, Atoxoplasma,* or some other genus. It is only recently that the relationship between *Isospora* and *Atoxoplasma* was recognized (see Levine, 1982). Further, some species of *Sarcocystis* have been named because their sarcocyst walls differ from those of species already named, yet it has been found (e.g., by Munday and Oberdorf, 1984) that the sarcocyst wall may change with age.

Our knowledge of hosts, too, is far from complete. A common fallacy is the DHDP (different host, different parasite) idea. It may or may not be true, depending on the parasite species. Actually many host species simply have not been examined, and certainly many more cross-transmission studies will be needed. Even if they are carried out, however, they may be misleading. The genetic composition of each strain must be known; Mayberry et al. (1982), for instance, found that the *E. separata* or *R. norvegicus* could be transmitted to certain genetic strains of *M. musculus,* but not to others, and the immunological status of the potential host must be taken into consideration; our recent understanding of the relationship of AIDS to susceptibility to *Cryptosporidium* is a case in point.

The following nomenclatural-taxonomic changes have been made: NEW CLASS: Conoidasida; NEW SPECIES: *Eimeria sericei;* NEW NAME: *Eimeria helenlevineae* Bray; NEW COMBINATIONS: *Babesia henryi, Dorisa bengalensis, Erhardovina bisphaera, E. carabodis, E. fuscozetis, E. nitida, E. oribatae, E. phthiracari, E. platynothri, E. postneri, Goussia iroquoiana, G. laureleus, Haemohormidium gallinarum, H. ptyodactyli, Plasmodium (Plasmodioides) neotropicalis, Pseudoklossia tellinorum;* EMENDATIONS: *Ascogregarina tripteroidis, Cephaloidophora japonici, Cnemidospora schizophylli, Echinomera erythrocephali, E. hispidi, E. horridi, Eimeria cabassi, E. choloepi, E. cyclopis, E. gouriae, E. himalayani, E. lutescensis, E. migratorii, E. nebulosi, E. nesokiae, E. ochetobii, E. ophiocephali, E. sphenocerci, E. sylvatici, Erhardovina carabodis, Ganapatiella odontotermitis, Gregarina acantholobi, G. arietuli, G. gryllodis, G. poeciloceri, G. scapsipedi, G. watsonae, Haemogregarina bicarinati, H. chironii, Haemoproteus palumbi, Hepatozoon dendromi, H. leimadophis, Hirmocystis speculitermitis, Hyalosporina cambolopsis, Monocystis octolasii, M. odontotermitis, M. pheretimae, M. saigonensis, Pileocephalus rhytinotae, Protomagalhaensia blaberi, Pyxinia myelophili, Quadruspinospora acridae, Q. chakravartyi, Q. indoaiolopi, Retractacephalus halticae, Schneideria quadrinotati, Sphaerocystis odontotermitis, Steinina alphitobii, S. coptotermitis, S. microgoni, Stenophora gongylorrhi, S. impressi, Stylocephaloides setenis, Tyzzeria typhlopis.*

Undoubtedly I have missed some species in this compendium. I would appreciate any additional named species being called to my attention.

Chapter 2

CLASS PERKINSASIDA

Phylum APICOMPLEXA Levine, 1970
Apical complex present at some stage, generally consisting of polar ring(s), rhoptries, micronemes, conoid and subpellicular microtubules; micropore(s) generally present at some stage; cilia absent; sexuality by syngamy; all species parasitic.

Class PERKINSASIDA Levine, 1978
With flagellated "zoospores" (sporozoites?); "zoospores" with anterior vacuole; conoid forms incomplete truncate cone; sexuality absent; homoxenous.

Order PERKINSORIDA Levine, 1978
With the characters of the class.

Family PERKINSIDAE *Levine, 1978*
With the characters of the order.

Genus *Perkinsus* Levine, 1978. Parasites of invertebrates; with the characters of the family. TYPE SPECIES *P. marinus* (Mackin, Owen, and Collier, 1950) Levine, 1978.

1.   *P. marinus* (Mackin, Owen, and Collier, 1950) Levine, 1978 (syns., *Dermocystidium marinum* Mackin, Owen, and Collier, 1950; *Labyrinthomyxa marina* Mackin, Owen, and Collier, 1950] Mackin and Ray, 1966 (TYPE SPECIES) in American oyster *Crassostrea virginica* (Molluska) all tissues.
2.   *P. olseni* Lester and Davis, 1981 in *Haliotis ruber* (Archaeogastropodorida) muscle and hemolymph.

Chapter 3

THE GREGARINES: ARCHIGREGARINES

Class CONOIDASIDA N. Cl.

New class. Apical complex well developed; conoid (formed by microtubules) a complete, hollow, truncate cone; reproduction generally both sexual and asexual; oocysts generally present, containing infective sporozoites resulting from sporogony; locomotion by body flexion, gliding, undulation of longitudinal ridges or flagellar lashing; flagella present only in microgametes of some groups; pseudopods ordinarily absent, and if present used for feeding, not locomotion; homoxenous or heteroxenous. (This class is named for the structure of its conoid; the term "Sporozoa" is obsolete.)

Subclass GREGARINASINA Dufour, 1828

Mature gamonts extracellular, large; mucron or epimerite in mature organism, the mucron being formed from the conoid; syzygy of gamonts generally occurs; gametes generally similar (isogamous) or nearly so; similar numbers of male and female gametes produced by gamonts; zygotes form oocysts within gametocysts; life cycle generally consists of gamogony and sporogony; parasites of digestive tract or body cavity of invertebrates or lower chordates; homoxenous.

Order ARCHIGREGARINORIDA Grassé, 1953

Life cycle apparently primitive, characteristically with three multiplications (merogony, gamogony, sporogony); gamonts ("trophozoites") aseptate; parasites of annelids, sipunculids, hemichordates, and ascidians.

Family SELENIDIOIDIDAE *Levine, 1971*

Foliaceous or cylindroid, with longitudinal striations ("myonemes"); body often curved or cresent-shaped; anterior end contains bacilliform chromatic bodies which are probably rhoptries; merogony in an epithelial cell; gamont develops extracellularly; syzygy generally tail to tail (caudal); oocysts sperical or ellipsoidal, containing four to n sporozoites.

Genus *Selenidioides* Levine, 1971. Oocysts contain four sporozoites; in gut of polychetes, hemichordates, and sipunculids. TYPE SPECIES *S. caulleryi* (Brasil, 1907) Levine, 1971.

1. *S. axiferens* (Fowell, 1936) Levine, 1971 (syns., *Selenidium axiferens* Fowell, 1936; *Selenidium* sp. Caullery and Mesnil, 1899) *Polydora flava* (Polychaetasida) intestine.
2. *S. caulleryi* (Brasil, 1907) Levine, 1971 (syn., *Selenidium caulleryi* Brasil, 1907) (TYPE SPECIES) in *Protula tubularia* and possibly *Apomatus similis* (Polychaetasida) intestine.
3. *S. fanthami* Levine, 1971 (syn., first *Selenidium* of Brasil and Fantham, 1907) in *Phascolosoma vulgare* and *P. elongatum* (Sipunculida) intestine.
4. *S. grassei* (Théodoridès and Desportes, 1968) Levine, 1984 (syn., *Selenidium grassei* Théoridès and Desportes, 1968) in *Balanoglossus clavigerus* (Hemichordata) intestine.
5. *S. hawesi* Levine, 1971 (syn., second *Selenidium* of Brasil and Fantham, 1907) in *Phascolosoma vulgare* and *P. elongatum* (Sipunculida) intestine.
6. *S. hollandei* (Vivier and Schrével, 1966) Levine, 1971 (syn., *Selenidium hollandei* Vivier and Schrével, 1966) in *Sabellaria alveolata* (Polychaetasida) intestine.
7. *S. intraepitheliale* (Reichenow, 1932) Levine, 1971 (syn., *Selenidium intraepitheliale* Reichenow, 1932) in *Scolelepis fuliginosa* (Polychaetasida) intestine.

8.   *S. mesnili* (Brasil, 1909) Levine, 1971 (syn., *Selenidium mesnili* Brasil, 1909) in *Myxicola infundibulum* (Polychaetasida) intestine.
9.   *S. metchnikovi* (Léger and Duboscq, 1917) Levine, 1971 (syn., *Selenidium metchnikovi* Léger and Duboscq, 1917) in *Glossobalanus minutus* (Hemichordata) intestine.
10. *S. potamillae* (Mackinnon and Ray, 1933) Levine, 1971 (syn., *Selenidium potamillae* Mackinnon and Ray, 1933) in *Potamilla reniformis* (Polychaetasida) intestine.
11. *S. sipunculi* (Dogiel, 1907) Levine, 1971 (syn., *Schizocystis sipunculi* Dogiel, 1907; *Selenidium sipunculi* [Dogiel, 1907] Fantham, 1908) in *Sipunculus nudus* (Sipunculida) intestine.

Genus *Meroselenidium* Mackinnon and Ray, 1933. Oocysts contain many sporozoites; in intestine of polychetes. TYPE SPECIES *M. keilini* Mackinnon and Ray, 1933.

1.   *M. keilini* Mackinnon and Ray, 1933 (TYPE SPECIES) in *Potamilla reniformis* (Polychaetasida) intestine.

Genus *Merogregarina* Porter, 1908. Gamonts similar to those of *Selenidioides*, but longitudinal striations confined to prenuclear region of body; oocysts ovoid or ellipsoidal, containing eight sporozoites. TYPE SPECIES *M. amaroucii* Porter, 1908.

1.   *M. amaroucii* Porter, 1908 (TYPE SPECIES) in *Amaroucium* sp. (Ascidiasida) intestine.

Family EXOSCHIZONIDAE *Levine, 1971*
Merogony present, extracellular, merozoites being formed as a cluster at the anterior end of the meront; gametocysts, gametes, and oocysts unknown.

Genus *Exoschizon* Hukui, 1939. With the characters of the family. TYPE SPECIES *E. siphonosomae* Hukui, 1939.

1.   *E. siphonosomae* Hukui, 1939 (TYPE SPECIES) in *Siphonosoma kumanense* (Sipunculida) intestine.

Chapter 4

THE GREGARINES: BLASTOGREGARINES

Order EUGREGARINORIDA Léger, 1900

Merogony absent; gamogony and sporogony present; typically parasites of annelids and arthropods but some species in other nonvertebrates; locomotion progressive, by gliding or undulation of longitudinal ridges, or nonprogressive.

Suborder BLASTOGREGARINORINA Chatton and Villeneuve, 1936 emend.

Gamogony by gamonts while still attached to the intestine, the gametes budding off of the gamonts; anisogamy present; syzygy absent; gametocysts absent; oocysts with 10 to 16 naked sporozoites; gamont composed of a single compartment without definite protomerite and deutomerite, but with a mucron; in marine polychetes.

Family SIEDLECKIIDAE *Chatton and Villeneuve, 1936*

With the characters of the suborder.

Genus *Siedleckia* Caullery and Mesnil, 1898. With the characters of the family; body elongate, with myonemes. TYPE SPECIES *S. nematoides* Caullery and Mesnil, 1898.

1.   *S. caulleryi* Chatton and Villeneuve, 1936 in *Theostoma oerstedi* (Polychaetasida) intestine.
2.   *S. dogieli* Chatton and Dehorne, 1929 in *Aricia foetida* (Polychaetasida) intestine.
3.   *S. mesnili* Chatton and Dehorne, 1929 in *Aricia latreillii* (Polychaetasida) intestine.
4.   *S. nematoides* Caullery and Mesnil, 1898 (TYPE SPECIES) in *Scoloplos armiger, S. norvegica, Scoloplos* sp., and *Aricia latreillii* (Polychaetasida) intestine.

Chapter 5

THE GREGARINES: ASEPTATE EUGREGARINES

Suborder ASEPTATORINA Chakravarty, 1960

Gamont composed of a single compartment, without definite protomerite and deutomerite, but with an epimerite or mucron in some species; syzygy present.

Family SELENIDIIDAE *Brasil, 1907*

Mucron present; motion nonprogressive, being pendulum-like, coiling or lashing.

Genus *Selenidium* Giard, 1884. Gamonts extracellular, foliaceous or cylindroid, with longitudinal striations (myonemes) and a body often curved or crescent-shaped; anterior end with bacilliform chromatic rods (rhoptries?); syzygy tail to tail (caudal), head to head (frontal), or lateral; oocysts spherical or ellipsoidal, with four sporozoites. TYPE SPECIES *S. pendula* Giard, 1884.

1. *S. alleni* Ray, 1930 in *Branchiomma vesiculosum* (Polychaetasida) intestine.
2. *S. amphinomi* Bhatia and Setna, 1938 in *Amphinome rostrata* (Polychaetasida) coelom.
3. *S. annulatum* (Greeff, 1885) Levine, 1984 (syns., *Gregarina annulata* Greeff, 1885; *Polyrabdina annulata* [Greeff, 1885] Mingazzini, 1893; *Polyrhabdina annulata* [Greeff, 1885] Mingazzini, 1893 emend. Labbé, 1899) in *Rhynchonerella fulgens* (Polychaetasida) intestine.
4. *S. branchiommatis* Ray, 1930 in *Branchiomma vesiculosum* (Polychaetasida) intestine.
5. *S. brasili* Ray, 1930 in *Pomatoceros triqueter, Serpula vermicularis*, and *Spirographis spallanzanii* (Polychaetasida) intestine.
6. *S. cantoui* Ormières, 1979 in *Phascolosoma granulatum* (Sipunculida) intestine.
7. *S. cirratuli* (Lankester, 1866) Caullery and Mesnil, 1899 (syns., *Monocystis cirratuli* Lankester, 1866; *Platycystis cirratuli* [Lankester, 1866] Léger, 1892; *Polyrhabdina cirratuli* Mingazzini, 1893 of Labbé [1899] in part) in *Cirratulus cirratus* (Polychaetasida) intestine.
8. *S. cometomorpha* Schrével, 1971 (syn., *S. "en point et virgule"* Caullery and Mesnil, 1899) in *Cirratulus cirratus* (Polychaetasida) intestine.
9. *S. costatum* Siedlecki, 1903 in *Polymnia nebulosa* (Polychaetasida) intestine, rarely coelom.
10. *S. cruzi* de Faria, da Cunha, and da Fonseca, 1917 in *Polydora socialis* (Polychaetasida) intestine.
11. *S. curvicollum* Bogolepova, 1953 in *Stylarioides plumosus* (Polychaetasida) intestine.
12. *S. echinatum* Caullery and Mesnil, 1899 in *Dodecaceria concharum* (Polychaetasida) intestine.
13. *S. fallax* MacGregor and Thomasson, 1965 (syns., *Polyrabdina cirratuli* [Lankester] of Mingazzini [1893]; *Polyrhabdina cirratuli* [Lankester] of Labbé [1899] in part; *S. cirratuli* [R. Lank.] Mingazzini of Caullery and Mesnil [1919]; *Selenidium* B of Mackinnon and Hawes, 1961; [non] *S. cirratuli* [Lankester, 1866] Caullery and Mesnil, 1899) in *Cirriformia tentaculata* (Polychaetasida) intestine.
14. *S. fauchaldi* Levine, 1974 in *Phragmatopoma californica* (Polychaetasida) intestine.
15. *S. filiformis* Schrével, 1971 in *Cirratulus cirratus* (Polychaetasida) intestine.
16. *S. flabelligerae* Bogolepova, 1953 in *Flabelligera* sp. (Polychaetasida) intestine.
17. *S. folium* Hukui, 1939 in *Siphonosoma kumanense* (Sipunculida) intestine.
18. *S. francianum* (Arvy, 1952) Tuzet and Ormières, 1965 emend. Levine, 1985 (syns.,

Lecudina franciana Arvy, 1952; *S. franciana* [Arvy, 1952] Tuzet and Ormières, 1965) in *Phascolion strombi* (Sipunculida) rectum.

19. *S. halteroide* Hasselmann, 1926 in *Nereis* sp. (Polychaetasida) intestine.

20. *S. hermellae* Hasselmann, 1926 in *Sabellaria* (syn., Hermella) sp. (Polychaetasida) intestine.

21. *S. mackinnonae* Levine, 1971 (syns., *S. cirratuli* [Mingazzini, 1893] of Reichenow [1932]; *S. cirratuli* [Ray Lank.] Mingazzini, 1893 of Schrével [1971]; *Selenidium* A of Mackinnon and Hawes, 1961; [non] *S. cirratuli* [Lankester, 1866] Caullery and Mesnil, 1899) in *Cirriformia tentaculata* (Polychaetasida) intestine.

22. *S. martinensis* Levine, 1971 (syn., *Selenidium* sp. Caullery and Mesnil, 1900) in *Spio martinensis* (Polychaetasida) intestine.

23. *S. melinnae* Schrével, 1971 in *Melinna palmata* (Polychaetasida) intestine.

24. *S. mercierellae* Théodoridès and Laubier, 1962 in *Mercierella enigmatica* (Polychaetasida) intestine.

25. *S. orientale* Bogolepova, 1953 in *Phascolosoma japonicum* (Sipunculida) intestine.

26. *S. parvum* Woodcock and Lodge, 1921 in an undetermined polychete (perhaps a sabellid) (Polychaetasida) intestine.

27. *S. pendula* Giard, 1884 (TYPE SPECIES) in *Nerine cirratulus* (Polychaetasida) intestine.

28. *S. plicatum* Ray, 1930 in *Cirratulus cirratus* (Polychaetasida) intestine.

29. *S. polydorae* Ganapati, 1946 in *Polydora ciliata* (Polychaetasida) intestine.

30. *S. productum* Tuzet and Ormières, 1964 (syn., *S. flabelligerae* Tuzet and Ormières, 1958; *[non] S. flabelligerae* Bogolepova, 1953) in *Flabelligera diplochaitos* (Polychaetasida) intestine.

31. *S. rayi* Levine, 1971 (syn., *S. cirratuli* Ganapati, 1946; *[non] S. cirratuli* [Lankester, 1866] Caullery and Mesnil, 1899) in *Cirratulus filiformis* (Polychaetasida) intestine.

32. *S. sabellae* (Lankester, 1863) Ray, 1930 (syns., *Monocystis sabellae* Lankester, 1863; *Polyrhabdina sabellae* [Lankester, 1863] Labbé, 1899) in *Sabella pavonia, S. infundibulum, S. alveolata, S. bombyx,* and *S. impendiculum* (Polychaetasida) intestine.

33. *S. sabellariae* Schrével, 1971 in *Sabellaria alveolata* (Polychaetasida) intestine.

34. *S. serpulae* (Lankester, 1863) Caullery and Mesnil, 1899 (syns., *Monocystis serpulae* Lankester, 1863; *Polyrabdina serpulae* [Lankester, 1863] Mingazzini, 1893; *Polyrhabdina serpulae* [Lankester, 1863] Mingazzini, 1893 emend. Labbé, 1899) in *Serpula contortuplicata* (Polychaetasida) intestine.

35. *S. spatulatum* Hasselmann, 1926 in *Nereis* sp. (Polychaetasida) intestine.

36. *S. spinosis* Ganapati, 1946 in *Prionospio cirrifera* (Polychaetasida) intestine.

37. *S. spionis* (von Kölliker, 1845) Ray, 1930 (syns., *Gregarina spionis* von Kölliker, 1845 in part; *Monocystis spionis* [von Kölliker, 1845] Lankester, 1863 in part; *Polyrabdina spionis* Mingazzini, 1891 in part; *Selenidium* sp. No. 1 Caullery and Mesnil, 1899) in *Scolelepis fuliginosa* (Polychaetasida) intestine.

38. *S. stellatum* Tuzet and Ormières, 1965 in *Aspidosiphon clavatus* (Sipunculida) intestine.

39. *S. synaptae* (Mingazzini, 1893) Levine, 1971 (syn., *Esarabdina synaptae* Mingazzini, 1893) in *Synapta* sp. (Holothurasida) intestine.

40. *S. telepsavi* (Stuart, 1871) Levine, 1974 (syn., *Monocystis telepsavi* Stuart, 1871) in *Telepsavus castarum* (Polychaetasida) intestinal appendages.

41. *S. terebellae* (von Kölliker, 1845) Ray, 1930 (syns., *Gregarina terebellae* von Kölliker, 1845; *Monocystis terebellae* [von Kölliker, 1845] Lankester, 1863; *Esarabdina terebellae* [von Kölliker, 1845] Mingazzini, 1891) in *Terebella lapidaris* and *Eulopymnia nebulosa* (Polychaetasida) intestine.

42. *S. virgula* Caullery and Mesnil, 1919 (syn., *S. "en virgule"* of Caullery and Mesnil [1899]) in *Cirratulus cirratus* (Polychaetasida) intestine.

Genus *Selenocystis* Dibb, 1938. Similar to *Selenidium*, but gametocysts attached to host intestinal cell by a bifid or multifid foot-like organelle formed by modification of the mucron. TYPE SPECIES *S. foliata* (Ray, 1930) Dibb, 1938.

1.   *S. foliata* (Ray, 1930) Dibb, 1938 (syns., *Selenidium foliatum* (Ray, 1930; *Haplozoon* sp. Mesnil, 1917; *Selenidium* sp. No. 2 of Caullery and Mesnil [1899]) (TYPE SPECIES) in *Scolelepis fuliginosa* (Polychaetasida) intestine.

Genus *Ditrypanocystis* Burt, Denny, and Thomasson, 1963. Gamonts vermiform, but broader anteriorly; with pellicular longitudinal ridges two to four of which are expanded and extended to form undulating membranes which originate just posterior to the mucron and extend posteriorly along part of the body; on each side of the undulating membranes is a less conspicuous ridge intermediate in size between that of the undulating membranes and that of the other pellicular ridges. TYPE SPECIES *D. cirratuli* Burt, Denny, and Thomasson, 1963.

1.   *D. cirratuli* Burt, Denny, and Thomasson, 1963 (TYPE SPECIES) in *Cirriformia tentaculata* (Polychaetasida) intestine.
2.   *D. coxi* Levine, 1971 (syn., *D. cirratuli* Burt, Denny, and Thomasson of Cox [1965] in *Cirratulus cirratus;* *[non] D. cirratuli* Burt, Denny, and Thomasson, 1963) in *Cirratulus cirratus* (Polychaetasida) intestine.

Genus *Heterospora* de Saint Joseph, 1907. Gamonts, syzygy, and gametocysts unknown; oocysts ellipsoidal, without thickenings at the ends but with a short, hair-like projection at each end and four hair-like projections from the middle of the sides; oocysts with four sporozoites, without residuum. TYPE SPECIES *H. eulaliae* de Saint Joseph, 1907.

1.   *H. eulaliae* de Saint Joseph, 1907 (TYPE SPECIES) in *Eulalia (Eumida) parva* (Polychaetasida) coelom.

Family LECUDINIDAE KAMM, *1922*
Gamonts elongate, their motion not pendulum-like or coiling but gliding; first stage of development generally intracellular; syzygy lateral of frontal; oocysts ellipsoidal or ovoid, with wall slightly thickened at one end; intestinal parasites of annelids, sipunculids, or echiurids.

Genus *Lecudina* Mingazzini, 1891. Mucron simple, without hooks or exfoliations; gamonts without longitudinal striations (myonemes); oocysts ovoid, thickened at one end; in polychetes and other marine invertebrates. TYPE SPECIES *L. pellucida* (von Kölliker, 1848) Mingazzini, 1891.

1.   *L. amphora* H. Hoshide, 1958 in *Glycera rouxi* and *Glycera* sp. (Polychaetasida) intestine.
2.   *L. aphroditae* (Lankester, 1863) Kamm, 1922 (syns., *Monocystis aphroditae* Lankester, 1863; *Doliocystis aphroditae* [Lankester, 1863] Labbé, 1899) in *Aphrodite aculeata* (Polychaetasida) intestine.
3.   *L. arabellae* H. Hoshide, 1958 (syn., *L. arabellae* Schrével, 1963) in *Arabella iricolor* and *tricolor* (Polychaetasida) intestine.
4.   *L. arrhyncha* Bogolepova, 1953 in *Nereis* sp. (Polychaetasida) intestine.

5. *L. attenuata* (Bogolepova, 1953) Théodoridès, 1969 (syn., *Cygnicollum attenuatum* Bogolepova, 1953) in *Lumbrineris* (syn., *Lumbriconereis*) *japonica* (Polychaetasida) presumably intestine.

6. *L. bhatiai* Levine, 1976 (syn., *L. eunicae* Bhatia and Setna, 1938) in *Eunice siciliensis* (Polychaetasida) intestine.

7. *L. bogolepovae* (Levine, 1971) Levine, 1976 (syns., *Polyrhabdina cirratuli* Bogolepova, 1953; [*non*] *P. cirratuli* Mingazzini, 1893 of Labbé [1899]; *P. bogolepovae* Levine, 1971) in *Cirratulus cirratus* (Polychaetasida) intestine.

8. *L. brasili* Ganapati and Aiyar, 1937 in *Lumbrineris* (syn., *Lumbriconereis* sp. (Polychaetasida) intestine.

9. *L. capitellae* Hasselmann, 1926 in *Capitella* sp. (Polychaetasida) intestine.

10. *L. catalinensis* Levine, 1974 in *Lumbrineris inflata* (Polychaetasida) intestine.

11. *L. caudata* K. Hoshide, 1977 in *Perinereis brevicirrus* (Polychaetasida) intestine.

12. *L. criodrilii* (Sciacchitano, 1931) Levine, 1976 (syn., *Doliocystis criodrilii* Sciacchitano, 1931) in *Criodrilus lacuum* (Oligochaetasida) intestine.

13. *L. danielae* Corbel, Desportes, and Théodoridès, 1979 in *Vanadis crystallina* (Polychaetasida) intestine.

14. *L. defretini* Schrével, 1969 in *Nereis irrorata* (Polychaetasida) intestine.

15. *L. elongata* (Mingazzini, 1891) Kamm, 1922 (syns., *Ophioidina elongata* Mingazzini, 1891; *Doliocystis elongata* [Mingazzini, 1891] Labbé, 1899) in *Lumbrineris impatiens, L. latreillei, L. tingens,* and *L. fragilis* (Polychaetasida) intestine.

16. *L. eunicae* (Lankester, 1866) Levine, 1976 (syns., *Monocystis eunicae* Lankester, 1866; *Polyrhabdina eunicae* [Lankester, 1866] Labbé, 1899; [*non*] *L. eunicae* Bhatia and Setna, 1938) in *Eunice harassii* (Polychaetasida) intestine.

17. *L. euphrosynes* Bogolepova, 1953 in *Euphrosyne* sp. (Polychaetasida) intestine.

18. *L. fluctus* Iitsuka, 1931 (syn., *L. fluktus* Iitsuka of H. Hoshide [1958] *lapsus calami*) in *Urechis unicinctus* (Echiurida) intestine.

19. *L. ganapatii* Vivier, Ormières, and Tuzet, 1964 (syn., *L. pellucida* [von Kölliker] Mingazzini, 1891 of Ganapati [1946]; [*non*] *L. pellucida* [von Kölliker, 1848] Mingazzini, 1891) in *Perinereis chilkaensis* (Polychaetasida) intestine.

20. *L. hesionis* Théodoridès, 1969 in *Hesione pantherina* (Polychaetasida) intestine.

21. *L. heterocephala* (Mingazzini, 1891) Kamm, 1922 (syns., *Ophioidina heterocephala* Mingazzini, 1891; *Doliocystis heterocephala* [Mingazzini, 1891] Labbé, 1899) in *Nephtys scolopendroides* and *N. hombergii* (Polychaetasida) intestine.

22. *L. indica* Ganapati, 1946 in *Lycastis indica* (Polychaetasida) intestine.

23. *L. krusadiensis* Ganapati, 1946 in *Platynereis abnormis* (Polychaetasida) intestine.

24. *L. laubieri* Théodoridès, 1969 (syn., *L. [Cygnicollum] laubieri* Théodoridès, 1969) in *Lumbrineris coccinea* and *Lumbrineris* sp. (Polychaetasida) intestine.

25. *L. legeri* (Brasil, 1909) Kamm, 1922 (syn., *Doliocystis legeri* Brasil, 1909) in *Glycera convoluta* and *G. alba* (Polychaetasida) intestine.

26. *L. leptonereidis* Schrével, 1963 in *Leptonereis glauca* (Polychaetasida) intestine.

27. *L. leuckartii* Mingazzini, 1891 in *Sagitta* sp. (Chaetognatha) intestine.

28. *L. linei* Vinckier, 1972 in *Lineus viridis* (Nemertini) intestine.

29. *L. longicephala* Ganapati, 1946 in *Pisionidens indica* (Polychaetasida) intestine.

30. *L. longissima* Tugawa [H. Hoshide], 1944 in *Lumbrineris japonica* and *L. zonata* (Polychaetasida) intestine.

31. *L. lysidicae* Bhatia and Setna, 1938 (syn., *L. lisidicae* Bhatia and Setna, 1938 of Bogolepova [1953] *lapsus calami*) in *Lysidice collaris* (Polychaetasida) intestine.

32. *L. mammilata* Tugawa [H. Hoshide] 1944 in *Nereis japonica* and *N. microdonta* (Polychaetasida) intestine.

33. *L. nereicola* (Bogolepova, 1953) Levine, 1976 (syn., *Polyrhabdina nereicola* Bogolepova, 1953) in *Nereis* spp. (Polychaetasida) intestine.

34. *L. pellucida* (von Kölliker, 1848) Mingazzini, 1891 (syns., *Gregarina pellucida* von Kölliker, 1848; *Doliocystis pellucida* [von Kölliker, 1848] Léger, 1893; *Monocystis pellucida* [von Kölliker, 1848] Lankester, 1863; *G. nereidis* Leidy, 1856; *M. nereidis* [Leidy, 1856] Lankester, 1863) (TYPE SPECIES) in *Nereis beaucoudrayi*, *N. pelagica*, *N. virens*, *Perinereis cultrifera*, and *Platynereis dumerilii* (?) (Polychaetasida) intestine.

35. *L. pelmatomorpha* Schrével, 1969 in *Perinereis marionii* (Polychaetasida) intestine.

36. *L. pherusae* Levine, 1974 in *Pherusa capulata* (Polychaetasida) intestine.

37. *L. phyllodocis* Théodoridès, 1969 in *Phyllodoce laminosa* and *Phyllodoce* sp. (Polychaetasida) intestine.

38. *L. platynereidis* Schrével, 1969 in *Platynereis massiliensis* and *P. dumerilii* (Polychaetasida) intestine.

39. *L. polydorae* (Léger, 1893) Kamm, 1922 (syns., *Doliocystis polydorae* Léger, 1893; *Polyrhabdina polydorae* [Léger, 1893] Caullery and Mesnil, 1914) in *Polydora agassizi*, *P. ciliata*, and *P. flava* (Polychaetasida) intestine.

40. *L. polymorpha* Schrével, 1963 in *Lumbrineris latreillei* (Polychaetasida) intestine.

41. *L. postae* Théodoridès and Desportes, 1978 in *Hyalinoecia tubicola* (Polychaetasida) intestine.

42. *L. pyriformis* Bogolepova, 1953 in *Nereis cyclurus* and *N. agassizi* (Polychaetasida) intestine.

43. *L. staurocephali* (Mingazzini, 1891) Levine, 1976 (syns., *Koellikeria staurocephali* Mingazzini, 1891; *Koellikerella staurocephali* [Mingazzini, 1891] Labbé, 1899) in *Staurocephalus rudolphi* (Polychaetasida) intestine.

44. *L. tuzetae* Schrével, 1963 in *Nereis diversicolor* (Polychaetasida) intestine.

45. *L. zimmeri* Levine, 1974 in *Pherusa capulata* (Polychaetasida) intestine.

Genus *Polyrhabdina* Mingazzini, 1891. Mucron with hooks; gamonts with longitudinal striations (myonemes); in polychetes. TYPE SPECIES *P. spionis* (von Kölliker, 1848) Mingazzini, 1891.

1. *P. bifurcata* Mackinnon and Ray, 1931 emend. Reichenow, 1932 (syn., *P. spionis* var. *bifurcata* Mackinnon and Ray, 1931) in *Scolelepis fuliginosa* (Polychaetasida) intestine.

2. *P. brasili* Caullery and Mesnil, 1914 in *Spio martinensis* (Polychaetasida) intestine.

3. *P. minuta* Ganapati, 1946 in *Prionospio cirrifera* (Polychaetasida) intestine.

4. *P. spionis* (von Kölliker, 1848) Mingazzini, 1891 (syns., *Gregarina spionis* von Kölliker, 1848 in part; *Monocystis spionis* [von Kölliker, 1848] Lankester, 1863 in part) (TYPE SPECIES) in *Scolelepis fuliginosa* and perhaps also *S. ciliata* (Polychaetasida) intestine.

Genus *Ulivina* Mingazzini, 1891. Mucron simple, without hooks or exfoliations; gamonts divided by an incomplete septum into a pseudoprotomerite and pseudodeutomerite whose contents differ; syzygy and oocysts unknown; in polychetes. TYPE SPECIES *U. elliptica* Mingazzini, 1891.

1. *U. acrobata* Hasselmann, 1926 in *Nereis* sp. (Polychaetasida) presumably intestine.

2. *U. dirigibilis* Hasselmann, 1926 in *Nereis* sp. (Polychaetasida) presumably intestine.

3. *U. elliptica* Mingazzini, 1891 (TYPE SPECIES) in *Cirratulus* (syn., *Audouinia) filigerus* and possibly other polychetes (Polychaetasida) intestine.

4. *U. eunicae* Bhatia and Setna, 1938 in *Nereis siciliensis* (Polychaetasida) intestine.

5. *U. macrocephala* Hasselmann, 1926 in *Nereis* sp. (Polychaetasida) presumably intestine.

6.   *U. rhyncoboli* (Crawley, 1903) Kamm, 1922 (syn., *Doliocystis rhyncoboli* Crawley, 1903) in *Glycera* (syn., *Rhynchobolus) americanus* (Polychaetasida) intestine.

Genus *Sycia* Léger, 1892. Mucron a simple knob with a fleshy ring at its base; gamont divided by an incomplete septum into a pseudoprotomerite and a pseudodeutomerite whose contents differ; syzygy, gametocysts, and oocysts unknown; in polychetes. TYPE SPECIES *S. inopinata* Léger, 1892.

1.   *S. cirratuli* H. Hoshide, 1958 in *Cirratulus cirratus* (Polychaetasida) intestine.
2.   *S. inopinata* Léger, 1892 (TYPE SPECIES) in *Cirriformia tentaculata* (syn., *Audouinia lamarki* (Polychaetasida) intestine.
3.   *S. legeri* Ganapati, 1946 in *Cirratulus filiformis* (Polychaetasida) intestine.
4.   *S. polydorae* Ganapati, 1946 in *Polydora ciliata* (Polychaetasida) intestine.

Genus *Pontesia* Hasselmann, 1926. Gamonts described as septate, elongate, cylindrical, tapering to a blunt pointed end; mucron (epimerite) a row of short spines the free upper extremity of the "protomerite"; in marine annelids. TYPE SPECIES *P. kelschi* Hasselmann, 1926.

1.   *P. kelschi* Hasselmann, 1926 (TYPE SPECIES) in *Nereis* sp. (Polychaetasida) intestine.

Genus *Bhatiella* Setna, 1931. Mucron a long, slender style with a small bulb at its apex; gamonts generally piriform; syzygy unknown; gametocyst without sporoducts; in polychetes. TYPE SPECIES *B. marphysae* Setna, 1931 emend. Levine, 1977.

1.   *B. iwamusii* Tugawa [H. Hoshide], 1944) Levine, 1977 (syns., *Cotyloepimeritus iwamusii* Tugawa [H. Hoshide], 1944; *Ferraria cornucephala iwamusi* Tugawa [H. Hoshide], 1944 [?]) *Marphysa iwamusi* (Polychaetasida) intestine.
2.   *B. marphysae* Setna, 1931 emend. Levine, 1977 (syns., *B. morphysae* Setna, 1931 lapsus calami; *Ferraria cornucephali* Setna, 1931) (TYPE SPECIES) in *Marphysa sanguinea* (syn., *M. furcellata*) (Polychaetasida) intestine.

Genus *Viviera* Schrével, 1963. Mucron a small, spherical ball; gamont divided into a broad, racquet-shaped anterior part and a long, narrow, isosceles triangle-shaped posterior part; other stages unknown; in polychetes. TYPE SPECIES *V. marphysae* Schrével, 1963.

1.   *V. marphysae* Schrével, 1963 (TYPE SPECIES) in *Marphysa sanguinea* (Polychaetasida) intestine.

Genus *Cochleomeritus* Tugawa [H. Hoshide], 1944. Mucron variable in shape, but usually elongate, with or without hairs; gamont broad anteriorly, flat in cross section, more or less folded over dorsum; gametocysts and oocysts unknown, in polychetes. TYPE SPECIES *C. lysidici* Tugawa [H. Hoshide], 1944.

1.   *C. emersoni* Levine, 1973 in *Diopatra ornata* (Polychaetasida) intestine.
2.   *C. lysidici* Tugawa [H. Hoshide], 1944 (TYPE SPECIES) in *Lysidice punctata* (Polychaetasida) intestine.

Genus *Ancora* Labbé, 1899. Mucron apparently simple; gamont with two lateral processes giving it the appearance of an anchor; in annelids. TYPE SPECIES *A. sagittata* (Leuckart, 1861) Labbé, 1899.

1. *A. criodrilii* (Sciacchitano, 1931) Levine, 1977 (syn., *Anchorina criodrilii* Sciacchitano, 1931) in *Criodrilus lacuum* (Oligochaetasida) coelom.
2. *A. lutzi* Hasselmann, 1918 in *Capitella capitata* (Polychaetasida) intestine.
3. *A. sagittata* (Leuckart, 1861) Labbé, 1899 (syns., *Gregarina sagittata* Leuckart, 1861; *Anc[h]orina sagittata* [Leuckart, 1861] Mingazzini, 1891; *Monocystis sagittata* [Leuckart, 1861] Lankester, 1863) (TYPE SPECIES) in *Capitella capitata* and *Capitella* sp. (Polychaetasida) intestine.

Genus *Hentschelia* Mackinnon and Ray, 1931. Mucron umbrella-like, with margin divided into four or five fluted lobes; gamonts with fine, inconspicuous, longitudinal striations; syzygy lateral, with the gamonts usually head to tail; in echiurids. TYPE SPECIES *H. thalassemae* Mackinnon and Ray, 1931.

1. *H. thalassemae* Mackinnon and Ray, 1931 (TYPE SPECIES) in *Thalassema neptuni* (Echiurida) intestine.

Genus *Lecythion* Mackinnon and Ray, 1931. Mucron long, cylindrical, with 14 or 15 petal-shaped lobes at its base; gamont spirally striated, with oblique rows of denticles; syzygy lateral, head to tail; gametocysts passed in feces; in echiurids. TYPE SPECIES *L. thalassemae* Mackinnon and Ray, 1931.

1. *L. thalassemae* Mackinnon and Ray, 1931 (TYPE SPECIES) in *Thalasssema neptuni* (Echiurida) intestine.

Genus *Hyperidion* Mackinnon and Ray, 1931. Mucron short, broad, simple; gamont with longitudinal myonemes; gamonts contractile, being either elongate or spherical; in echiurids. TYPE SPECIES *H. thalassemae* Mackinnon and Ray, 1931.

1. *H. thalassemae* Mackinnon and Ray, 1931 (syn., *H. neptuni* Reichenow, 1932) (TYPE SPECIES) in *Thalassema neptuni* (Echiurida) intestine.

Genus *Zygosoma* Labbé, 1899. Mucron short, broad, simple; gamonts more or less piriform, covered with nipple-like projections; syzygy lateral; gametocysts spherical or ovoid, with about 1250 oocysts; gametes spherical to piriform, isogamous; oocysts with 4 naked sporozoites; in echiurids. TYPE SPECIES *Z. gibbosum* (Greeff, 1880) Labbé, 1899.

1. *Z. gibbosum* (Greeff, 1880) Labbé, 1899 (syns., *Conorhynchus gibbosus* Greeff, 1880 in part; *C. echiuri* Greeff, 1880 in part) (TYPE SPECIES) in *Echiurus pallasi* (Echiurida) intestine.
2. *Z. globosum* Noble, 1938 in *Urechis caupo* (Echiurida) intestine.

Genus *Filipodium* Hukui, 1939. Mucron broadly funnel-shaped, with papillae around its rim; gamonts elongate, with longitudinal striations and with many protrusible filaments emerging from beneath the pellicle; gametocysts with many oocysts; gametes dissimilar, but male gametes not flagellated; oocysts ellipsoidal or ovoid, with eight sporozoites; in sipunculids. TYPE SPECIES *F. ozakai* Hukui, 1939.

1. *F. aspidosiphoni* Tuzet and Ormières, 1965 in *Aspidosiphon clavatus* (Sipunculida) intestine.
2. *F. ozakai* Hukui, 1939 (syn., *Lithocystis parquerae* Jones, 1971) (TYPE SPECIES) in *Siphonosoma kumanense* (Sipunculida) intestine.

Genus *Diplauxis* Vivier, Ormières, and Tuzet, 1964. Gamont elongate, more or less pointed at one end; ectoderm of gamont densely hairy; syzygy tail to tail, occurring very early; gametocysts elongate, crescent- to sausage-shaped, preserving the form of the coupled gamonts; dehiscence by simple rupture; oocysts ovoid, with a thickening at one end, emitted in chains; in polychetes. TYPE SPECIES *D. hatti* Vivier, Ormières, and Tuzet, 1964.

1. *D. hatti* Vivier, Ormières, and Tuzet, 1964 (TYPE SPECIES) in *Perinereis cultrifera* (Polychaetasida) coelom.
2. *D. schreveli* Porchet-Henneré and Fischer, 1973 in *Platynereis dumerilii* (Polychaetasida) coelom.

Genus *Chlamydocystis* Dogel', 1910. Mucron presumably simple; gamonts elongate, within host amebocytes; locomotion gliding; gametocysts and oocysts unknown; in polychetes. TYPE SPECIES *C. captiva* Dogel', 1910.

1. *C. captiva* Dogel', 1910 (TYPE SPECIES) in *Telephus circinnatus* (Polychaetasida) amebocytes in coelom.

Genus *Contortiocorpa* Bhatia and Setna, 1938. Gamonts solitary; body twisted spirally upon itself. TYPE SPECIES *C. prashadi* Bhatia and Setna, 1938.

1. *C. prashadi* Bhatia and Setna, 1938 (TYPE SPECIES) in *Eunice siciliensis* (Polychaetasida) intestine.

Genus *Lankesteria* Mingazzini, 1891. Mucron present although not always apparent, usually simple but sometimes complex; gamonts more or less spatulate; syzygy head to head or scissors-like; gametocysts spherical; anisogamy present; oocysts ellipsoidal, often with a plug at each end; in ascidians. TYPE SPECIES *L. ascidiae* (Lankester, 1872) Mingazzini, 1891.

1. *L. abbotti* Levine, 1981 in *Clavelina huntsmani* (Ascidiasida) intestine and/or stomach.
2. *L. acutissima* Ormières, 1965 in *Ascidiella aspersa* (Ascidiasida) intestine.
3. *L. amaroucii* (Giard, 1873) Labbé, 1899 (syns., *Monocystis amaroecii* Giard, 1873 *lapsus calami; L. giardi* Mingazzini, 1893) in *Aplidium (Amaroucium) punctum* (Ascidiasida) intestine.
4. *L. aplidii* Levine, 1981 in *Aplidium solidum* (Ascidiasida) intestine and/or stomach.
5. *L. ascidiae* (Lankester, 1872) Mingazzini, 1891 (syns., *Monocystis ascidiae* Lankester, 1872; *Gregarina cionae* Frenzel, 1885; *Urospora cionae* Parona, 1886 and also Gruber, 1886) (TYPE SPECIES) in *Ciona intestinalis* (Ascidiasida) stomach and intestine.
6. *L. ascidiellae* Duboscq and Harant, 1923 in *Ascidiella (Ascidia) aspersa* and *A. scabra* (Ascidiasida) stomach and intestine.
7. *L. botrylli* Ormières, 1965 in *Botryllus schlosseri* and *B. leachi* (Ascidiasida) intestine.
8. *L. buetschlii* (Mingazzini, 1891) Ormières, 1965 emend. Levine, 1977 (syns., *Pleurozyga buetschlii* Mingazzini, 1891; *L. butschlii [sic]*[Mingazzini, 1893] Ormières, 1965) in *Phallusia mammillata* (Ascidiasida) intestine.
9. *L. clavellinae* (von Kölliker, 1848) Ormières, 1965 (syns., *Gregarina clavellinae* von Kölliker, 1848; *Monocystis clavellinae* [von Kölliker, 1848] Lankester, 1863; *Pleurozyga clavellinae* [von Kölliker, 1848] Labbé, 1899) in *Clavellina producta* and *C. lepadiformis* (Ascidiasida) intestine.
10. *L. diaphanis* Levine, 1981 in *Archidistoma diaphanes* (Ascidiasida) intestine and/or stomach.

11. *L. diazonae* (Mingazzini, 1891) Labbé, 1899 (syn., *Cytomorpha diazonae* Mingazzini, 1891) in *Diazona violacea* (Ascidiasida) intestine.
12. *L. distapliae* (Mingazzini, 1891) Ormières, 1965 (syns., *Pleurozyga distapliae* Mingazzini, 1891; *Gregarina distapliae* Bogoyavlensky, 1926) in *Distaplia magnilarva, D. stelligera,* and *D. rosea* (Ascidiasida) intestine.
13. *L. euherdmaniae* Levine, 1981 in *Euherdmania claviformis* (Ascidiasida) intestine and/or stomach.
14. *L. gigantea* Ormières, 1965 in *Pyura microcosmus* (Ascidiasida) intestine.
15. *L. globosa* Ormières, 1965 in *Distomus variolosus* (Ascidiasida) intestine.
16. *L. gracilis* Ormières, 1965 in *Polycarpa gracilis* (Ascidiasida) intestine.
17. *L. gyriniformis* Ormières, 1965 in *Rhopalea neapolitana* (Ascidiasida) intestine.
18. *L. maculata* Ormières, 1965 (syns., *Selenidium* [gen.?] *giganteum* Harant, 1931 in part [?]; *Selenidioides giganteum* [Harant, 1931] Levine, 1971 in part [?]) in *Polycarpa pomaria* (Ascidiasida) digestive tract.
19. *L. molgulidarum* Ormières, 1965 in *Ctenicella appendiculata* (Ascidiasida) intestine.
20. *L. monstrosa* Ormières, 1965 in *Diazona violacea* (Ascidiasida) intestine.
21. *L. montereyensis* Levine, 1981 in *Archidistoma molle* (Ascidiasida) intestine and/or stomach.
22. *L. morchellii* Ormières, 1965 in *Morchellium argus* (Ascidiasida) intestine.
23. *L. ormieresi* Levine, 1977 in *Polyclinum aurantium* (Ascidiasida) intestine.
24. *L. parascidiae* Duboscq and Harant, 1923 in *Sidnyum* (syn., *Parascidia) elegans* (Ascidiasida) intestine.
25. *L. perophoropsis* Ormières, 1965 in *Perophoropsis herdmanni* (Ascidiasida) intestine.
26. *L. pescaderoensis* Levine, 1981 in *Ritterella rubra* (Ascidiasida) intestine and/or stomach.
27. *L. pittendrighi* Levine, 1981 in *Ascidia ceratodes* (Ascidiasida) intestine and/or stomach.
28. *L. psammii* Levine, 1981 in *Archidistoma psammion* (Ascidiasida) intestine and/or stomach.
29. *L. ritterellae* Levine, 1981 in *Ritterella pulchra* (Ascidiasida) intestine and/or stomach.
30. *L. ritterii* Levine, 1981 in *Archidistoma ritteri* (Aschdiasida) intestine and/or stomach.
31. *L. siedleckii* Duboscq and Harant, 1923 (syn., *L. siedlickii* Duboscq and Harant, 1923 *lapsus calami)* in *Ascidia* (syn., *Phallusia) mentula* (Ascidiasida) intestine and sperm duct.
32. *L. striata* Ormières, 1965 in *Aplidium (Amaroucium) nordmanni* (Ascidiasida) intestine.
33. *L. styelae* Ormières, 1965 (syn., *L. ascidiae* [Lankester] of Harant [1931]; [*non] L. ascidiae* [Lankester, 1872] Mingazzini, 1891) in *Styela partita* (Ascidiasida) intestine.
34. *L. synoici* Levine, 1981 in *Synoicum parfustis* (Ascidiasida) intestine and/or stomach.
35. *L. tethyi* Bogolepova, 1953 in *Tethyum aurantium* (Ascidiasida) intestine.
36. *L. tuzetae* Ormières, 1965 in *Aplidium (Amaroucium) nordmanni* and *A. pallidum* (Ascidiasida) intestine.
37. *L. zonata* Ormières, 1965 in *Dendrodoa grossularia* (Ascidiasida) intestine.

Genus *Monocystella* Valkanov, 1935. Similar to *Lankesteria* but in turbellarians. TYPE SPECIES *M. arndti* (Valkanov, 1931) Valkanov, 1935.

1. *M. arndti* (Valkanov, 1931) Valkanov, 1935 (syn., *Lankesteria arndti* Valkanov, 1931) (TYPE SPECIES) in *Planaria albissima* and *Fonticola ochridana* (Turbellariasida) intestine.
2. *M. compacta* (Geus, 1967) Levine, 1977 (syn., *Lankesteria compacta* Geus, 1967) in *Neodendrocoelum sanctinaumi* (Turbellariasida) intestine and parenchyma.
3. *M. cyclopori* (Poisson, 1921) Levine, 1977 (syns., *Lankesteria cyclopori* Poisson,

1921; *Pseudolankesteria cyclopori* [Poisson, 1921] Ormières, 1965) in *Cycloporus maculatus* (Turbellariasida).

4. *M. discocelidis* (Mingazzini, 1893) in *Discocelis tigrina* (Turbellariasida) intestine.

5. *M. leptoplanae* (Bhatia and Setna, 1924) Levine, 1977 (syn., *Leidyana leptoplanae* Bhatia and Setna, 1924) in *Leptoplana* sp. (Turbellariasida) parenchyma.

6. *M. neodendrocoeli* (Georgévitch, 1951) de Puytorac and Grain, 1960 (syns., *Monocystis neodendrocoeli* Georgévitch, 1951; *Lankesteria claviformis* [Georgévitch] of Geus [1967]) in *Neodendricoelum* sp., *N. sanctinaumi, N. maculatum,* and *N. ochridense* (Turbellariasida) intestine, parenchyma, and gastrovascular system.

7. *M. plagiostomae* (Georgévitch, 1951) Levine, 1977 (syns., *Monocystis plagiostomae* Georgévitch, 1951; *Lankesteria plagiostomae* [Georgévitch, 1951] Geus, 1967) in *Plagiostomum lemani* (Turbellariasida) intestine and parenchyma.

8. *M. planariae* (Mingazzini, 1893) Levine, 1977 (syns., *Pleurozyga planariae* Mingazzini, 1893; *Lankesteria planariae* [Mingazzini, 1893] Labbé, 1899; *Monocystis planariae* [?] [Schultze] Lankester, 1863; *Pseudolankesteria planariae* [Mingazzini, 1893] Ormières, 1965) in *Planaria torva, P. fusca, Planaria* sp., *Dugesia polychroa, D. lugubris,* and *Sorocoelis* sp. (Turbellariasida) intestine and parenchyma.

9. *M. spelaea* R. Codreanu and Balcescu, 1971 in *Dendrocoelum tismanae* Turbellariasida) parenchyma.

10. *M. swarczewskyi* R. Codreanu and Balcescu, 1971 (syn., *Lankesteria* sp. Swarczewsky, 1910) in *Planaria* sp. and *Sorocoelis* sp. (Turbellariasida) intestine and parenchyma.

Genus *Ascogregarina* Ward, Levine, and Craig, 1982. Similar to *Lankesteria,* but in insects. TYPE SPECIES *A. culicis* (Ross, 1898) Ward, Levine, and Craig, 1982.

1. *A. armigerei* (Lien and Levine, 1980) Ward, Levine, and Craig, 1982 (syn., *Ascocystis armigerei* Lien and Levine, 1980) in *Armigeres subalbatus* (Dipterorida) midgut and Malpighian tubules.

2. *A. barretti* (Vavra, 1969) Ward, Levine, and Craig, 1982 (syns., *Lankesteria barretti* Vavra, 1969; *Ascocystis barretti* [Vavra, 1969] Levine, 1977) in *Aedes triseriatus* and experimentally *A. geniculati* (Dipterorida) midgut and Malpighian tubules.

3. *A. brachyceri* (Purrini, 1980) Levine, 1985 (syn., *Ascocystis brachyceri* Purrini, 1980) in *Megaselia subnitida* (Dipterorida) ovaries and eggs.

4. *A. chagasi* (Adler and Mayrink, 1961) Ward, Levine, and Craig, 1982 (syns., *Monocystis chagasi* Adler and Mayrink, 1961; *M. shagasi* Adler and Mayrink, 1961 *lapsus calami; Ascocystis chagasi* [Adler and Mayrink, 1961] Tuzet and Rioux, 1965) in *Lutzomyia* (syn., *Phlebotomus) longipalpis, L. townsendi,* and possibly *L. vexatrix occidentalis, L. flaviscotellata,* and other species of *Lutzomyia* (Dipterorida) ovary, accessory glands, hemocoele.

5. *A. clarki* (Sanders and Poinar, 1973) Ward, Levine, and Craig, 1982 (syns., *Lankesteria clarki* Sanders and Poinar, 1973; *Ascocystis clarki* [Sanders and Poinar, 1973] Levine, 1977) in *Aedes sierrensis* (Dipterorida) intestine and Malpighian tubules.

6. *A. culicis* [Ross, 1898] Ward, Levine and Craig, 1982 (syns., *Gregarina culicidis* Ross, 1895 *lapsus calami; G. culicis* Ross, 1898; *Lankesteria culicis* [Ross, 1898] Wenyon, 1911; *Ascocystis culicis* [Ross, 1898] Grassé, 1953) (TYPE SPECIES) in *Aedes (Stegomyia) aegypti* (Dipterorida) midgut and Malpighian tubules.

7. *A. galliardi* (Garnham, 1973) Ward, Levine, and Craig, 1982 (syns., *Lankesteria galliardi* Garnham, 1973; *Ascocystis galliardi* [Garnham, 1973] Garnham in Levine, 1977) in *Nycteribia dentata* (Dipterorida) location not given.

8. *A. geniculati* Munstermann and Levine, 1983 in *Aedes geniculatus* (Dipterorida) intestine and Malpighian tubules.

9. *A. lanyuensis* (Lien and Levine, 1980) Ward, Levine, and Craig, 1982 (syn., *Ascocystis lanyuensis* Lien and Levine, 1980) in *Aedes alcasidi* (Dipterorida) midgut and Malpighian tubules.
10. *A. legeri* (Blanchard, 1902) Ward, Levine, and Craig, 1982 (syns., *Monocystis legeri* Blanchard, 1902; *Ascocystis legeri* [Blanchard, 1902] Levine, 1977) in *Carabus auratus, Chrysocarabus punctatoauratus, C. splendens, Pterostichus niger,* and *P. vulgaris* (Coleopterorida) coelom.
11. *A. mackiei* (Shortt and Swaminath, 1927) Ward, Levine, and Craig, 1982 (syns., *Monocystis mackiei* Shortt and Swaminath, 1927; *Lankesteria mackiei* [Shortt and Swaminath, 1927] Bhatia, 1938; *L. phlebotomi mackiei* Missiroli, 1932; *Diplocystis* sp. [?] of Lisova [1962]; *Ascocystis mackiei* [Shortt and Swaminath, 1927] Tuzet and Rioux, 1965; *A. mackiei* [Shortt and Swaminath, 1927] Ormières, 1965) in *Phlebotomus papatasii* and *P. argentipes* (Dipterorida) intestine.
12. *A. polynesiensis* Levine, 1985 (syn., *Lankesteria culicis* of Pillai, Neill, and Sone [1976]) in *Aedes polynesiensis* (Dipterorida) intestine and Malpighian tubules.
13. *A. taiwanensis* (Lien and Levine, 1980) Ward, Levine, and Craig, 1982 (syn., *Ascocystis taiwanensis* Lien and Levine, 1980) in *Aedes albopictus* (Dipterorida) midgut and Malpighian tubules.
14. *A. tripteroidis* (Bhatia, 1938) Ward, Levine, and Craig, 1982 emend. (syns., *Lankesteria tripteroidesi* Bhatia, 1938; *Ascocystis tripteroidesi* [Bhatia, 1938] Ormières, 1965) in *Tripteroides dofleini* (Dipterorida) larval coelom, trachea, and anal gills.

Genus *Kofoidina* Henry, 1933. Gamonts in syzygy in chains of 2 to 14; in insects. TYPE SPECIES *K. ovata* Henry, 1933.

1. *K. ovata* Henry, 1933 (TYPE SPECIES) in *Zootermopsis angusticollis* and *Z. nevadensis* (Isopterorida) midgut.

Genus *Ophioidina* Mingazzini, 1891. Gamont body very elongated, cylindrical, of the same thickness throughout; anterior end more or less truncated; mucron absent; gametocysts and oocysts unknown, in echiurids. TYPE SPECIES *O. bonelliae* (Frenzel, 1885) Labbé, 1899.

1. *O. bonelliae* (Frenzel, 1885) Labbé, 1899 (syn., *Gregarina bonelliae* Frenzel, 1885) (TYPE SPECIES) in *Bonellia viridis* (Echiurida) intestine.

Genus *Paraophioidina* Levine, 1977. Gamont body elongated, cylindrical, of the same thickness throughout except at the anterior end, which is swollen in the mature forms (forming a mucron?) but not in the young ones; gametocysts and oocysts unknown; in crustacea. TYPE SPECIES *P. haeckeli* (Mingazzini, 1891) Levine, 1977.

1. *P. apsteini* (Théodoridès and Desportes, 1972) Levine, 1977 (syn., *Ganymedes apsteini* Théodoridès and Desportes, 1972) in *Calanus gracilis* and *Clausocalanus arcuicornis* (Copepodasina) intestine.
2. *P. copiliae* (Rose, 1933) Levine, 1977 (syn., *Monocystis copiliae* Rose, 1933) in *Copilia vitrea* (Copepodasina) alimentary tract.
3. *P. eucopiae* (Théodoridès and Desportes, 1975) Levine, 1984 (syn., *Ganymedes eucopiae* Théodoridès and Desportes, 1975) in *Eucopia hanseni* (Mysidae) alimentary canal.
4. *P. haeckeli* (Mingazzini, 1891) Levine, 1977 (syns., *Ophioidina haeckelii* Mingazzini, 1891; *Ganymedes haeckeli* [Mingazzini, 1893] Théodoridès and Desportes, 1972) (TYPE SPECIES) in *Sapphirina* spp. (Copepodasina) intestine.

5. *P. korotneffi* (Théodoridès and Desportes, 1975) Levine, 1984 (syn., *Ganymedes korotneffi* Théodoridès and Desportes, 1975) in *Sergestes robustus* (Decapodasida) presumably intestine.
6. *P. lacrima* (Vejdovsky, 1882) Levine, 1977 (syns., *Monocystis lacrima* Vejdovsky, 1882; *M. lacryma* Vejdovsky, 1882 of Labbé [1899]) in *Canthocamptus minutus* (Copepodasina) digestive tract.
7. *P. oaklandi* (Jones, 1968) Levine, 1977 (syn., *Ganymedes oaklandi* Jones, 1968) in *Gammarus fasciatus* (Amphipodasida) digestive tract.
8. *P. pisae* Léger and Duboscq, 1911) Levine, 1977 (syns., *Porospora pisae* Léger and Duboscq, 1911; *Thiriota pisae* [Léger and Duboscq, 1911] Desportes, Vivares, and Théodoridès, 1977) in *Pisa gibbsii, P. armata, P. nodipes, P. tetraodon,* and *Herbstia condyliata* (Decapodorida) intestine.
9. *P. portuni* (Frenzel, 1885) Levine, 1977 (syn., *Gregarina portuni* Frenzel, 1885) in *Portunus arcuatus* (Decapodasina) midgut and anterior hindgut.
10. *P. putanea* (Lachmann, 1859) Levine, 1977 (syns., *Zygocystis putanea* Lachmann, 1859; *Monocystis putanea* (?) [Lachmann, 1859] Lankester, 1863) in *Niphargus subterraneous* (syn., *Gammarus putaneus*) *(Amphipodasida)* intestine.
11. *P. vibiliae* (Théodoridès and Desportes, 1972) Levine, 1977 (syn., *Ganymedes vibiliae* Théodoridès and Desportes, 1972) in *Vibilia armata* and *V. propinqua* (Amphipodasida) intestine.

Genus *Lateroprotomeritus* Théodoridès and Desportes, 1975. Trophozoites said to be separate, with lateral mucron ("protomerite"); development solitary; syzygy laterofrontral, the trophozoites joining mucron to mucron. TYPE SPECIES *L. conicus* Théodoridès and Desportes, 1975.

1. *L. conicus* Théodoridès and Desportes, 1975 (TYPE SPECIES) in *Nematoscelis megalops* (Decapodasida) stomach.

Genus *Extremocystis* Setna, 1931. Gamont cylindrical, resembling a nematode; syzygy end to end, the tapering end of one individual fitting into a regular hemispherical depression in the other, but not forming a ball-and-socket joint as in *Ganymedes*; gametocysts not seen; oocysts spindle-shaped, with both ends finely pointed; no sporozoites seen; in sipunculids. TYPE SPECIES *E. dendrostomi* Setna, 1931.

1. *E. dendrostomi* Setna, 1931 (TYPE SPECIES) in *Dendrostoma signifer* (Sipunculida) coelom.

Family UROSPORIDAE *Léger, 1892*

Mucron more or less marked; syzygy frontal or lateral; anisogamy little accentuated; dehiscence of gametocysts by simple rupture; oocysts with dissimilar ends, with an anterior neck and a more or less marked posterior prolongation; oocyst with a well-differentiated wall ("epispore"); oocysts with eight sporozoites; in coelom and rarely intestine; in polychetes, oligochetes, nemerteans, mollusks, echinoderms, probably ascidians, and perhaps nematodes.

Genus *Urospora* Schneider, 1875. Mucron simple, small; gamont large (up to 1 mm), longitudinally striated; syzygy lateral, often between gamonts of unequal size; oocysts ovoid, with an appendage ("tail") at one end; some species covered with bristles; oocysts with six to eight (actually eight?) sporozoites. In annelids, sipunculids, nemertines, and echinoderms. TYPE SPECIES *U. nemertis* (von Kölliker, 1845) Schneider, 1875.

1. *U. chiridotae* (Dogiel, 1906) Goodrich, 1925 (syn., *Cystobia chiridotae* Dogel', 1906 in *Chiridota laevis* (syn., *C. pellucida) (Holothurasida)* blood vessels.
2. *U. echinocardii* Pixell-Goodrich, 1915 in *Echinocardium* spp. and *Spatangus* spp. (Echinidasida) coelom.
3. *U. grassei* (Changeux, 1961) Levine, 1977 (syn., *Cystobia grassei* Changeux, 1961) in *Holothuria tubulosa* and *H. stellati* (Holothurasida) hemal sinus.
4. *U. hardyi* Goodrich, 1950 in *Sipunculus nudus* (Sipunculida) coelom.
5. *U. holothuriae* (Schneider, 1858) Trégouboff, 1918 (syns., *Gregarina holothuriae* Schneider, 1858; *Monocystis holothuriae* [Schneider, 1858] Lankester, 1863; *Syncystis holothuriae* [Schneider, 1858] Cuénot, 1891; *Cystobia holothuriae* [Schneider, 1858] Mingazzini, 1891; *Lithocystis chiajii* Giard, 1886) in *Holothuria tubulosa* and *H. stellati* (Holothurasida) blood vessels, coelom, and intestine.
6. *U. intestinalis* Bogolepova, 1953 in *Cucumaria japonica* (Holothurasida) intestine.
7. *U. lagidis* de Saint Joseph, 1898 in *Lagis* (syn., *Pectinaria) koreni* (Polychaetasida) coelom.
8. *U. legeri* Goodrich, 1950 in *Sipunculus nudus* (Sipunculida) coelom.
9. *U. longicauda* Mingazzini, 1891 in *Cirratulus filigerus* (Polychaetasida) location not given.
10. *U. longissima* (Caullery and Mesnil, 1898) Schrével, 1965 (syn., *Gonospora longissima* Caullery and Mesnil, 1898) in *Dodecaria caulleryi* (syn., *D. concharum* form B) (Polychaetasida) coelom.
11. *U. muelleri* (Cuénot, 1891) Cuénot, 1892 (syn., *Syncystis muelleri* Cuénot, 1891) in *Synapta digitata* (Holothurasida) coelom.
12. *U. nemertis* von Kölliker, 1845) Schneider, 1875 (syns., *Gregarina nemertis* von Kölliker, 1845; *Monocystis nemertis* [von Kölliker, 1845] Lankester, 1863) (TYPE SPECIES) in *Nemertes delineatus* (syn., *Polia delineata)* and possibly *Valenciennia* sp. (Nemertini) intestine.
13. *U. neopolitana* Pixell-Goodrich, 1915 in *Echinocardium cordatum* (Echinidasida) coelom.
14. *U. ovalis* Dogiel, 1910 in *Travisia forbesi* (Polychaetasida) coelom.
15. *U. pulmonalis* Bogolepova, 1953 in *Cucumaria japonica* (Holothurasida) gills.
16. *U. rhyacodrili* Gabriel, 1929 in *Rhyacodrilus coccineus, Tubifex albicola,* and *T. barbatus* (Oligochaetasida) seminal vesicles and coelom.
17. *U. schneideri* (Mingazzini, 1891) Levine, 1977 (syn., *Cystobia schneideri* Mingazzini, 1891) in *Holothuria polii, H. impatiens, H. tubulosa,* and *H. stellati* (Holothurasida) blood vessels and coelom.
18. *U. sipunculi (von Kölliker, 1845) Léger, 1892 (syns., Gregarina sipunculi* von Kölliker, 1845; *Monocystis sipunculi* [von Kölliker, 1845] Lankester, 1863; *Pachysoma sipunculi* [von Kölliker, 1845] Mingazzini, 1891; *Zygocystis sipunculi* von Stein, 1867) in *Sipunculus nudus* (Sipunculida) coelom.
19. *U. synaptae* (Cuénot, 1891) Léger, 1892 (syns., *Syncystis synaptae* Cuénot, 1891; *Esarabdina synaptae* [Cuénot, 1891] Mingazzini, 1893; *Lithocystis muelleri* Giard, 1886) in *Synapta digitata, Leptosynapta* (syn., *Synapta) inhaerens,* and *L. galliennei* (Holothurasida) coelom.
20. *U. travisiae* Dogiel, 1910 in *Travisia forbesi* (Polychaetasida) coelom.
21. *U. tubificis* Hesse, 1909 (syns., *Gregarina saenuridis* von Kölliker, 1848 of Lankester [1872] and other authors; *U. saenuridis* [von Kölliker, 1848] Nasse, 1922; *[non] G. saenuridis* von Kölliker, 1848) in *Tubifex tubifex* (Oligochaetasida) seminal vesicles and coelom.

Genus *Gonospora* Schneider, 1875. Gamonts polymorphic, ovoid, piriform, or vermiform;

gametocysts spherical; oocysts ovoid, with dissimilar ends, rounded at base and with a simple or denticulate point or funnel at the other end. In coelom and rarely intestine of polychetes; testes of prosobranch mollusks. coelom, respiratory tree, and blood vessels of holothurians. TYPE SPECIES *G. varia* Léger, 1892.

1. *G. arenicolae* (Cunningham, 1907) Trégouboff, 1918 (syns., *Kalpidorhynchus arenicolae* Cunningham, 1907; *Cystobia arenicolae* [Cunningham, 1907] Dogel', 1909) in *Arenicola ecaudata* (Polychaetasida) coelom.
2. *G. belenoides* (Mingazzini, 1891) Corbel, Desportes, and Théodoridès, 1979 (syns., *Lobianchella belenoides* Mingazzini, 1891; *Lobiancoella belenoides* Mingazzini, 1891 emend. Labbé, 1899) in *Alciope* sp. and *Naiaides cantrainii* (syn., *Alciopa cantrainii*) (Polychaetasida) coelom.
3. *G. duboscqui* Tuzet, 1931 in *Bithynia tentaculata* (Molluska) testes.
4. *G. glycerae* Pixell-Goodrich, 1916 in *Glycera siphonostoma* (Polychaetasida) coelom.
5. *G. gonadipertha* (Djakonov, 1923) Levine, 1977 (syn., *Diplodina gonadipertha* Djakonov, 1923) in *Cucumaria frondosa* (Holothurasida) gonads.
6. *G. goodrichae* Levine, 1977 (syn., *G. minchinii* Goodrich and Goodrich, 1920; [non] *G. minchinii* [Woodstock, 1904] Trégouboff, 1918) in *Arenicola ecaudata* (Polychaetasida) coelom and egg.
7. *G. intestinalis* (Ssokoloff, 1913) Trégouboff, 1918 (syn., *Cystobia intestinalis* Ssokoloff, 1913) in *Glycera* (syn., *Rhynchobolus*) *siphonostoma* (Polychaetasida) intestine.
8. *G. irregularis* (Minchin, 1893) Trégouboff, 1918 (syns., *Gregarina irregularis* Minchin, 1893; *Cystobia irregularis* [Minchin, 1893] Labbé, 1899; *Diplodina irregularis* [Minchin, 1893] Woodcock, 1904) in *Holothuria nigra (impatiens[?])* and *H. forskali* (Holothurasida) blood vessels.
9. *G. isogokaii* Tugawa [H. Hoshide], 1944 in *Nereis microdonta* (Polychaetasida) coelom.
10. *G. mercieri* Cuénot, 1912 in *Synapta digitata* (Holothurasida) coelom.
11. *G. minchini* (Woodcock, 1904) Trégouboff, 1918 (syns., *Cystobia minchinii* Woodcock, 1904; *Diplodina minchinii* [Woodcock, 1904] Woodcock, 1906; *Lithocystis minchini* [Woodcock, 1904] Reichenow, 1932; [non] *G. minchinii* Goodrich and Goodrich, 1920) in *Cucumaria saxicola* (syn., *C. pentactes)* and *C. normani* (syn., *C. planci* (Holothurasida) blood vessels.
12. *G. ormieri* Porchet, 1978 in *Notomastus latericeus* (Polychaetasida) coelom.
13. *G. ozakii* Tugawa [H. Hoshide], 1944 in *Nereis japonica* (Polychaetasida) coelom.
14. *G. sparsa* Léger, 1892 in *Phyllodoce* sp. and *Glycera* sp. (Polychaetasida) coelom.
15. *G. stichopi* (Lützen, 1967) Levine, 1977 (syn., *Cystobia stichopi* Lützen, 1967) in *Stichopus tremulus* (Holothurasida) along muscle bands.
16. *G. testiculi* (Trégouboff, 1916) Trégouboff, 1918 (syn., *Cystobia testiculi* Trégouboff, 1916) in *Cerithium vulgatum* (Molluska) testis.
17. *G. varia* Léger, 1892 (syn., *Gonospora terebellae* [von Koelliker] Schneider, 1875) (TYPE SPECIES) in *Cirriformia tentaculata* (syns., *Audouinia tentaculata, A. lamarkii)* and *Terebella* sp. (Polychaetasida) coelom.
18. *G. vertiginosa* Tugawa [H. Hoshide], 1944 in *Cirratulus vertiginosa* (Polychaetasida) coelom.

Genus *Lithocystis* Giard, 1876. Gamonts large, ovoid or cylindrical, attached for a long time to the host tissue; syzygy in the form of an X; gametocysts with a sphere of calcium oxalate derived from the endoplasm of the gamonts; oocysts truncate at one end, with a more or less long process at the other; in echinoderms and sipunculids. TYPE SPECIES *L. schneideri* Giard, 1876.

1. *L. brachycercus* Goodrich, 1925 in *Chiridota laevis* (Holothurasida) intestinal wall, coelom.
2. *L. cucumariae* Goodrich, 1930 in *Cucumaria sacicola* (Holothurasida) respiratory tree.
3. *L. foliacea* Pixell-Goodrich, 1915 in *Echinocardium cordatum* (Echinoidasida) coelom.
4. *L. lankesteri* Goodrich, 1950 in *Sipunculus nudus* (Sipunculida) coelom.
5. *L. latifronsi* Brownwell and McCauley, 1971 in *Brisaster latifrons* (Echinoidasida) coelom.
6. *L. microspora* Pixell-Goodrich, 1915 in *Spatangus purpureus* (Echinoidasida) coelom.
7. *L. oregonensis* Brownwell and McCauley, 1971 in *Brisaster latifrons* (Echinoidasida) coelom.
8. *L. schneideri* Giard, 1876 (TYPE SPECIES) in *Echinocardium cordatum* (Echinoidasida) coelom.

Genus *Ceratospora* Léger, 1892. Gamonts elongate conical; syzygy head to head, without encystment; oocysts ovoid, with dissimilar ends, with a small collar at one end and two divergent elongate filaments at the other; in polychetes. TYPE SPECIES *C. mirabilis* Léger, 1892.

1. *C. mirabilis* Léger, 1892 (TYPE SPECIES) in *Glycera* sp. (Polychaetasida) coelom.

Genus *Pterospora* Labbé and Racovitza, 1897. Gamonts in syzygy or solitary; posterior end drawn out into four bifurcated processes; syzygy head to head; gametocysts spherical or ovoid; oocysts with wall drawn out into three lateral processes; mostly in coelom of polychetes. TYPE SPECIES *P. maldaneorum* Labbé and Racovitza, 1897.

1. *P. clymenellae* (Porter, 1897) Dogel', 1909 (syns., *Monocystis clymenellae* Porter, 1897; *P. clymenellae* Théodoridès and Laird, 1970) in *Clymenella torquata* (Polychaetasida) coelom.
2. *P. maldaneorum* Labbé and Racovitza, 1897 (TYPE SPECIES) in *Leiochone* (syn., *Leiocephalus*) *leiopygos, L. clypeata,* and probably *Clymene lumbriocoides (lumbricalis)* [?] (Polychaetasida) coelom.
3. *P. petaloprocti* Ormières, 1977 in *Petaloproctus terricola* (Polychaetasida) location not given.
4. *P. ramificata* Dogel', 1910 in *Nicomache lumbricalis* (Polychaetasida) coelom.

Genus *Paragonospora* Lang, 1954. Mucron simple; gamonts with sparse, poorly developed myonemes, with ellipsoidal paramylon (amylopectin?) bodies; gamont locomotion by gliding; gamont nucleus with one or more endosomes; syzygy head to head; oocysts with dissimilar ends, having a funnel-shaped opening at one end and six small tubular formations at the other; oocysts with eight sporozoites and a very small residuum; in polychetes. TYPE SPECIES *P. typica* Lang, 1954.

1. *P. typica* Lang, 1954 (TYPE SPECIES) in *Terebellides stroemi* (Polychaetasida) coelom.

Family AIKINETOCYSTIDAE *Bhatia, 1930*
Gamont sacciform, continued at one end by 2 branches which divide dichotomously, forming 8 to 16 secondary branches bearing groups of suckers; oocysts biconical (fusiform).

Genus *Aikinetocystis* Gates, 1926. Syzygy in pairs, tail to tail, by the nonramified ends; in coelom of oligochetes. TYPE SPECIES *A. singularis* Gates, 1926.

1.   *A. singularis* Gates, 1926 (TYPE SPECIES) in *Eutyphaeus foveatus, E. peguanus, E. rarus,* and *E. spinulosus* (Oligochaetasida) coelom.

Genus *Nellocystis* Gates, 1933. Gamonts solitary or in syzygy tail to tail in pairs or threes; anterior end ramified; binucleate (?); in oligochetes. TYPE SPECIES *N. birmanica* Gates, 1933.

1.   *N. birmanica* Gates, 1933 (TYPE SPECIES) in *Pheretima compta* (Oligochaetasida) coelom.

Family MONOCYSTIDAE *Bütschli, 1882*

Gamonts spherical to cylindrical, with anterior end little differentiated if at all; oocysts biconical or navicular; mostly coelomic; the great majority parasites of oligochetes.

Subfamily MONOCYSTINAE *Bhatia, 1930*

Gamonts nematoid, cylindroid, or ovoid, without mucron or with an inconspicuous one; syzygy late; ends of oocysts thickened; ordinary habitat seminal vesicles of oligochetes, although some live in the coelom.

Genus *Monocystis* von Stein, 1848. Mucron not marked; gamonts ovoid, short or elongate, solitary; oocysts bioconical, symmetrical. TYPE SPECIES *M. agilis* von Stein, 1848.

1.   *M. abegbei* Segun, 1978 in *Libyodrilus violaceus* (Oligochaetasida) seminal vesicles and coelom.
2.   *M. acuta* Berlin, 1924 in *Lumbricus rubellus* and *L. castaneus* (Oligochaetasida) seminal vesicles.
3.   *M. agilis* von Stein, 1848 (syns., *Gregarina agilis* [von Stein, 1848] Diesing, 1849; *M. tenax* [Dujardin, 1835] Labbé, 1899; [*non*] *Proteus tenax* Dujardin, 1835) (TYPE SPECIES) *Lumbricus terrestris, L. rubellus, L. castaneus, Allolobophora longa (?), Pheretima hypeiensis,* and *Eisenia foetida* (Oligochaetasida) seminal vesicles.
4.   *M. arcuata* Boldt, 1910 in *Lumbricus castaneus, L. rubellus,* and *Eisenia foetida* (Oligochaetasida) seminal vesicles.
5.   *M. banyulensis* Tuzet and Loubatières, 1946 in *Octolasium complanatum* and *Dichrogaster baeri* (Oligochaetasida) seminal vesicles.
6.   *M. beddardi* Ghosh, 1923 in *Eutyphaeus nicholsoni* (Oligochaetasida) seminal vesicles.
7.   *M. bengalensis* Ghosh, 1923 in *Pheretima posthuma* (Oligochaetasida) seminal vesicles.
8.   *M. berlini* Levine, 1977 (syn., *M. turbo* Hesse, 1909 var. *suecica* Berlin, 1923) in *Lumbricus rubellus* (Oligochaetasida) seminal vesicles.
9.   *M. biacuminata* Boisson, 1957 in *Glossoscolex corethrurus* (Oligochaetasida) seminal vesicles.
10. *M. boissoni* Levine, 1977 (syn., *M. minima* Boisson, 1957; [*non*] *M. minima* Konsuloff, 1916) in *Pheretima posthuma* (Oligochaetasida) seminal vesicles.
11. *M. bretscheri* Hesse, 1909 in *Fridericia polycheta* (Oligochaetasida) coelom.
12. *M. buccalis* Tuzet and Loubatières, 1948 (syn., *M. buccalis* Loubatières, 1955) in *Allolobophora gigas* (Oligochaetasida) seminal vesicles.
13. *M. cambrensis* Rees, 1961 in *Eisenia foetida* (Oligochaetasida) coelom.
14. *M. capillata* Tuzet and Vogeli, 1956 in *Millsonia anomala* (Oligochaetasida) seminal vesicles.
15. *M. carlgrenii* Berlin, 1924 in *Lumbricus terrestris* and *L. rubellus* (Oligochaetasida) seminal vesicles and coelom.
16. *M. catenata* Mulsow, 1911 in *Lumbricus terrestris* and *L. rubellus* (Oligochaetasida) coelom.

17. *M. caudata* Berlin, 1924 in *Lumbricus rubellus, L. castaneus,* and *Allolobophora longa* (Oligochaetasida) seminal vesicles.
18. *M. cazali* Tuzet and Loubatières, 1946 in *Octolasium complanatum* (Oligochaetasida) seminal vesicles.
19. *M. ciliata* Drzhevetskii, 1907 in *Allolobophora longa* (Oligochaetasida) seminal vesicles and coelom.
20. *M. crenulata* Hesse, 1909 in *Allolobophora longa, A. caliginosa,* and *A. caliginosa* var. *trapezoides* (Oligochaetasida) seminal vesicles.
21. *M. criodrilii* Sciacchitano, 1931 in *Criodrilus lacuum* (Oligochaetasida) seminal vesicles.
22. *M. densa* Berlin, 1924 in *Lumbricus terrestris, L. rubellus, L. castaneus, Eisenia foetida, Allolobophora longa, A. caliginosa,* and *A. chlorotica* (Oligochaetasida) seminal vesicles.
23. *M. dichogasteri* Tuzet and Zuber-Vogeli, 1955 in *Dichogaster inermis* (Oligochaetasida) seminal vesicles.
24. *M. duboscqi* Hesse, 1909 in *Lumbricus variegatus* (Oligochaetasida) seminal vesicles.
25. *M. eiseniae* Levine, 1977 (syn., *M. cuneiformis* Loubatières, 1955; [*non*] *M. cuneiformis* Ruschhaupt, 1885) in *Eisenia foetida* (Oligochaetasida) seminal vesicles.
26. *M. eudrilii* Tuzet and Vogeli, 1956 in *Eudrilus eugeniae* (Oligochaetasida) seminal vesicles.
27. *M. hederacea* Loubatières, 1955 in *Allolobophora rosea* (Oligochaetasida) seminal vesicles.
28. *M. hessei* Berlin, 1924 in *Lumbricus terrestris* and *L. rubellus* (Oligochaetasida) seminal vesicles.
29. *M. hirsuta* Hesse, 1909 in *Lumbricus castaneus* (Oligochaetasida) seminal vesicles.
30. *M. krin* Righi, 1974 in *Guaranidrilus sawayai* (Oligochaetasida) coelom and seminal vesicles.
31. *M. lanceata* Rees, 1961 in *Allolobophora caliginosa* (Oligochaetasida) coelom.
32. *M. lememei* Hesse, 1909 in *Allolobophora caliginosa* and *Octolasium complanatum* (Oligochaetasida) seminal vesicles.
33. *M. libyodrili* Segun, 1978 in *Libyodrilus violaceus* (Oligochaetasida) coelom.
34. *M. lloydi* Ghosh, 1923 emend. Loubatières, 1955 (syn., *M. lloidi* Ghosh, 1923) in *Pheretima posthuma* (Oligochaetasida) seminal vesicles.
35. *M. lobosa* Tuzet and Vogeli, 1956 in *Millsonia anomala* (Oligochaetasida) seminal vesicles.
36. *M. longispora* Boisson, 1957 in *Perionyx excavatus* (Oligochaetasida) seminal vesicles.
37. *M. lopadiformis* Loubatières, 1955 in *Eisenia foetida* (Oligochaetasida) seminal vesicles.
38. *M. loubatieresi* Levine, 1977 (syn., *M. cristata* Loubatières, 1955; [*non*] *M. cristata* Schmidt, 1854) in *Eisenia foetida* (Oligochaetasida) seminal vesicles.
39. *M. lumbrici* (Henle, 1845) Cuénot, 1901 (syns., *Gregarina lumbrici* Henle, 1845; *G. nematoides* [Schmidt, 1854] Diesing, 1859; *Proteus tenax* Dujardin, 1835; *Sablier proteiforme* Suriray, 1836; *M. cristata* Schmidt, 1854; *M. tenax* var. *cristata* Schmidt, 1854 of Labbé [1899]; *M. nematoides* Schmidt, 1854; *M. rostrata* Mulsow, 1911; *Zygocystis wenrichi* Troisi, 1933) in *Lumbricus terrestris, L. rubellus, L. castaneus,* and *Eisenia foetida* (Oligochaetasida) seminal vesicles.
40. *M. lumbricilli* Giere, 1971 in *Lumbricillus lineatus* (Oligochaetasida) seminal vesicles and coelom.
41. *M. lumbricoides* (Hesse, 1909) Meier, 1956 (syn., *Nematocystis lumbricoides* Hesse, 1909) in *Allolobophora c. caliginosa, A. longa, A. r. rosea, Dendrobaena tenuis, Eisenia foetida,* and *Pheretima heterochaeta* (Oligochaetasida) seminal vesicles and coelom.
42. *M. macrospora* Hesse, 1909 in *Pheretima hawayana* and *P. rodericensis* (Oligochaetasida) coelom.

43. *M. mammaliae* Segun, 1968 in *Dendrobaena mammalis* (Oligochaetasida) seminal vesicles.
44. *M. minima* Konsuloff, 1916 in *Euchlanis dilatata* and *Salpina mucronata* (Rotifera) intestine and ovary.
45. *M. minor* Boisson, 1957 in *Pheretima peguana* (Oligochaetasida) seminal vesicles.
46. *M. mollis* Bereczky, 1967 in *Dendrobaena platyura* var. *montana* (Oligochaetasida) seminal vesicles.
47. *M. mrazeki* Hahn, 1928 in *Rhynchelmis limosella* and *R. komareki* (Oligochaetasida) seminal vesicles and coelom.
48. *M. naidis* von Voss, 1921 in *Nais elinguis* or *N. obtusa* (Oligochaetasida) coelom.
49. *M. nidata* Boisson, 1957 in *Pontodrilus ephippiger* (Oligochaetasida) seminal vesicles.
50. *M. oblonga* Berlin, 1924 in *Lumbricus terrestris, L. rubellus, L. castaneus,* and *Allolobophora longa* (Oligochaetasida) seminal vesicles.
51. *M. octavi* Righi, 1974 in *Pristina minuta* (Oligochaetasida) intestine.
52. *M. octolasii* Tuzet and Loubatières, 1946 emend. (syn., *M. octolasiae* Tuzet and Loubatières, 1946) in *Octolasium complanatum* (Oligochaetasida) seminal vesicles.
53. *M. odontotermitis* Kalavati, 1979 emend. (syn., *M. odontotermi* Kalavati, 1979) in *Odontotermes obesus* (Isopterorida) hemocoele.
54. *M. omodeoi* Tuzet and Vogeli, 1956 in *Dichogaster baeri* (Oligochaetasida) seminal vesicles.
55. *M. perforans* Pinto, 1918 in *Glossoscolex wiengreeni* (Oligochaetasida) testis.
56. *M. perichaetae* (Beddard, 1889) Labbé, 1899 (syn., *Gregarina perichaetae* Beddard, 1888) in *Megascolex* (syn., *Perichaeta*) *novaezealandiae, Diporochaeta intermedia, M. mauritii* (?), and *M. armatus* (?) (Oligochaetasida) seminal vesicles and coelom.
57. *M. pheretimae* Bhatia and Chatterjee, 1925 emend. (syn., *M. pheretimi* Bhatia and Chatterjee, 1925) in *Pheretima posthuma* (Oligochaetasida) seminal vesicles.
58. *M. piriformis* Boldt, 1910 in *Octolasium complanatum, O. transpadanum, Fridericia galba,* and *F. hegemon* (Oligochaetasida) seminal vesicles and coelom.
59. *M. polymorpha* Berlin, 1924 in *Lumbricus rubellus* (Oligochaetasida) seminal vesicles and coelom.
60. *M. pontodrili* Subbarao, Kalavati, and Narasimhamurti, 1979 in *Pontodrilus bermudensis* (Oligochaetasida) coelom.
61. *M. radiata* Boisson, 1957 in *Glossoscolex corethrurus* (Oligochaetasida) seminal vesicles.
62. *M. rhabdota* Giere, 1971 in *Lumbricillus lineatus* (Oligochaetasida) coelom.
63. *M. saigonensis* Boisson, 1957 emend. Levine, 1977 (syn., *M. minor saigonensis* Boisson, 1957) in *Pheretima saigonensis* (Oligochaetasida) location not given.
64. *M. securiformis* Berlin, 1924 in *Allolobophora caliginosa* (Oligochaetasida) seminal vesicles.
65. *M. setosa* Tuzet and Loubatières, 1946 in *Allolobophora gigas* (Oligochaetasida) seminal vesicles.
66. *M. striata* Hesse, 1909 in *Lumbricus terrestris, L. rubellus,* and *L. castaneus* (Oligochaetasida) seminal vesicles.
67. *M. suecica* Berlin, 1923 in *Eisenia foetida, Lumbricus terrestris,* and *L. rubellus* (Oligochaetasida) seminal vesicles.
68. *M. thamnodrili* Cognetti de Martiis, 1911 in *Thamnodrilus incertus* (Oligochaetasida) coelom.
69. *M. tubiformis* Berlin, 1923 in *Lumbricus rubellus* and *L. castaneus* (Oligochaetasida) seminal vesicles and coelom.
70. *M. tupi* Righi, 1974 in *Tupidrilus lacteus* and *Guaranidrilus oiepe* (Oligochaetasida) coelom.

71. *M. turbo* Hesse, 1909 in *Octolasium lacteum* and *Eisenia foetida* (Oligochaetasida) seminal vesicles.
72. *M. tuzetae* Levine, 1977 (syn., *M. legeri* Tuzet and Loubatières, 1946; [*non*] *M. legeri* Blanchard, 1902) in *Octolasium complanatum* (Oligochaetasida) seminal vesicles.
73. *M. ventrosa* Berlin, 1924 (syn., *M. agilis* Forma B Hesse, 1909) in *Lumbricus rubellus*, *L. castaneus*, and *Eisenia foetida* (Oligochaetasida) seminal vesicles.
74. *M. wallengrenii* Berlin, 1924 (syn., *M. hispida* Loubatières, 1955) in *Allolobophora c. caliginosa, A. longa*, and *Lumbricus rubellus* (Oligochaetasida) seminal vesicles and coelom.

Genus *Nematocystis* Hesse, 1909. Gamonts large, cylindroid, nematoid, often with mucron at anterior end, solitary; oocysts biconical. TYPE SPECIES *N. magna* (Schmidt, 1854) Hesse, 1909.

1. *N. almae* Cognetii de Martiis, 1921 in *Alma emini* var. *aloysii-sabaudii* (Oligochaetasida) coelom.
2. *N. anguillula* Hesse, 1909 in *Pheretima rodericensis* and *P. hawayana* (Oligochaetasida) seminal vesicles.
3. *N. bunmii* Segun, 1978 in *Heliodrilus lagosensis* (Oligochaetasida) seminal vesicles.
4. *N. caerenis* Mohammed and Ramadan, 1972 in *Alma nilotica* (Oligochaetasida) seminal vesicles.
5. *N. cecconii* Cognetii de Martiis, 1918 in *Pheretima (Parapheretima) wendessiana* (Oligochaetasida) coelom.
6. *N. claviformis* Loubatières, 1955 in *Allolobophora chlorotica* (Oligochaetasida) seminal vesicles.
7. *N. clipeiformis* Loubatières, 1955 in *Octolasium complanatum* (Oligochaetasida) seminal vesicles.
8. *N. criodrilii* Sciacchitano, 1931 in *Criodrilus lacuum* (Oligochaetasida) coelom.
9. *N. cylindroides* (Georgèvitch, 1951) Levine, 1977 (syn., *Monocystis cylindroides* Georgévitch, 1951) in *Criodrilus ochridensis* (Oligochaetasida) seminal vesicles.
10. *N. dendrobaenae* Segun, 1968 in *Dendrobaena rubida* f. *subrubicunda* and *D. mammalis* (Oligochaetasida) coelom and seminal vesicles.
11. *N. elmassiani* (Hesse, 1909) Bhatia, 1929 (syns., *Monocystis elmassiani* Hesse, 1909; *M. proteiformis* Loubatières, 1955) in *Dendrobaena subrubicunda, D. tenuis, Lumbricus terrestris, L. rubellus, Allolobophora r. rosea, A. longa*, and *Eisenia foetida* (Oligochaetasida) seminal vesicles.
12. *N. glossoscoli* Boisson, 1957 in *Glossoscolex corethrurus* (Oligochaetasida) seminal vesicles.
13. *N. goliatti* (Georgévitch, 1951) Levine, 1977 (syn., *Monocystis goliatti* (Georgévitch, 1951) in *Criodrilus ochridensis* (Oligochaetasida) seminal vesicles.
14. *N. gracilis* Berlin, 1924 emend. Segun, 1968 (syn., *N. anguillula* var. *gracilis* Berlin, 1924) in *Lumbricus terrestris, L. rubellus*, and *L. castaneus* (Oligochaetasida) seminal vesicles.
15. *N. hessei* Bhatia and Chatterjee, 1925 in *Pheretima heterochaeta* (Oligochaetasida) seminal vesicles.
16. *N. magna* (Schmidt, 1854) Hesse, 1909 (syns., *Monocystis magna* Schmidt, 1854; *Gregarina magna* Diesing, 1859; *N. caudata* Loubatières, 1955) (TYPE SPECIES) in *Lumbricus terrestris* (Oligochaetasida) testis.
17. *N. meierae* Levine, 1977 (syns., *N. lumbricoides* var. *pilosa* Meier of Segun [1968, 1971]; *Monocystis lumbricoides* [Hesse, 1909] Meier, 1956; [*non*] *N. lumbricoides* Hesse, 1909; [*non*] *N. pilosa* Tuzet and Loubatières, 1946) in *Allolobophora longa* (Oligochaetasida) seminal vesicles and coelom.

18. *N. navicula* Loubatières, 1955 in *Octolasium complanatum* (Oligochaetasida) seminal vesicles.
19. *N. pilosa* Tuzet and Loubatières, 1955 in *Octolasium complanatum, O. transpadanum,* and *Dichogaster baeri* (Oligochaetasida) seminal vesicles.
20. *N. pistilliformis* Loubatières, 1955 in *Allolobophora gigas* (Oligochaetasida) location not given.
21. *N. plurikaryosomata* Bhatia and Chatterjee, 1924 in *Eisenia foetida* (Oligochaetasida) seminal vesicles.
22. *N. sinuosa* Loubatières, 1955 in *Allolobophora rosea* (Oligochaetasida) seminal vesicles.
23. *N. stephensoni* Bhatia and Setna, 1926 in *Eutyphaeus incommodus* (Oligochaetasida) seminal vesicles.
24. *N. testiculi* Tuzet and Loubatières, 1946 in *Octolasium complanatum* (Oligochaetasida) seminal vesicles.
25. *N. tuzetae* Loubatières, 1947 emend. Levine, 1977 (syn., *N. tuzeti* Loubatières, 1947) in *Octolasium complanatum* and *O. transpadanum* (Oligochaetasida) testes and coelom.
26. *N. variabilis* Sciacchitano, 1931 in *Criodrilus lacuum* (Oligochaetasida) coelom.
27. *N. vermicularis* Hesse, 1909 in *Allolobophora longa, Lumbricus terrestris, L. rubellus,* and *Pheretima barbadensis* (Oligochaetasida) seminal vesicles.

Genus *Rhabdocystis* Boldt, 1910. Gamonts long and narrow, with club-shaped anterior end, solitary; sarcocyte partly foamy; gametocyst biscuit-shaped; oocysts biconical, without appendages. TYPE SPECIES *R. claviformis* Boldt, 1910.

1. *R. claviformis* Boldt, 1910 (TYPE SPECIES) in *Octolasium complanatum* (Oligochaetasida) seminal vesicles.
2. *R. gyriniformis* Loubatières, 1955 in *Allolobophora caliginosa* (Oligochaetasida) seminal vesicles.
3. *R. pilosa* Meier, 1956 in *Lumbricus rubellus* (Oligochaetasida) seminal vesicles.

Genus *Apolocystis* Cognetii de Martiis, 1923. Gamonts spherical, solitary; oocysts biconical. TYPE SPECIES *A. lumbriciolidi* (Schmidt, 1854) Cognetti de Martiis, 1923 emend. Levine, 1977.

1. *A. beaufortii* (Cognetti de Martiis, 1918) Cognetti de Martiis, 1923 (syns., *Monocystis beaufortii* Cognetti de Martiis, 1918; *M. ambigua* Cognetti de Martiis, 1918; *M. tricingulata* Cognetti de Martiis, 1918; *A. ambigua* [Cognetti de Martiis, 1918] Cognetti de Martiis, 1923) in *Pheretima (Parapheretima) beaufortii* (Oligochaetasida) seminal vesicles.
2. *A. catenata* (Mulsow, 1911) Cognetti de Martiis, 1923 (syn., *Monocystic catenata* Mulsow, 1911) in *Lumbricus terrestris* (Oligochaetasida) coelom.
3. *A. chattoni* Tuzet and Loubatières, 1946 in *Allolobophora gigas* (Oligochaetasida) seminal vesicles.
4. *A. dichogasteri* Tuzet and Vogeli, 1956 in *Dichogaster baeri* (Oligochaetasida) seminal vesicles.
5. *A. dudichi* Bereczky, 1967 in *Dendrobaena platyura* var. *depressa* (Oligochaetasida) seminal vesicles.
6. *A. gigantea* Troisi, 1933 in *Eisenia foetida* and *Lumbricus rubellus* (Oligochaetasida) seminal vesicles.
7. *A. gigas* Tuzet and Loubatières, 1946 in *Octolasium complanatum* (Oligochaetasida) seminal vesicles.
8. *A. granulata* Tuzet and Loubatières, 1946 in *Allolobophora gigas* and *A. chlorotica* (Oligochaetasida) seminal vesicles.

9. *A. herculea* (Bosanquet, 1894) Meier, 1956 (syn., *Monocystis herculea* Bosanquet, 1894) in *Lumbricus terrestris, L. rubellus, L. castaneus, Octolasium lacteum, O. cyaneum, Allolobophora caliginosa, A. chlorotica, Dendrobaena rubida, D. tenuis,* and *Libyodrilus violaceus* (Oligochaetasida) seminal vesicles and coelom.

10. *A. iridodrili* Segun, 1978 in *Iridodrilus preussi* and *I. roseus* (Oligochaetasida) seminal vesicles.

11. *A. lavernensis* Rees, 1963 in *Allolobophora longa* (Oligochaetasida) seminal vesicles.

12. *A. libyodrili* Segun, 1978 in *Libyodrilus violaceus* (Oligochaetasida) seminal vesicles.

13. *A. lumbriciolidi* (Schmidt, 1854) Cognetti de Martiis, 1923 emend. Levine, 1977 (syns., *Monocystis lumbrici olidi* Schmidt, 1854; *A. elongata* Phillips and Mackinnon, 1946) (TYPE SPECIES) in *Eisenia foetida, Dendrobaena rubida, D. subrubicunda,* and *D. tenuis* (Oligochaetasida) seminal vesicles.

14. *A. mattheii* (Bathia and Setna, 1926) Loubatières, 1955 (syn., *Monocystis matthei* Bhatia and Setna, 1926) in *Megascolex trilobatus* (Oligochaetasida) seminal vesicles.

15. *A. megagranulata* Segun, 1971 in *Dendrobaena rubida* f. *subrubicunda* (Oligochaetasida) seminal vesicles.

16. *A. michaelseni* (Hesse, 1909) Cognetti de Martiis, 1923 (syn., *Monocystis michaelseni* Hesse, 1909) in *Pheretima hawayana* (Oligochaetasida) coelom.

17. *A. minuta* Troisi, 1933 in *Lumbricus terrestris, L. castaneus,* and *L. rubellus* (Oligochaetasida) seminal vesicles.

18. *A. pareudrili* (Cognetti de Martiis, 1911) Cognetti de Martiis, 1923 (syns., *Monocystis pareudrili* Cognetti de Martiis, 1911; *A. parendrili* Cognetti of Loubatières [1955] *lapsus calami*) in *Pareudrilus pallidus* (Oligochaetasida) seminal vesicles.

19. *A. perfida* Rees, 1963 in *Allolobophora chlorotica* (Oligochaetasida) coelom.

20. *A. pertusa* Loubatières, 1955 in *Allolobophora chlorotica, A. gigas,* and *A. rosea* (Oligochaetasida) seminal vesicles.

21. *A. pilosa* Meier, 1956 in *Lumbricus terrestris, L. rubellus, L. festivus,* and *L. castaneus* (Oligochaetasida) seminal vesicles.

22. *A. rotaria* Rees, 1963 in *Octolasium cyaneum* (Oligochaetasida) coelom.

23. *A. spinosa* Rees, 1963 in *Allolobophora chlorotica* (Oligochaetasida) seminal vesicles.

24. *A. stammeri* Meier, 1956 in *Fridericia striata, F. perrieri,* and *F. ratzeli* (Oligochaetasida) coelom.

25. *A. villosa* (Hesse, 1909) Cognetti de Martiis, 1923 (syn., *Monocystis villosa* Hesse, 1909) in *Octolasium lacteum* (Oligochaetasida) seminal vesicles.

26. *A. vivax* (Berlin, 1924) Meier, 1956 (syn., *Monocystis vivax* Berlin, 1924) in *Eiseniella t. tetraedra* (Oligochaetasida) seminal vesicles.

Genus *Cephalocystis* Rees, 1962. Gamonts elongate, irregularly cylindrical, with a well-differentiated striated mucron with two posteriolaterally directed processes; solitary. TYPE SPECIES *C. singularis* Rees, 1962.

1. *C. singularis* Rees, 1962 (TYPE SPECIES) in *Eisenia foetida* (Oligochaetasida) coelom.

Genus *Trigonepimerus* Sciacchitano, 1932. Gamonts cylindrical; mucron (epimerite) triangular, separated from rest of gamont by a filament. TYPE SPECIES *T. criodrilii* (Sciacchitano, 1931) Sciacchitano, 1932.

1. *T. criodrilii* (Sciacchitano, 1931) Sciacchitano, 1932 (syn., *Trigonocephalus criodrilii* Sciacchitano, 1931) (TYPE SPECIES) in *Criodrilus lacuum* (Oligochaetasida) coelom.

Genus *Mastocystis* Boisson, 1957. Form varies; with small peripheral pseudopodic lobes due to a radiating peripheral system of myofibrils. TYPE SPECIES *M. denticulata* Boisson, 1957.

1.   *M. denticulata* Boisson, 1957 (TYPE SPECIES) in *Pheretima peguana* (Oligochaetasida) seminal vesicles.
2.   *M. tuzeti* Boisson, 1957 in *Pheretima peguana saigonensis* (Oligochaetasida) seminal vesicles.

Subfamily ZYGOCYSTINAE *Bhatia, 1930*
     Syzygy extremely early or permanent; oocyst navicular or biconical, with peculiar thickenings at the ends.

Genus *Zygocystis* von Stein, 1848. Gamonts piriform, always in frontal (head-to-head) syzygy; oocysts navicular. TYPE SPECIES *Z. cometa* von Stein, 1848.

1.   *Z. aegyptica* Mohammed and Ramadan, 1971 in *Pheretima californica* and *P. hawayana* (Oligochaetasida) seminal vesicles.
2.   *Z. aster* Bereczky, 1968 in *Allolobophora dubiosa* (Oligochaetasida) coelom.
3.   *Z. cometa* von Stein, 1848 (syn., *Gregarina cometa* [von Stein, 1848] Diesing, 1859) (TYPE SPECIES) in *Lumbricus terrestris, Allolobophora longa, A. caliginosa, A. caliginosa* var. *trapezoides, A. chloroticum, A. dubiosa,* and *Eisenia foetida* (Oligochaetasida) seminal vesicles.
4.   *Z. cordiformis* Loubatières, 1955 in *Allolobophora gigas* (Oligochaetasida) seminal vesicles.
5.   *Z. eiseniae* Loubatières, 1955 in *Eisenia foetida* (Oligochaetasida) seminal vesicles.
6.   *Z. grassei* Loubatières, 1955 in *Allolobophora caliginosa* (Oligochaetasida) amebocytes.
7.   *Z. henleae* Meier, 1956 in *Henlea ventriculosa* (Oligochaetasida) coelom.
8.   *Z. ictericae* Segun, 1968 in *Allolobophora icterica* f. *typica* (Oligochaetasida) seminal vesicles.
9.   *Z. legeri* Hesse, 1909 (syn., *Monocystis cognettii* Hesse, 1909) in *Allolobophora chlorotica* and *Pheretima diffringens* (Oligochaetasida) seminal vesicles.
10. *Z. limnodrili* Janiszewska, 1968 in *Limnodrilus hoffmeisteri* (Oligochaetasida) seminal vesicles.
11. *Z. pagesi* Tuzet and Loubatières, 1946 in *Allolobophora gigas* (Oligochaetasida) seminal vesicles.
12. *Z. pheretimae* de Puytorac and Tourret, 1963 in *Pheretima schmardae, P. houlleti,* and *Pheretima* sp. (Oligochaetasida) coelom.
13. *Z. pilosa* Hesse, 1909 in *Allolobophora longa* (Oligochaetasida) seminal vesicles.
14. *Z. suecia* Berlin, 1924 in *Eisenia foetida* (Oligochaetasida) seminal vesicles.
15. *Z. violacea* Segun, 1978 in *Libyodrilus violaceus* (Oligochaetasida) coelom.

Genus *Adelphocystis* Cox, 1967. Gamonts very elongate, always in head-to-head syzygy. TYPE SPECIES *A. aeikineta* Cox, 1967.

1.   *A. aeikineta* Cox, 1967 (TYPE SPECIES) in *Keffia variabilis* (Oligochaetasida) coelom.

Genus *Pleurocystis* Hesse, 1909. Gamonts cylindroid, always in lateral syzygy; oocysts biconical; in funnel of seminal vesicles of oligochetes, never in the sac itself. TYPE SPECIES *P. cuenoti* Hesse, 1909.

1.   *P. cuenoti* Hesse, 1909 (TYPE SPECIES) in *Allolobophora longa terrestris* and *A. caliginosa* (Oligochaetasida) seminal vesicles.

2.   *P. eiseniella* Ormières, 1977 in *Eiseniella tetraedra* (Oligochaetasida) seminal vesicles.

Subfamily RHYNCHOCYSTINAE *Bhatia, 1930*
   Gamonts ovoid, spherical, or elongated, with a conical or cylindroconical trunk at the anterior end, solitary; oocysts biconical, with similar nonappendiculate ends, with eight sporozoites.

Genus *Rhynchocystis* Hesse, 1909. Rostrum of gamont metabolic, most often elongated into a conical or cylindroconical trunk. TYPE SPECIES *R. cuneiformis* (Ruschhaupt, 1885) Levine, 1977.

1.   *R. awatii* Bhatia and Setna, 1926 (syn., *R. hawatii* Bhatia and Setna of Loubatières [1955] *lapsus calami*) in *Pheretima elongata* (Oligochaetasida) seminal vesicles.
2.   *R. cognetti* Bhatia and Chatterjee, 1925 in *Allolobophora caliginosa* (Oligochaetasida) seminal vesicles.
3.   *R. cuneiformis* (Ruschhaupt, 1885) Levine, 1977 (syns., *Monocystis cuneiformis* Ruschhaupt, 1885; *M. pilosa* Cuénot, 1901; *M. minuta* Ruschhaupt, 1885; *M. tenax* var. *cuneiformis* Ruschhaupt, 1885 of Labbé [1899]; *M. tenax* var. *minuta* Ruschhaupt, 1885 of Labbé [1899]; *R. pilosa* [Cuénot, 1901] Hesse, 1909) (TYPE SPECIES) in *Lumbricus terrestris, L. rubellus, L. castaneus, L. festivus,* and *Eisenia foetida* (Oligochaetasida) seminal vesicles.
4.   *R. hessei* Cognetti de Martiis, 1911 in *Pareudrilus pallidus, Lumbricus rubellus,* and *L. terrestris* (Oligochaetasida) seminal vesicles.
5.   *R. mamillata* Bhatia and Setna, 1926 in *Pheretima elongata* (Oligochaetasida) seminal vesicles.
6.   *R. oculata* Berlin, 1924 emend. Levine, 1977 (syn., *R. pilosa* var. *oculata* Berlin, 1924) in *Lumbricus terrestris, L. rubellus,* and *L. castaneus* (Oligochaetasida) seminal vesicles.
7.   *R. ovata* Loubatières, 1955 in *Allolobophora rosea* (Oligochaetasida) seminal vesicles.
8.   *R. pessoai* Righi, 1974 in *Pristina minuta* (Oligochaetasida) seminal vesicles.
9.   *R. piriformis* Berlin, 1924 in *Lumbricus terrestris, L. rubellus,* and *Eisenia foetida* (Oligochaetasida) seminal vesicles.
10.   *R. porrecta* (Schmidt, 1854) Hesse, 1909 (syns., *Monocystis porrecta* Schmidt, 1854; *M. tenax* var. *porrecta* Schmidt, 1854 of Labbé [1899]; *Gregarina porrecta* [Schmidt, 1854] Diesing, 1859) in *Lumbricus terrestris, L. rubellus, L. castaneus, Eisenia foetida,* and *Allolobophora caliginosa* (Oligochaetasida) seminal vesicles.

Genus *Dirhynchocystis* Cognetti de Martiis, 1921. Rostrum of gamont metabolic, most often elongated into a conical or cylindroconical trunk; gamont with projections at both ends. TYPE SPECIES *D. brasiliensis* Cognetti de Martiis, 1921.

1.   *D. brasiliensis* Cognetti de Martiis, 1921 (TYPE SPECIES) in *Fimoscolex inurus* (Oligochaetasida) seminal vesicles.
2.   *D. elongata* Loubatières, 1955 (syn., *Echinocystis elongata* Loubatières, 1955) in *Allolobophora rosea* and *A. chlorotica* (Oligochaetasida) seminal vesicles.
3.   *D. eudrilii* Tuzet and Vogeli, 1956 in *Eudrilus eugeniae* (Oligochaetasida) seminal vesicles.
4.   *D. globosa* (Bhatia and Chatterjee, 1925) Tuzet and Loubatières, 1946 (syn., *Echinocystis globosa* Bhatia and Chatterjee, 1925) in *Pheretima heterochaeta* (Oligochaetasida) seminal vesicles.
5.   *D. minuta* Ruston, 1959 in *Lumbricus terrestris* (Oligochaetasida) seminal vesicles.

6.   *D. oblonga* Tuzet and Loubatières, 1946 in *Octolasium complanatum* (Oligochaetasida) seminal vesicles.

7.   *D. sacciformis* Boisson, 1957 in *Pheretima peguana* (Oligochaetasida) seminal vesicles.

Genus *Grayallia* Setna, 1927. Gamonts long and slender, with four spines at each end. TYPE SPECIES *G. quadrispina* Setna, 1927.

1.   *G. quadrispina* Setna, 1927 (TYPE SPECIES) in *Pheretima heterochaeta* (Oligochaetasida) coelom.

Subfamily STOMATOPHORINAE *Bhatia, 1930*
   Gamonts highly variable in shape (cylindrical, globular, stellate, etc.), solitary; mucron transformed into a sucker, myocyte complex; oocysts navicular, with truncate ends, with eight sporozoites.

Genus *Stomatophora* Drzhevetskii, 1907. Gamont spherical or ovoid; sucker petaloid, with radiating sides, oocysts biconical, with a flattened button at each end, attached to each other end to end in long chains inside gametocysts. TYPE SPECIES *S. coronata* (Hesse, 1904) Drzhevetskii, 1907.

1.   *S. bulbifera* Bhatia and Setna, 1926 in *Pheretima elongata* (Oligochaetasida) seminal vesicles.

2.   *S. coronata* (Hesse, 1904) Drzhevetskii, 1907 (syn., *Monocystis coronatus* Hesse, 1904) (TYPE SPECIES) in *Pheretima rodericensis, P. hawayana, P. barbadensis, P. peguana,* and *P. elongata* (Oligochaetasida) seminal vesicles.

3.   *S. diadema* Hesse, 1909 in *Pheretima hawayana, P. barbadensis, P. elongata,* and *Pheretima* sp. (Oligochaetasida) seminal vesicles.

4.   *S. primitiva* Bhatia and Setna, 1938 in *Eunice siciliensis* (Polychaetasida) intestine.

5.   *S. simplex* Bhatia, 1924 in *Pheretima rodericensis* Oligochaetasida) seminal vesicles.

Genus *Craterocystis* Cognetti de Martiis, 1918. Gamont globular, covered laterally with "hairs"; deep myocyte wall well developed; with strong myonemes going from sucker to lateral face; gamonts solitary. TYPE SPECIES *C. papua* Cognetti de Martiis, 1918.

1.   *C. myonemata* Boisson, 1957 in *Pheretima posthuma* (Oligochaetasida) free among the spermatogenic elements.

2.   *C. papua* Cognetti de Martiis, 1918 (TYPE SPECIES) in *Pheretima (Parapheretima) wendessiana* (Oligochaetasida) lymphocytes of the prostate gland.

Genus *Astrocystella* Cognetti de Martiis, 1918. Gamont body flattened, stellate, cut into five to nine deep lobes arising from a central region which contains the nucleus and forms a sucker; gamonts solitary. TYPE SPECIES *A. lobosa* Cognetti de Martiis, 1918.

1.   *A. lobosa* Cognetti de Martiis, 1918 (TYPE SPECIES) in *Pheretima (Parapheretima) beaufortii* (Oligochaetasida) seminal vesicles.

Genus *Albertisella* Cognetti de Martiis, 1926. Gamont body a thick cupule, with a central sucker having smooth walls; myocyte deep, comprising myonemes running from the base of the sucker to the opposite face. TYPE SPECIES *A. crater* Cognetti de Martiis, 1926.

1.   *A. crater* Cognetti de Martiis, 1926 (syn., *A. cratesae* Cognetti of Loubatières [1955])

(TYPE SPECIES) in *Pheretima (Parapheretima) sermowaiana* (Oligochaetasida) seminal vesicles.

Genus *Beccaricystis* Cognetti de Martiis, 1926. Gamont body cylindrical, lobed laterally; sucker anterior; axial cluster of myonemes present. TYPE SPECIES *B. loriai* Cognetti de Martiis, 1926.

1. *B. loriai* Cognetti de Martiis, 1926 (TYPE SPECIES) in *Pheretima (Parapheretima) sermowaiana* (Oligochaetasida) seminal vesicles.

Genus *Parachoanocystoides* De Saedeleer, 1930. Gamont body ovoid, with anterior mobile sucker limited by an annular sphincter of myonemes; a retractile, pseudopod-like tentacle bearing cytoplasmic "hairs" arises from the base of the sucker; gamonts solitary. TYPE SPECIES *P. tentaculata* (Cognetti de Martiis, 1918) De Saedeleer, 1930.

1. *P. tentaculata* (Cognetti de Martiis, 1918) De Saedeleer, 1930 (syns., *Choanocystis tentaculata* Cognetti de Martiis, 1918; *Choanocystella tentaculata* [Cognetti de Martiis, 1918] de Mello, 1931) (TYPE SPECIES) in *Pheretima (Parapheretima) beaufortii* (Oligochaetasida) seminal vesicles.

Genus *Choanocystoides* Cognetti de Martiis, 1925. Similar to *Parachoanocystoides,* but tentacle arises from center of sucker, which is borderd by cytoplasmic filaments; gamonts solitary, rounded or cup-shaped. TYPE SPECIES *C. costaricensis* Cognetti de Martiis, 1925.

1. *C. costaricensis* Cognetti de Martiis, 1925 (TYPE SPECIES) in *Pheretima heterochaeta* and *P. indica* (Oligochaetasida) seminal vesicles.

Genus *Zeylanocystis* Dissanaike, 1953. Mature gamonts bordered by papillae and filaments that may be absorbed to form a rim; gamonts rounded when in syzygy; oocysts navicular, with truncate ends. TYPE SPECIES *Z. burti* Dissanaike, 1953.

1. *Z. burti* Dissanaike, 1953 (TYPE SPECIES) in *Pheretima peguana* (Oligochaetasida) seminal vesicles.
2. *Z. fernandoi* Dissanaike, 1953 in *Pheretima peguana* (Oligochaetasida) seminal vesicles.

Genus *Arborocystis* Rees and Howell, 1966. Gamonts solitary, with a circle of contractile processes anteriorly forming the mucron. TYPE SPECIES *A. piriformis* (Rees, 1962) Rees and Howell, 1966.

1. *A. piriformis* (Rees, 1962) Rees and Howell, 1966 (syn., *Dendrocystis piriformis* Rees, 1962) (TYPE SPECIES) in *Allolobophora caliginosa* (Oligochaetasida) coelom.

Genus *Chakravartiella* Misra and Raychaudhury, 1973. Gamonts solitary, elongated and cylindrical, with an anterior immobile, circular, sucker-like organelle bearing tooth-like structures; sucker on a short neck; myonemes present; in intestine of millipedes. TYPE SPECIES *C. sugereiformes* Misra and Raychaudhury, 1973.

1. *C. sugereiformes* Misra and Raychaudhury, 1973 (syn., *C. suckeriformes* Misra and Raychaudhury, 1973 *lapsus calami*) (TYPE SPECIES) in *Trigoniulus goesii* (Diplopodasida) intestine.

Subfamily OLIGOCHAETOCYSTINAE *Meier, 1956 emend. Levine, 1977*
Oocysts without thickenings at ends.

Genus *Oligochaetocystis* Meier, 1956. Gamonts club-shaped, solitary or in syzygy; syzygy head to head. TYPE SPECIES *O. pachydrili* (Lankester, 1863) Meier, 1956 emend. Levine, 1977.

1. *O. mesenchytraei* Meier, 1956 in *Mesenchytraeus flavidus* (Oligochaetasida) coelom.
2. *O. pachydrili* (Lankester, 1863) Meier, 1956 emend. Levine, 1977 (syns., *Monocystis pachydrili* Lankester, 1863; *M. packydrili* [Claparède, 1861] Labbé, 1899; *Gonospora pachydrili* Vejdovsky, 1879; *Gregarina saenuridis* von Kölliker, 1848 of Delphy [1922]; [*non*] *G. saenuridis* von Kölliker, 1848) (TYPE SPECIES) in *Lumbricillus* (syns., *Pachydrilus, Marionina) semifuscus, L. pagenstecheri,* and *L. lineatus* (Oligochaetasida) coelom.
3. *O. saenuridis* (von Kölliker, 1848) Levine, 1977 (syns., *Gregarina saenuridis* von Kölliker, 1848; *Monocystis saenuridis* [von Kölliker, 1848] Hesse, 1909; *Zygocystis saenuridis* von Stein, 1848) in *Tubifex tubifex* (syns., *Saenuris variegata; Lumbricus variegatus; T. rivulorum*) (Oligochaetasida) coelom and seminal vesicles.

Genus *Echiurocystis* Levine, 1977. Gamonts spindle-shaped, with bulging middle; syzygy lateral (side by side); in echiurids. TYPE SPECIES *E. yumushii* (Iitsuka, 1933) Levine, 1977.

1. *E. bullis* (Noble, 1938) Levine, 1977 (syn., *Enterocystis bullis* Noble, 1938) in *Urechis caupo* (Echiurida) intestine.
2. *E. greeffi* (Noble, 1938) Levine, 1977 (syns., *Enterocystis greeffi* Noble, 1938; *Conorynchus gibbosus* Greeff, 1880 in part) in *Echiurus pallasii* (Echiurida) intestine.
3. *E. yumushii* (Iitsuka, 1933) Levine, 1977 (syns., *Enterocystis yumushii* Iitsuka, 1933; *E. umushii* Iitsuka, 1933 *lapsus calami*) (TYPE SPECIES) in *Urechis unicinctus* (Echiurida) intestine.

Genus *Neomonocystis* Geus, 1969. Gamonts spindle-shaped, not deformed, always solitary; oocysts with flattened ends; mucron absent; in chilopods. TYPE SPECIES *N. lithobii* Rauchalles in Geus, 1969.

1. *N. lithobii* Rauchalles in Geus, 1969 (TYPE SPECIES) in *Lithobius mutabilis, L. muticus,* and *L. piceus* (Chilopodasida) coelom.

Genus *Acarogregarina* Levine, 1977. Gamonts with a crown of spines at anterior end; gametocysts and oocysts unknown; in mites. TYPE SPECIES *A. corolla* (Erhardova, 1955) Levine, 1977.

1. *A. corolla* (Erhardova, 1955) Levine, 1977 (syn., *Gregarina corolla* Erhardova, 1955) in *Scutovertex minutus* (Acarinorida) location in host not given.

Family DIPLOCYSTIDAE *Bhatia, 1930*
Syzygy precocious, with fusion of gamonts; oocysts ellipsoidal, spherical, or ovoid, with eight sporozoites; parasites of insects.

Genus *Diplocystis* Künstler, 1887. Gametocysts in hemocoele; oocysts ellipsoidal, open in host midgut; sporozoites penetrate between columnar cells of epithelium, where they develop under the intestinal serosa and then enter hemocoele; infection by cannibalism, at least in cockroaches. TYPE SPECIES *D. schneideri* Künstler, 1887.

1. *D. clerci* Léger, 1904 (syn., *D. clercki* Léger of Dumbleton [1949] *lapsus calami*) in *Haploembia* (syn., *Embia*) *solieri* (Embiopterorida) coelom.
2. *D. horni* Kalavati, 1977 in *Odontotermes horni* (Isopterorida) fat body and hemocoele.
3. *D. johnsoni* van Thiel, 1954 in *Anopheles maculipennis* (Dipterorida) outer wall of midgut.
4. *D. major* Cuénot, 1897 in *Gryllus domesticus* (Orthopterorida) coelom.
5. *D. metselaari* van Thiel, 1954 in *Anopheles farauti* (Dipterorida) midgut wall.
6. *D. minor* Cuénot, 1897 in *Gryllus domesticus* (Orthopterorida) coelom.
7. *D. oxycani* Dumbleton, 1949 in *Oxycanus cervinatus* (Lepidopterorida) coelom.
8. *D. schneideri* Künstler, 1887 (TYPE SPECIES) in *Periplaneta americana* and *Blatella orientalis* (Orthopterorida) coelom.
9. *D. tipulae* Sherlock, 1979 in *Tipula paludosa* and (experimentally) *T. oleracea* (Dipterorida) larval midgut.
10. *D. zootermopsidis* Desportes, 1963 in *Zootermopsis nevadensis* (Isopterorida) hemoceoele.

Family ALLANTOCYSTIDAE *Bhatia, 1930*
Gamonts elongate; gametocysts elongate, sausage-like; oocysts spindle-shaped, with dissimilar sides.

Genus *Allantocystis* Keilin, 1920. Syzygy head to head. TYPE SPECIES *A. dasyhelei* Keilin, 1920.

1. *A. dasyhelei* Keilin 1920 (TYPE SPECIES) in *Dasyhelea obscura* (Dipterorida) midgut.

Family SCHAUDINNELLIDAE *Poche, 1913*
Gamonts do not encyst; male gamonts produce microgametes; females form macrogametes; oocysts spherical.

Genus *Schaudinnella* Nusbaum, 1903. Microgametes fusiform; macrogametes spherical; in digestive tract of oligochetes. TYPE SPECIES *S. enchytraei* (von Kölliker, 1848) Meier, 1956.

1. *S. enchytraei* (von Kölliker, 1848) Meier, 1956 (syns., *Gregarina enchytraei* von Kölliker, 1848; *G. pachydrili* Claparède, 1861; *Monocystis mitis* Leidy, 1882; *M. enchytraei* [von Kölliker, 1848] Labbé, 1899; *S. henleae* Nusbaum, 1903) (TYPE SPECIES) in *Henlea nasuta, H. ventriculosa, Bryodrilus ehlersi, Lumbricillus* (syn.,*Pachydrilus) semifuscus, L. pagenstecheri, Mesenchytraeus flavidus, Fridericia perrieri, F. leydigi, F. galba, F. polycheta, F. ratzeli, Guaranidrilus oiepe, Pristina minuta, Tupidrilus lacteus,* and *Distichopus silvestris* (Oligochaetasida) intestine.

Family ENTEROCYSTIDAE *M. Codreanu, 1940*
Development intracellular at first; syzygy early, the primite enlarging at its base and the satellite remaining more or less cylindrical so that the two cells have a sword or dagger shape; gametocysts without sporoducts, opening by simple rupture; oocysts ellipsoidal.

Genus *Enterocystis* Tsvetkov, 1926. With the characters of the family. TYPE SPECIES *E. ensis* Tsvetkov, 1926.

1. *E. ensis* Tsvetkov, 1926 (TYPE SPECIES) in *Caenis* sp. nymph and *Baetis rhodani* (Ephemeropterorida) intestine.
2. *E. ephemerae* (von Frantzius, 1848) Desportes, 1963 (syns., *Zygocystis ephemerae* von Frantzius, 1848; *Gamocystis francisci* Schneider, 1882; *G. ephemerae* [von Frantzius, 1848] Labbé, 1899; *Gregarina ephemerae* [von Frantzius, 1848] Diesing, 1859; *G. clavata* von Kölliker, 1848) in *Ephemera vulgata* nymph and *Ephemerella ignita* (Ephemeropterorida) intestine.
3. *E. fungoides* M. Codreanu, 1940 in *Baetis vernus* nymph and *B. rhodani* (Ephemeropterorida) intestine.
4. *E. grassei* Desportes, 1963 in *Beatis vernus, Epeorus torrentium, Heptagenia flava,* and *Ecdyonurus* sp. nymphs (Ephemeropterorida) intestine.
5. *E. hydrophili* (Foerster, 1938) Baudoin and Maillard, 1972 (syn., *Sphaerocystis hydrophili* Foerster, 1938) in *Hydrochara* (syn., *Hydrophilus) caraboides* (Coleopterorida) intestine.
6. *E. palmata* M. Codreanu, 1940 in *Baetis buceratus* nymph (Ephemeropterorida) intestine.
7. *E. racovitza* M. Codreanu, 1940 in *Baetis vernus* and *B. rhodani* nymphs (Ephemeropterorida) intestine.
8. *E. rhithrogenae* M. Codreanu, 1940 in *Rhithrogena semicolorata* nymph (Ephemeropterorida) intestine.

Family GANYMEDIDAE *Huxley, 1910*

Syzygy caudofrontal (head to tail), with the posterior end of the primite forming a cuplike depression into which the mucron of the satellite fits (this is a ball-and-socket arrangement); cytoplasm of gamonts fuses; gametocysts spherical; oocysts unknown; in intestine of crustacea.

Genus *Ganymedes* Huxley, 1910. With characters of the family. TYPE SPECIES *G. anaspidis* Huxley, 1910.

1. *G. anaspidis* Huxley, 1910 (TYPE SPECIES) in *Anaspides tasmaniae* (Crustaceasida) intestine and liver.

Chapter 6

THE GREGARINES: SEPTATE EUGREGARINES

Suborder SEPTATORINA Lankester, 1885
Gamont divided into protomerite and deutomerite by a septum; with epimerite, in invertebrates, especially arthropods.

Superfamily POROSPORICAE *Chakravarty, 1960*
Heteroxenous; two host species involved, one crustacean and the other molluskan.

Family POROSPORIDAE *Labbé, 1899*
Vegetative development in digestive tract of decapod crustacea and sporogony in connective tissue of lamellibranch mollusks; an enormous number of gymnospores produced. (One or the other host type unknown for some porosporid species.)

Genus *Porospora* Schneider, 1875. Oocysts absent, the sporozoites occurring in the host leukocytes; gymnospores develop into naked sporozoites rather than resistant oocysts in molluskan host; mature gamonts relatively long, with a tendency to remain isolated. TYPE SPECIES *P. gigantea* (van Beneden, 1869) Schneider, 1875.

1. *P. gigantea* (van Beneden, 1869) Schneider, 1875 (syn., *Gregarina gigantea* van Beneden, 1869) (TYPE SPECIES) in *Homarus gammarus* and *H. americanus* (Decapodorida) intestine and *Trochocochlea mutabilis* (Gastropodasida) gills and other tissues.
2. *P. nephropis* Léger and Duboscq, 1915 in *Nephrops norvegicus* (Decapodorida) intestine; other host unknown.

Genus *Nematopsis* Schneider, 1892. Oocysts with a single sporozoite; sporozoites in a double envelope; gymnospores develop into monozoic resistant oocysts in molluskan host; syzygy precocious, typically with several individuals in straight or forked chains; in older associations primite and satellite may be enclosed in a common epicyte, while satellite(s) become(s) a single (often multinucleate) compartment; protomerite of primite with a muscular collar. TYPE SPECIES *N. portunidarum* (Frenzel, 1885) Hatt, 1931.

1. *N. brasiliensis* Feigenbaum, 1975 in *Penaeus brasiliensis* (Decapodorida); other host unknown.
2. *N. calappae* (Ball, 1951) Sprague and Couch, 1971 (syn., *Carcinoecetes calappae* Ball, 1951) in *Calappa flammea* (Decapodorida) intestine; other host unknown.
3. *N. clausii* (Frenzel, 1885) Levine, 1980 (syn., *Gregarina clausii* Frenzel, 1885 in part) in *Phronima* sp. (Amphipodorida) intestine; other host unknown.
4. *N. dorippe* Bogolepova, 1953 in *Dorippe granulata* (Decapodorida) intestine; other host unknown.
5. *N. duorari* Kruse, 1966 in *Panaeus duorarum* (Decapodorida) intestine; oocysts in *Aequipecten irradians, Cardita floridana, Chione cancellata,* and *Macrocallista nimbosa* (Bivalvasida).
6. *N. foresti* (Théodoridès, 1967) Levine, 1984 (syns., *Cephaloidophora foresti* Théodoridès, 1967; *Porospora foresti* [Théodoridès, 1967] Théodoridès, 1979 in part) in *Xantho poressa* (Decapodorida) intestine; other host unknown.
7. *N. goneplaxi* Tuzet and Ormières, 1961 (syn., *Porospora petiti* Théodoridès, 1962 in

part) in *Goneplax angulata* and *G. rhomboides* (Decapodorida) intestine; other host unknown.

8. *N. grassei* (Théodoridès, 1962) Vivares and Rubio, 1969 (syn., *Porospora grassei* Théodoridès, 1962) in *Calappa granulata* (Decapodorida) intestine; other host unknown.

9. *N. hesperus* (Ball, 1938) Tuzet and Ormières, 1961 (syns., *Carcinoecetes hesperus* Ball, 1938; *Porospora hesperus* [Ball, 1938] Théodoridès, 1961) in *Pachygrapsus crassipes* (Decapodorida) intestine; other host unknown.

10. *N. lamellaris* Bogolepova, 1953 in an unidentified shoreline crab (Decapodorida) intestine; other host unknown.

11. *N. legeri* (de Beauchamp, 1910) Hatt, 1931 (syns., *N. mediterranen* Léger, 1905 *nomen oblitum; Porospora legeri* de Beauchamp, 1910; *P. galloprovincialis* Léger and Duboscq, 1925) in *Eriphia spinifrons* and *E. verrucosa* (Decapodorida) intestine and *Mytilus galloprovincialis* (Bivalvasida).

12. *N. maraisi* (Léger and Duboscq, 1911) Sprague, 1954 (syn., *Porospora maraisi* Léger and Duboscq, 1911) in *Macropipus* (= *Portunus) depurator, M. vernalis, Portumnus latipes, Pirimela denticulata*, and *Carcinus mediterraneus* (Decapodorida) intestine; other host unknown.

13. *N. matutae* (Ball, 1959) Tuzet and Ormières, 1961 (syn., *Carcinoecetes matutae* Ball, 1959) in *Matuta lunaris* (Decapodorida) intestine; other host unknown.

14. *N. mizoulei* (Théodoridès, 1964) Sprague and Couch, 1971 (syn., *Porospora mizoulei* Théodoridès, 1964) in *Solenocera membranacea* (Decapodorida) intestine; other host unknown.

15. *N. ormieresi* Vivares, 1973 in *Macropipus corrugatus* (Decapodorida) intestine; other host unknown.

16. *N. ostrearum* Prytherch, 1938 in *Panopeus herbstii, Eurypanopeus depressus, Neopanope texana sayi*, and *Eurytium limosum* (Decapodorida) intestine; and *Crassostrea virginica* (Bivalvasida).

17. *N. panopei* Ball, 1951 in *Panopeus occidentalis* and *P. herbstii* (Decapodorida) intestine; other host unknown.

18. *N. parapeneopsisi* (Setna and Bhatia, 1934) Sprague and Couch, 1971 (syns., *Hirmocystis* (?) *parapeneopsisi* Setna and Bhatia, 1934; *Protomagalhensia* (?) *attenuata* Setna and Bhatia, 1934) in *Parapeneopsis sculptilis* (Decapodorida) intestine; other host unknown.

19. *N. pectinis* (Léger and Duboscq, 1925) Sprague, 1970 (syn., *Porospora pectinis* Léger and Duboscq, 1925) in *Chlamys varia* (Bivalvasida) gills; other host unknown.

20. *N. penaeus* Sprague, 1954 (syn., *Porospora penaeus* [Sprague, 1954] Théodoridès, 1965) in *Penaeus aztecus* and *P. duorarum* (Decapodorida) intestine; other host unknown.

21. *N. petiti* (Théodoridès, 1962) Tuzet and Ormières, 1962 (syn., *Porospora petiti* Théodoridès, 1962 in part) in *Goneplax rhomboides* (Decapodorida) intestine; other host unknown.

22. *N. portunidarum* (Frenzel, 1885) Hatt, 1931 (syns., *Aggregata portunidarum* Frenzel, 1885 in part; *Zygocystis portunidarum* [Frenzel, 1885] Labbé, 1899; *N. schneideri* Léger, 1903; *Frenzelina portunidarum* [Frenzel, 1885] Léger and Duboscq, 1907; *Porospora portunidarum* [Frenzel, 1885] Léger and Duboscq, 1911) (TYPE SPECIES) in *Carcinus maenas, C. mediterraneus*, and also *Macropipus* (syn., *Portunus) arcuatus* (Decapodorida) intestine; and *Cardium edule, Scrobularia plana*, and *Solen vagina* (Bivalvasida) gills and mantle.

23. *N. prytherchi* Sprague, 1949 (syn., *N. ostrearum* Prytherch, 1938 in part) in *Menippe mercenaria* (Decapodorida) intestine and *Crassostrea virginica* (Bivalvasida).

24. *N. raouadi* Vivares, 1970 (syn., *Porospora raouadi* [Vivares, 1970] Théodoridès, 1979) in *Portumnus latipes* (Decapodorida) intestine; other host unknown.

25. *N. sinaloensis* Feigenbaum, 1975 in *Panaeus vannamei* (Decapodorida); other host unknown.
26. *N. soyeri* (Théodoridès, 1965) Sprague and Couch, 1971 (syn., *Porospora soyeri* Théodoridès, 1965) in *Aristeus antennatus* (Decapodorida) intestine; other host unknown.
27. *N. theodori* Vivares, 1972 (syn., *Porospora theodori* [Vivares, 1972] Théodoridès, 1979) in *Ethusa mascarone* (Decapodorida) intestine; other host unknown.
28. *N. tuzetae* Vivares, 1972 (syn., *Porospora tuzetae* [Vivares, 1972] Théodoridès, 1979) in *Carcinus mediterraneus* (Decapodorida) intestine; other host unknown.
29. *N. vannamei* Feigenbaum, 1975 in *Penaeus vannamei* (Decapodorida); other host unknown.
30. *N. veneris* (Léger and Duboscq, 1925) Sprague, 1970 (syn., *Porospora veneris* Léger and Duboscq, 1925) in *Venus fasciata* (Bivalvasida) gills; other host unknown.

Genus *Pachyporospora* Théodoridès, 1961. Presumably only individuals in syzygy known; at any rate, they have two or three nuclei and no septa, and Théodoridès (1961) believed that the septa disappeared completely after syzygy; with many transverse folds which go deep into the cytoplasm; epicyte thick; mucron (epimerite?) apparently present; in intestine of crabs; no other stages known, the position of this genus in the family Porosporidae being speculative. TYPE SPECIES *P. laubieri* Théodoridès, 1961.

1. *P. lamellaris* (Bogolepova, 1953) Ormières, 1968 (syns., *Nematopsis lamellaris* Bogolepova, 1953; *Tricystis plicata* Polyanskii and Kheisin, 1965) in unnamed crab (Decapodorida) intestine.
2. *P. laubieri* Théodoridès, 1961 (TYPE SPECIES) in *Atelecyclus septemdentatus* and *A. rotundatus* (Decapodorida) intestine.
3. *P. retorta* Ormières, 1968 in *Macropipus holsatus, M. vernalis, Pirimela denticulata,* and *Xaiva biguttata* (Decapodorida) intestine.
3a. *P. retorta pacifica* Théodoridès, 1977 in *Cancer productus* (Decapodorida) intestine.

Superfamily GREGARINICAE *Chakravarty, 1960*
Homoxenous; syzygy early.

Family CEPHALOIDOPHORIDAE *Kamm, 1922*
    Mucron present; early development intracellular; syzygy head to tail (caudofrontal), early, with primite different from satellite; marked anisogamy; gametocysts open by simple rupture; oocysts ovoid or spherical, with protruding equatorial ridge; no distinct epispore; in intestine of crustacea and other relatively primitive arthropods.

Genus *Cephaloidophora* Mavrodiadi, 1908. Mucron small, lenticular; oocysts ellipsoidal or spherical, with an inconspicuous equatorial ring, expelled in chains or singly; in intestine of cirripedes, decapods, and amphipods. TYPE SPECIES *C. communis* Mavrodiadi, 1908.

1. *C. akayebi* H. Hoshide, 1958 in *Penaeopsis akayebi* (Decapodorida) intestine.
2. *C. alii* Théodoridès and Desportes, 1975 in *Sergestes arcticus* (Decapodorida) intestine.
3. *C. ampelisca* (Nowlin and Smith, 1917) Kamm, 1922 (syn., *Frenzelina ampelisca* Nowlin and Smith, 1917) in *Ampelisca spinipes* (Amphipodorida) intestine.
4. *C. anisogammari* K. Hoshide, 1971 in *Anisogammarus pugettensis* (Amphipodorida) intestine.
5. *C. apsteini* Théodoridès and Desportes, 1975 in *Meganyctiphanes norvegica* (Decapodorida) degestive tract.

6. *C. baicalensis* (Tsvetkov, 1928) Théodoridès, 1962 (syns., *Gregarina baicalensis* Tsvetkov, 1928; *Rotundula baicalensis* [Tsvetkov, 1928] Lipa, 1968) in *Pallasea brandti* (Amphipodorida) intestine.

7. *C. bermudensis* (Ball, 1951) Sprague and Couch, 1971 (syn., *Carcinoecetes bermudensis* Ball, 1951) in *Pachygrapsus transversus* (Decapodorida) intestine.

8. *C. brasili* Poisson, 1920 in *Orchestia gammarella* (syn., *O. littorea*) (Amphipodorida) intestine.

9. *C. caprellae* (Frenzel, 1885) Levine, 1980 (syns., *Gregarina caprellae* Frenzel, 1885; *Aggregata caprellae* [Frenzel, 1885] Labbé, 1899) in *Caprella* sp. (Amphipodorida) intestine.

10. *C. carpilodei* Ball, 1963 in *Carpilodes rugatus* (Decapodorida) intestine.

11. *C. chthamalicola* Bogolepova, 1953 in *Chthamalus challengeri* (Cirripedasina) intestine.

12. *C. clausii* (Frenzel, 1885) Théodoridès and Desportes, 1975 (syn., *Gregarina clausii* Frenzel, 1885 in part) in *Phronima sedentaria* and *Phronima* sp. (Amphipodorida) intestine.

13. *C. communis* Mavrodiadi, 1908 (syn., *Pyxinioides chthamali* of Tuzet and Ormières [1956]) (TYPE SPECIES) in *Balanus improvisus, B. eburneus, B. amphitrite* var. *pallidus, B. amphitrite albicostatus, B. tintinnabulum, B. perforatus, B. crenatus, B. glandula, B. cariosus,* and *Chthamalus* sp. (Cirripedasina) intestine.

14. *C. conformis* (Diesing, 1851) Léger and Duboscq, 1911 (syns., *Gregarina conformis* Diesing, 1851; *Aggregata conformis* [Diesing, 1851] Labbé, 1899; *Frenzelina conformis* [Diesing, 1851] Léger and Duboscq, 1907; *Carcinoecetes conformis* [Diesing, 1851] Ball, 1938) in *Pachygrapsus marmoratus* (Decapodorida) intestine.

15. *C. delphinia* (Watson, 1916) Kamm, 1922 (syn., *Frenzelina delphinia* Watson, 1916) in *Telorchestia longicornis* (Amphipodorida) intestine.

16. *C. drachi* Théodoridès, 1962 in *Xantho rivulosus, X. granulocarpus,* and *X. poressa* (Decapodorida) gastric ceca.

17. *C. dromiae* (Frenzel, 1885) Vivares and Rubio, 1969 (syns., *Gregarina dromiae* Frenzel, 1885; *Frenzelina dromiae* [Frenzel, 1885] Léger and Duboscq, 1907; *Aggregata dromiae* [Frenzel, 1885] Labbé, 1899) in *Dromia personata* (syn., *D. vulgaris*) (Decapodorida) intestine.

18. *C. duboscqi* Poisson, 1924 in *Athanas nitescens* (Decapodorida) intestine.

19. *C. elongata* K. Hoshide, 1971 in *Hyale schmidti* (Amphipodorida) intestine.

20. *C. etisi* (Ball, 1959) Sprague and Couch, 1971 (syn., *Carcinoecetes etisi* Ball, 1959) in *Etisus laevimanus* (Decapodorida) intestine.

21. *C. florencae* Vivares, 1978 in *Inachus dorsettensis* (Decapodorida) intestine.

22. *C. fossor* (Léger and Duboscq, 1907) Trégouboff, 1912 (syns., *Aggregata coelomica* Léger, 1901 in part; *Frenzelina fossor* Léger and Duboscq, 1907) in *Pinnotheres pisum* (Decapodorida) intestine.

23. *C. gershensoni* Lipa, 1968 in *Gmelinoides fasciatus* (Amphipodorida) intestine.

24. *C. guinotae* Théodoridès, 1967 in *Xantho incisus granulocarpus, Pilumnus hirtellus spinifer,* and *P. spinifer* (Decapodorida) intestine.

25. *C. japonici* H. Hoshide, 1958 emend. Levine, 1980 (syn., *Carcinoecetes japonicus* H. Hoshide, 1958) in *Penaeus japonicus* (Decapodorida) intestine.

26. *C. knoepffleri* Théodoridès, 1962 in *Pinnotheres pinnotheres* (Decapodorida) gastric ceca.

27. *C. lata* H. Hoshide, 1958 in *Penaeopsis akayebi* (Decapodorida) intestine.

28. *C. macropodiae* Vivares, 1978 in *Macropodia longirostris* and *M. rostrata* (Decapodorida) GI tract.

29. *C. magna* Henry, 1938 in *Balanus nubilis* (Cirripedasina) intestine.

30. *C. margaretae* Balcescu, 1972 in *Synurella ambulans* (Amphipodorida) intestine.

31. *C. metaplaxi* (Pearse, 1932) Levine, 1980 (syn., *Steinina metaplaxi* Pearse, 1932) in *Metaplax dentipes* (Decapodorida) intestine.
32. *C. mithraxi* (Ball, 1951) Sprague and Couch, 1971 (syn., *Carcinoecetes mithraxi* Ball, 1951) in *Mithrax forceps* (Decapodorida) intestine.
33. *C. multiplex* Henry, 1938 in *Balanus balanus pugetensis* and *B. rostratus heteropus* (Cirripedasina) intestine.
34. *C. nephropis* Tuzet and Ormières, 1961 (syn., *Porospora nephropis* Léger and Duboscq, 1915 in part) in *Nephrops norvegicus* (Decapodorida) intestine.
35. *C. nicaeae* (Frenzel, 1885) Levine, 1980 (syns., *Gregarina nicaeae* Frenzel, 1885; *Aggregata nicaeae* [Frenzel, 1885] Labbé, 1899) in *Hyale pontica* (syn., *Nicaea nilsonii*) (Amphipodorida) intestine.
36. *C. nigrofusca* (Watson, 1916) Kamm, 1922 (syn., *Frenzelina nigrofusca* Watson, 1916) in *Uca pugnax* and *U. pugilator* (Decapodorida) intestine.
37. *C. obatakeensis* H. Hoshide, 1958 in *Telorchestia* sp. (Amphipodorida) intestine.
38. *C. ocellata* (Léger and Duboscq, 1907) Kamm, 1922 (syns., *Aggregata vagans* Léger and Duboscq, 1903 in part; *Frenzelina ocellata* Léger and Duboscq, 1907) in *Eupagurus prideauxi* (Decapodorida) intestine.
39. *C. olivia* (Watson, 1916) Kamm, 1922 (syn., *Frenzelina olivia* Watson, 1916) in *Libinia dubia* and *Ocypode cursor* (Decapodorida) intestine.
40. *C. orchestiae* Poisson and Remy, 1925 in *Orchestia bottae* (Amphipodorida) intestine.
41. *C. ozakii* (H. Hoshide, 1958) Levine, 1980 (syn., *Carcinoecetes ozakii* H. Hoshide, 1958) in *Penaeopsis akayebi* (Decapodorida) intestine.
42. *C. pagri* H. Hoshide, 1958 in *Eupagurus samuelis* and *Pagurus* sp. (Decapodorida) intestine.
43. *C. petiti* Gobillard, 1964 in *Candacia longimana* and *C. aethiopica* (Copepodasina) intestine.
44. *C. phrosinae* Théodoridès and Desportes, 1975 in *Phrosina semilunata* (Amphipodorida) presumably intestine.
45. *C. pinguis* Ball, 1963 in *Carpilodes rugatus* (Decapodorida) intestine.
46. *C. poissoni* Théodoridès, 1967 in *Alpheus ruber* and *A. dentipes* (Decapodorida) intestine.
47. *C. poltevi* Lipa, 1968 in *Baicalogammarus pullus*, *Gmelinoides fasciatus*, and *Microrupus v. vortex* (Amphipodorida) intestine.
48. *C. punctata* H. Hoshide, 1958 in *Amphithoe japonica* (Amphipodorida) intestine.
49. *C. setoutiensis* H. Hoshide, 1958 in *Orchestia platensis* (Amphipodorida) intestine.
49a. *C. setoutiensis minor* K. Hoshide, 1969 in *Orchestia ochotensis* (Amphipodorida) intestine.
50. *C. synurellae* Balcescu, 1972 in *Synurella ambulans* (Amphipodorida) intestine.
51. *C. talitri* Mercier, 1912 in *Talitrus saltator*, *Gammarus pulex*, and *Pontoporeia affinis* (Amphipodorida) intestine.
52. *C. tregouboffi* Théodoridès and Desportes, 1975 in *Meganyctiphanes norvegica* (Decapodorida) ceca of anterior midgut.
53. *C. ucae* (Pearse, 1933) Levine, 1980 (syn., *Gregarina ucae* Pearse, 1933) in *Uca dussumieri* (Decapodorida) intestine.
54. *C. vibiliae* Thæeodoridès and Desportes, 1975 in *Vibilia armata* and *V. propinqua* (Amphipodorida) intestine.
55. *C. vivieri* Théodoridès and Desportes, 1975 in *Stylocheiron abbreviatum* (Decapodorida) stomach.
56. *C. warekara* H. Hoshide, 1969 in *Caprella* sp. (Amphipodorida) intestine.

Genus *Caridohabitans* Ball, 1959. Mucron functional, transparent, crescent-shaped, concave anteriorly; epicyte thick; nucleus with granules distributed irregularly along membrane, without endosome; gametocysts unknown; in digestive tract of crustacea. TYPE SPECIES *C. setnai* Ball, 1959.

1. *C. dobsoni* Janardanan, 1978 in *Metapenaeus dobsoni* (Decapodorida) intestine.
2. *C. indicus* Janardanan, 1978 in *Penaeus indicus, Metapenaeus affinis*, and *M. dobsoni* (Decapodorida) intestine.
3. *C. setnai* Ball, 1959 (TYPE SPECIES) in *Penaeus simisulcatus* (Decapodorida) midgut.
4. *C. stylifera* Janardanan, 1978 in *Parapenaeopsis stylifera* (Decapodorida) intestine.

Genus *Rotundula* Goodrich, 1949. Mucron button-like; gamont rotund; oocysts small, spherical or subspherical, with equatorial suture; in amphipods. TYPE SPECIES *R. maculata* (Léger and Duboscq, 1911) Goodrich, 1949.

1. *R. dybowskii* Lipa, 1968 in *Brandtia l. lata, Gmeilinoides fasciatus, Pallasea cancellus, P. kessleri*, and *P. viridis* (Amphipodorida) intestine.
2. *R. gammari* (Diesing, 1859) Goodrich, 1949 (syns., *Gregarina gammari* Diesing, 1859; *G. gammari* von Siebold, 1839 of Uspenskaja [1960] [?]; *G. gammari* Georgévitch, 1951 [?]; *Cephaloidophora echinogammari* Poisson, 1921; *C. gammari* [von Frantzius-Seibold, 1848] Théodoridès, 1967) in *Gammarus pulex, G. olivii, G. ochridensis, G. roeselii, Gammarus* sp., *Paramaera walkerii, Echinogammarus berilloni, E. pungens*, and probably *Amphithoe rubricata* (Amphipodorida) intestine.
3. *R. godlewskii* Lipa, 1968 in *Brandtia l. lata, Pallasea kessleri*, and *P. viridis* (Amphipodorida) intestine.
4. *R. maculata* (Léger and Duboscq, 1911) Goodrich, 1949 (syn., *Cephaloidophora maculata* Léger and Duboscq, 1911) (TYPE SPECIES) in *Gammarus marinus, G. duebeni, G. locusta*, and *Pallasea quadrispinosa* (Amphipodorida) intestine.

Family CEPHALOLOBIDAE *Théodoridès and Desportes, 1975*
Gamonts fixed to stomach epithelium by a differentiation from the protomerite; this protoepimerite is dilated into a sucker and forms lobes which adhere closely to the microvillosities of the stomach epithelium; association precocious, composed of a primite and one or two satellites, these last being placed side by side.

Genus *Cephalolobus* Kruse, 1959. Development extracellular; syzygy caudofrontal (head to tail), occurring while attached to host gut, with one, two, or three satellites (smaller than the primite); protomerite of satellite without specialized holdfast; in gut of crustacea. TYPE SPECIES *C. penaeus* Kruse, 1959.

1. *C. lavali* Théodoridès and Desportes, 1975 in *Phronima sedentaria* (Amphipodorida) stomach.
2. *C. penaeus* Kruse, 1959 (TYPE SPECIES) in *Penaeus aztecus* and *P. duorarum* (Decapodorida) stomach.
3. *C. petiti* Théodoridès, 1964 in *Solenocera membranacea* (Decapodorida) stomach and hindgut.

Genus *Callynthrochlamys* Frenzel, 1885. Gamont with tubules radiating out into the cytoplasm from the nucleus; syzygy head to tail; gametocysts and oocysts unknown; in intestine of amphipods. TYPE SPECIES *C. phronimae* Frenzel, 1885.

1. *C. phronimae* Frenzel, 1885 (TYPE SPECIES) in *Phronima elongata, P. sedentaria*, and *P. atlantica* (Amphipodorida) intestine.

Family URADIOPHORIDAE *Grassé, 1953*

Mucron simple and cylindrical; development extracellular; syzygy caudofrontal (head to tail), precocious, with protomerite of satellite compressing deutomerite of primite; anisogamous; gametocyst opens by simple rupture; oocysts spherical, isolated, with fine equatorial ridge or radial processes.

Genus *Uradiophora* Mercier, 1912. Mucron simple; gametocysts ovoid; oocysts emitted in chains; in crustacea. TYPE SPECIES *U. cuenoti* (Mercier, 1911) Mercier, 1912.

1. *U. athanasi* Poisson, 1924 in *Athanas nitescens* (Decapodorida) intestine.
2. *U. cuenoti* (Mercier, 1911) Mercier, 1912 (syn., *Cephaloidophora cuenoti* Mercier,1911) (TYPE SPECIES) in *Atyaephyra desmaresti* and *Neocardina denticulata* (Decapodorida) intestine.
3. *U. denticulata* H. Hoshide, 1969 in *Neocardina denticulata* (Decapodorida) intestine.
4. *U. gammari* Poisson, 1924 in *Gammarus locusta, G. duebeni,* and *G. oceanicus* (Amphipodorida) intestine.
5. *U. mercieri* (Poisson, 1920) Poisson, 1921 (syns., *Didymophyes longissima* Poisson; *Frenzelina mercieri* Poisson, 1920) in *Orchestia gammarella* (syn., *O. littorea*) (Amphipodorida) intestine.
6. *U. ramosa* Balcescu-Codreanu, 1974 in *Pontogammarus robustoides* (Amphipodorida) midgut.

Genus *Heliospora* Goodrich, 1949. Similar to *Uradiophora,* but with six radiating processes formed by epispore at equator; in amphipods. TYPE SPECIES *H. longissima* (von Siebold in von Kölliker, 1848) Goodrich, 1949.

1. *H. acanthogammari* (Tsvetkov, 1928) Lipa, 1968 (syn., *Gregarina acanthogammari* Tsvetkov, 1928) in *Acanthogammarus godlevskii* var. *victori, Eulimnogammarus cruentus,* and *E. viridis* (Amphipodorida) intestine.
2. *H. eximia* Geus, 1967 in *Niphargus svetinaumi* (Amphipodorida) intestine.
3. *H. longissima* (von Siebold in von Kölliker, 1848) Goodrich, 1949 (syns., *Gregarina gammari* von Kölliker, 1847; *G. longissima* von Siebold in von Kölliker, 1848; *G. pediepiscopalis* von Wasielewski, 1896; *Uradiophora gammari* [von Kölliker, 1847] Poisson, 1924; *U. longissima* [von Siebold in von Kölliker, 1848] Poisson, 1921; *Monocystis gammari* Georgévitch, 1951; *Didymophyes longissima* [von Siebold, 1839] von Frantzius, 1848; *Gregarina millaria* [Zenker] Diesing, 1851) (TYPE SPECIES) in *Gammarus pulex, G. roeslii, Caprella aequalitra, Niphargus* sp., *Orchestia littorea, Pallasea quadrispinosa,* and *Pontoporeia affinis* (Amphipodorida) intestine.

Genus *Pyxinioides* Trégouboff, 1912. Mucron a ribbed button with 16 longitudinal furrows and a small cone at the end, or a cupule with a central trunk; gametocysts open by simple rupture; oocysts either unknown or ellipsoidal, with a large appendage at one end, often united in packets; in barnacles. TYPE SPECIES *P. balani* (von Kölliker, 1848) Trégouboff, 1912.

1. *P. balani* (von Kölliker, 1848) Trégouboff, 1912 (syn., *Gregarina balani* von Kölliker, 1848) (TYPE SPECIES) in *Balanus amphitrite, B. eburneus, B. balanoides,* and *B. pusillus* (Cirripedasina) intestine.
2. *P. bolitoides* Henry, 1938 emend. Levine, 1985 (syn., *Pyxinoides bolitoides* Henry, 1938) in *Balanus crenatus, B. cariosus,* and *B. glandula* (Cirripedasina) intestine.

3.   *P. chthamali* (Léger and Duboscq, 1909) Trégouboff, 1912 (syns., *Frenzelina chtham-ali* Léger and Duboscq, 1909; *Cephaloidophora chthamali* [Léger and Duboscq, 1909]) in *Chthamalus stellatus* (Cirripedasina) intestine.

4.   *P. fujitubo* H. Hoshide, 1951 in *Balanus amphitrite communis* (Cirripedasina) intestine.

5.   *P. japonicus* H. Hoshide, 1951 emend. Levine, 1985 (syn., *Pyxinoides japonicus* H. Hoshide, 1968) in *Chthamalus challengeri* (Cirripedasina) intestine.

6.   *P. kamenote* H. Hoshide, 1951 emend. Levine, 1985 (syn., *Pyxinoides kamenote* H. Hoshide, 1951) in *Mitella mitella* (Cirripedasina) intestine.

7.   *P. kurofuji* H. Hoshide, 1951 emend. Levine, 1985 (syn., *Pyxinoides kurofuji* H. Hoshide, 1968) in *Tetraclita squamosa japonica* (Cirripedasina) intestine.

8.   *P. oshoroensis* H. Hoshide, 1951 emend. Levine, 1985 (syn., *Pyxinoides oshoroensis* H. Hoshide, 1968) in *Balanus cariosus* (Cirripedasina) intestine.

9.   *P. pugetensis* Henry, 1938 emend. Levine, 1985 (syn., *Pyxinoides pugetensis* Henry, 1938) *Balanus balanus pugetensis* (Cirripedasina) intestine.

10. *P. valettei* (Nussbaum, 1890) Levine, 1980 (syn., *Gregarina valettei* Nussbaum, 1890) in *Polliceps* (syn., *Mitella) polymerus* (Cirripedasina) intestine.

Genus *Nematoides* Mingazzini, 1891. Gamonts vermiform; without septum; epimerite in the form of a fork or of pincers, on an elongate neck; in barnacles. TYPE SPECIES *N. fusiformis* Mingazzini, 1891.

1.   *N. fusiformis* Mingazzini, 1891 (TYPE SPECIES) in *Balanus perforatus* (Cirripedasina) intestine.

Genus *Bifilida* Tuzet and Ormières, 1964. Mucron (epimerite) unknown; oocysts cylindrical, with a long filament at each end, emitted in chains; in barnacles. TYPE SPECIES *B. rara* Tuzet and Ormières, 1964.

1.   *B. rara* Tuzet and Ormières, 1964 (TYPE SPECIES) in *Chthamalus stellatus* (Cirripedasina) intestine.

Family GREGARINIDAE *Labbé, 1899*
    Mucron always simple; early development extracellular; syzygy caudofrontal (head to tail), ordinarily early, even very precocious; anisogamy moderately marked; gametocysts with sporoducts; oocysts clearly elongate or cylindrical, symmetrical.

Genus *Gregarina* Dufour, 1828. Mucron conical, button-shaped, globular, or cylindrical; syzygy rather precocious; oocysts doliform (barrel-shaped), navicular, or spherical; in intestine of insects. TYPE SPECIES *G. ovata* Dufour, 1828.

1.   *G. acantholobi* H. Hoshide, 1952 emend. (syn., *G. acantholobae* H. Hoshide, 1952) in *Acantholobus japonicus* and *Acrydium japonicum* (Orthopterorida) intestine.

2.   *G. acinopi* Ormières, 1966 in *Acinopus picipes* (Coleopterorida) intestine.

3.   *G. acridiorum* (Léger, 1893) Labbé, 1899 (syn., *Clepsidrina acridiorum* Léger, 1893) in *Acanthacris ruficarnis, Accytera brevipennis, Acrotylus insubricus, A. patruelis, Anonconctus alpinus, Arphia sulphurea, Auloserpusia* sp., *Brachystola magna, Calliptamus italicus, C. itericus, C. wettenwylianus, Catantops quadratus, Chorthippus curtipennis, Chortophaga viridifasciata, Decticus verrucivorus, Dissosteira carolina, Ephippiger perforatus, E. terrestris, Ephippigerida nigromarginata, Glyptobothrus biguttulus, G. pullus, Hesperotettix pratensis, Homorocoryphus nitidulus, Melanoplus bivittatus, M. coloradensis, M. differentialis, M. femoratus, M. femorrubrum, M.*

luridus, M. oboratipennis, M. sanguinipes, M. mexicanus (= *M. atlantis), M. scudderi, Metrioptera grisea, Myrmeleotetrix maculatus, Occidentophena ruandensis, Oedalus decorus, Oedipoda coerulescens, O. fuscocineta* var. *coerulea, O. germanica, O. miniatra, Omocestus haemorrhoidalis, O. viridulus, Pamphagus,* sp., *Parapropacris rhodopterus, Pardalophora apiculata, Phaneroptera* sp., *Psophus stridulus, Saga pedo, Schistocerca americana, Serpusia lemarineli, Spharagemon bolli, S. collare, Sphingonotus caerulans, Sphingonotus* sp., *Stenobothrus biguttulus, S. nigromaculatus, Tetrix bipunctuata, Truxalis* sp. (Orthopterorida) intestine.

4. *G. acrydiinarum* Semans, 1939 (syn., *G. acrydiimarum* Semans, 1939 of Amoji and Rodgi [1976] *lapsus calami)* in *Acrydium arenosum angustum* and *Paratettix c. cucullatus* (Orthopterorida) midgut.

5. *G. acuta* (Léger, 1892) Labbé, 1899 (syn., *Clepsidrina acuta* Léger, 1892) in *Trox perlatus* and *T. hispidus* (Coleopterorida) digestive tract.

6. *G. aethiopica* Théodoridès, Desportes, and Jolivet, 1964 in *Gonocephalum simplex* and *Mesomorphus setosus* (Coleopterorida) intestine.

7. *G. africana* Théodoridès, Ormières, and Jolivet, 1958 in *Homorocyoryphus nitudulus, Conocephalus maculatus, C. iris,* and *Conocephalus* sp. (Orhtopterorida) intestine.

8. *G. alcidesii* Haldar and Chakraborty, 1978 in *Alcides* sp. nr. *leopardus* (Coleopterorida) midgut.

9. *G. alphitophagi* Foerster, 1938 in *Alphitophagus bifasciatus* (Coleopterorida) intestine.

10. *G. amarae* von Frantzius, 1848 (syn., *Clepsidrina ovata* Hammerschmidt, 1838) in *Amara familiaris, Amara* sp., *Harpalus aeneus, H. rufipes,* and *Pterostichus cupreus* (Coleopterorida) intestine.

11. *G. ambigua* Amoji and Rodgi, 1976 in *Forficula ambigua* (Dermapterorida) midgut.

12. *G. ampullaria* H. Hoshide and K. Hoshide, 1969 in *Altica caerulescens* (Coleopterorida) intestine.

13. *G. anaboliae* Tsvetkov, 1929 in *Anabolia nervosa, A. sororcula, Drusus annulatus,* and *D. plicatus* (Trichopterorida) intestine.

14. *G. antherophagi* Rauchalles in Geus, 1969 in *Antherophagus nigricornis* and *Philonthus* sp. (Coleopterorida) location not stated.

15. *G. anthici* Kamm, 1921 in *Anthicus* sp. and *Formicomus pedestris* (Coleopterorida) intestine.

16. *G. aragaoi* Pinto, 1918 in *Systena* sp. (Coleopterorida) intestine.

17. *G. arietuli* K. Hoshide, 1978 emend. (syn., *G. arietuliae* K. Hoshide, 1978) in *Loxoblemmus arietulus* (Orthopterorida) intestine.

18. *G. arthromacrae* Obata, 1953 in *Arthromacra sumptuosa* (Coleopterorida) intestine.

19. *G. atomariae* Foerster, 1938 in *Atomaria linearis* and *Atomaria* sp. (Coleopterorida) midgut.

20. *G. ausoniae* Ghidini and Moriggi, 1941 in *Reticulitermes lucifugus* and *Cubitermes* sp. (Isopterorida) midgut.

21. *G. bancoi* Gisler, 1967 in *Thoracotermes macrothorax, Anoplotermes* sp., *Alloghathotermes hypogeus,* and *Pericaptritermes urgens* (Isopterorida) intestine.

22. *G. barbarara* M. E. Watson, 1915 in *Coccinella bipunctata, Coccinella* sp., *Exochomus quadripustulatus,* and *Tytthaspis sedecipunctata* (Coleopterorida) intestine.

23. *G. bembidii* Rauchalles in Geus, 1969 in *Bembidium* sp. (Coleopterorida) midgut.

24. *G. bergi* Frenzel, 1892 (syn., *Pileocephalus bergi* [Frenzel, 1892] Labbé, 1899) in *Corynetes coeruleus, Corynetes* sp., *Necrobia ruficollis,* and *N. violaceus* (Coleopterorida) intestine.

25. *G. bidari* Patil, 1982 in tenebrionid beetle (Coleopterorida) gut.

26. *G. bilobosa* Kundu and Haldar, 1981 in *Longitarsus* sp. (Coleopterorida) midgut.

27. *G. blaberae* Frenzel, 1892 (syn., *Pileocephalus blaberae* [Frenzel, 1892] Labbé, 1899) in *Blabera claraziana, B. fusca,* and *B. discoidalis* (Orthopterorida) intestine.

28. *G. blattarum* von Siebold, 1839 (syns., *G. blattae orientalis* Leidy, 1853; *Clepsidrina blattarum* [von Siebold, 1839] Schneider, 1875) in *Blatta orientalis, Blattella germanica, Ischroptera pennsylvanica, Periplaneta americana,* and perhaps *Blaberus craniifer* (Orthopterorida) intestine.

29. *G. boevi* Corbel, 1964 in *Oedipoda germanica, O. miniata,* and *Sphingonotus coerulans* (Orthopterorida) intestine.

30. *G. boletophagi* Crawley, 1903 (syn., *Anthorhynchus boletophagi* [Crawley, 1903] Ellis, 1913) in *Boletophagus cornutus* (Coleopterorida) intestine.

31. *G. brachypteri* Tsvetkov, 1929 in *Brachypterus urticae* (Coleopterorida) location not given.

32. *G. brosci* Geus, 1969 in *Broscus cephalotes* (Coleopterorida) intestine.

33. *G. byrrhina* Foerster, 1938 in *Byrrhus fasciatus, B. pilula, Cytilus sericeus,* and *Simplocaria semistriata* (Coleopterorida) location not given.

34. *G. californica* Lipa, 1968 in *Coccinella californica* (Coleopterorida) intestine.

35. *G. capritermitis* Desai, 1961 in *Capritermes incola* (Isopterorida).

36. *G. caudata* Georgévitch, 1951 in Trichopterorida gen. spp. (Trichopterorida) intestine.

37. *G. cavalierina* Blanchard, 1905 (syn., *Protomagalhaensia marottai* Filipponi, 1952) in *Dendarus trisis, Scaurus striatus, S. unicinus,* and *Scaurus* sp. (Coleopterorida) midgut.

38. *G. ceropriae* Théodoridès and Desportes, 1967 in *Ceropria subocellata* (Coleopterorida) intestine.

39. *G. cerylonis* Geus, 1969 in *Cerylon ferrugineum* and *C. histeroides* (Coleopterorida) intestine.

40. *G. cestiforme* Foerster, 1938 in *Rhagio* sp. (Dipterorida) intestine.

41. *G. cetoniae* Foerster, 1938 in *Cetonia aurata, Osmoderma eremita,* and *Potosia cuprea* (Coleopterorida) intestine.

42. *G. ceuthophili* Semans, 1939 in *Ceuthophilus brevipes, C. divergens,* and *C. gracilipes* (Orthopterorida) intestine.

43. *G. chagasi* Pinto, 1918 in *Conocephalus frater* (Orthopterorida) intestine.

44. *G. chelidurellae* Geus, 1969 in *Chelidurella acanthopygia* (Dermapterorida) location not stated.

45. *G. chilocori* Obata, 1953 in *Chilocorus rubidus* (Coleopterorida) intestine.

46. *G. chironomus* Georgévitch, 1951 in *Chironomus* sp. (Dipterorida) intestine.

47. *G. chrysomelae* Lipa, 1967 in *Chrysomela polita* (Coleopterorida) gastric ceca.

48. *G. cis* Foerster, 1938 in *Cis boleti, C. micans,* and *Rhopalodontus fracticornis* (Coleopterorida) intestine.

49. *G. clavata* von Kölliker, 1848 (syns., *Sporadina clavata* [von Kölliker, 1848] von Frantzius, 1848; *Zygocystis clavata* [von Kölliker, 1848] Lankester, 1863; *Clepsidrina granulosa* Schneider, 1887; *G. granulosa* [Schneider, 1887] Labbé, 1899) in *Ephemera vulgata, Ephemera* sp., *Habrophlebia* sp., and *Pothamanthus* sp. (Ephemeropterorida) intestine.

50. *G. coccinellae* Lipa, 1967 in *Coccinella septempunctata, C. quinquepunctata, Hippodamia tridecipunctata,* and *Myrrha octodecimoguttata* (Coleopterorida) intestine.

51. *G. coelomica* Foerster, 1938 in *Pyrochroa coccinea* (Coleopterorida) intestine and coelom.

52. *G. columnata* Geus, 1968 in *Cis castaneus* (Coleopterorida) location not stated.

53. *G. compressa* Ormières, 1966 in *Diaperis boleti* (Coleopterorida) intestine.

54. *G. concava* H. Hoshide, 1952 in *Gampsocleis burgeri* (Orthopterorida) intestine and gastric ceca.

55. *G. conoducla* H. Hoshide, 1958 in *Mycetophaga* sp. (Colopterorida) intestine.

56. *G. consobrina* Ellis, 1913 in *Ceuthophilus valgus* and *Harposcepa* sp. (Orthopterorida) location not stated.

57. *G. coptotomi* Watson, 1916 in *Coptotomus interrogatus* (Coleopterorida) intestine.
58. *G. corbeli* Théodoridès, Desportes, and Jolivet, 1972 in *Panesthia javanica* (Orthopterorida) intestine.
59. *G. cousinea* Corbel, 1968 in *Gryllus campestris* (Orthopterorida) intestine.
60. *G. craspedonoti* Obata, 1953 in *Craspedonotus tibialis* (Coleopteorida) intestine.
61. *G. crassa* Ishii, 1914 in *Tribolium ferrugineum* (Coleopterorida) intestine.
62. *G. crawleyi* Levine, 1980 (syns., *Euspora lucani* Crawley, 1903; *G. lucani* [Crawley, 1903] Watson, 1916) in *Lucanus dama* (Coleopterorida) intestine.
63. *G. crenata* (Bhatia and Setna, 1914) Théodoridès and Jolivet, 1959 (syns., *Caulocephalus crenatus* Bhatia and Setna, 1924 in part; *C. japonicus* Hoshide, 1957) in *Aulacophora foveicollis, A. femoralis, A. africana, A. similis, Aulacophora* sp., *Chapuisia usambarica, C. nitida, Chrysomela aurichalces, C. menthastri, Chrysolina staphylea, Leptaulaca vinula, Monolepta pauperata, Morphosphaeroides africana, Nisotra delecta, Ootheca mutabilis, Paracantha vicina musavakii, P. multicolor, Phaedonia arcata, Pseudocophora* sp. apud *perplexa, Rhaphidopalpa similis,* and *Strobiderus aequatorialis* (Coleopterorida) intestine.
64. *G. crescentica* Haldar and Chakraborty, 1978 in *Amblyrrhinus* sp. (Coleopterorida) midgut.
65. *G. cubensis* Peregrine, 1970 in *Blaberus discoidalis* (Orthopterorida) intestine.
66. *G. cucumiformis* Gisler, 1967 in *Alloghathotermes hypogeus silvestri* (Isopterorida) intestine.
67. *G. culleata* Geus, 1969 in *Anabolia nervosa* (Trichopterorida) intestine.
68. *G. cuneata* von Stein, 1848 (syns., *Clepsidrina cuneata* [von Stein, 1848] Pfeffer, 1910; *C. polymorpha* Hammerschmidt, 1838 in part; *G. polymorpha* [Hammerschmidt, 1838] of Lankester [1863] in part; *G. polymorpha* var. *cuneata* [Hammerschmidt, 1838] Schneider, 1875; *C. trimosa* Schneider, 1875; *G. xylopini* Crawley, 1903) in *Alphitobium ovatum, Ceropria anthracina, C. romandi, Chiroscelis digitata, Stenosis angustata, Strongylium buettneri, Tenebrio molitor, T. nitidulus, T. obscrurus, Tribolium casteneum, T. ferrugineum* (Coleopterorida) intestine.
69. *G. curvata* (von Frantzius, 1848) Diesing, 1851 (syns., *Rhizinia* sp. Hammerschmidt, 1838; *Sporadina curvata* von Frantzius, 1848; *G. cetoniae* Foerster, 1938) in *Cetonia aurata, Cetonia* sp., *Osmoderma eremita,* and *Potosia cuprea* (Coleopterorida) intestine.
70. *G. cylindrica* Foerster, 1938 in Cucujidae gen. sp. and *Ditoma crenata* (Coleopterorida) intestine.
71. *G. cylindrosa* Haldar and Kundu, 1979 in *Supella superllectilium* (Orthopterorida) midgut.
72. *G. darchenae* Théodoridès, Desportes, and Mateu, 1976 in *Cubitermes* sp. (Isopterorida) intestine.
73. *G. davini* Léger and Duboscq, 1899 in *Gryllomorpha dalmatina* (Orthopterorida) intestine.
74. *G. decourti* Théodoridès, Desportes, and Jolivet, 1975 in *Pseudoblaps javana* (Coleopterorida) intestine.
75. *G. delmasi* Tuzet and Rambier, 1953 in *Arantia* sp., *Catoptopteryx apicalis, Conocephalus maculatus, Decticus verrucivorus* var. *monspeliensis, D. albifrons, Ephippiger terrestris, Harposcepa* sp., *Heterophorus obscurella, Phaneroptera nanasparsa, Platycleis affinis, Schulthessina* sp., *Tylopsis* sp., *Zeuneria biraminosa,* and *Zeuneria* sp. (Orthopterorida) intestine.
76. *G. derispiae* K. Hoshide, 1979 (syn., *G. displae* K. Hoshide, 1979 *lapsus calami*) in *Derispia maculipennis* (Coleopterorida) intestine.
77. *G. desaegeri* Théodoridès, Ormières, and Jolivet, 1958 in *Cyphocerastis* sp. (Orthopterorida) location not given.

78. *G. desmopterae* Théodoridès, Desportes, and Jolivet, 1972 in *Desmoptera novaeguineae* (Orthopterorida) intestine.
79. *G. dichirotrichi* Massot and Ormières, 1979 in *Dichirotrichus pallidus* (Coleopterorida) intestine.
80. *G. diestrammenae* H. Hoshide, 1953 in *Diestrammena japonica* (Orthopterorida) intestine.
81. *G. diforma* Nelson, 1970 in *Harpalus fraternus* (Coleopterorida) intestine.
82. *G. dimorpha* Filipponi, 1947 (syn., *G. ovosatellitis* Obata, 1953) in *Agosterus* (syn., *Chlaenius) vestitus, C. velutinus, C. nogushii, C. nigricons, C. circumdotus,* and *Chlaeniellus inops* (Coleopterorida) intestine.
83. *G. discocephala* Kundu and Haldar, 1981 in Blattellidae gen. sp. (Orthopterorida) gut.
84. *G. distinguenda* Tsvetkov, 1929 in *Limnophilus decipiens* and *Chaetopteryx villosa* (Trichopterorida) intestine.
85. *G. dorotheae* Geus, 1969 in *Niptus hololeucus* (Coleopterorida) location in host not given.
86. *G. dragescoi* Théodoridès and Desportes, 1966 in *Amarygmus laosensis* (Coleopterorida) intestine.
87a. *G. dragescoi* var. *novaguineae* Théodoridès, Desportes, and Jolivet, 1972 in *Amarygmus laniger* (Coleopterorida) intestine.
88. *G. eccrinifera* Geus, 1969 in *Stenophylax rotundipennis* (Trichopterorida) intestine.
89. *G. echinata* H. Hoshide, 1958 in *Lesticus magnus* (Coleopterorida) intestine.
90. *G. electae* (Ludwig, 1946) Ormières, 1968 (syn., *Didymophyes electae* Ludwig, 1946) in *Tipula abdominalis* (Dipterorida) hemocoele.
91. *G. embiae* Loubes, Ormières, and Bouix, 1971 in *Monotylota ramburi* and *Haploembia solieri* (Embiopterorida) intestine.
92. *G. endomychi* Foerster, 1938 in *Endomychus coccineus* and *Micrencaustes* sp. (Coleopterorida) intestine.
93. *G. epilachnae* K. Hoshide, 1980 in *Epilachna vigintioctomaculata pustolosa* (Coleopterorida) intestine.
94. *G. epilamprae* Théodoridès, Desportes, and Jolivet, 1972 in *Epilampra kerandrenii* (Orthopterorida) intestine.
95. *G. erecta* Wellmer, 1910 in *Broscus ecphalotes* (Coleopterorida) intestine.
96. *G. ernobii* Rauchalles in Geus, 1969 in *Ernobius abietis* and *Ernobius* sp. (Coleopterorida) intestine.
97. *G. erotylidarum* Rauchalles in Geus, 1969 in *Scaphidium quadrimaculatum* and *Tritoma bipustulata* (Coleopterorida) intestine.
98. *G. fallax* Ormières, 1975 in *Forficula auricularia* and *Anechura bipunctata* (Dermapterorida) intestine.
99. *G. fastidiosa* Harrison, 1955 in *Aptera cingulata* (Orthopterorida) intestine.
100. *G. fernandoi* Mano-Sabaratham, 1971 in *Pycnoscelus surinamensis* (Orthopterorida) intestine.
101. *G. fontanae* Théodoridès, Desportes, and Jolivet, 1965 in *Pogonobasis raffrayi* (Coleopterorida) intestine.
102. *G. fontinalis* Tsvetkov, 1929 (syn., *G. frontinalis* Zwetkow, 1929 of Amoji and Rodgi [1976] *lapsus calami)* in *Chaetopteryx* sp., *C. villosa, Drusus rectus,* and *Potamophylax nigricornis* (Trichopterorida) intestine.
103. *G. forficulae* Lipa, 1967 in *Forficula auricularis* (Dermapterorida) intestine.
104. *G. fragilis* M. E. Watson, 1915 in *Coccinella* sp. (Coleopterorida) intestine.
105. *G. fusipara* Baudoin and Maillard, 1972 (syn., *G. pusipara* Baudoin and Maillard, 1972 of Amoji and Rodgi [1976] *lapsus calami)* in *Hydrochara caraboides* (Coleopterorida) intestine.

106. *G. galliveri* M. E. Watson, 1915 in *Gryllus abbreviatus* (Orthopterorida) intestine.
107. *G. garnhami* Canning, 1956 in *Schistocerca gregaria, Serpusia lemarineli, Auloserpusia* sp., *Duronia* sp., *Locusta migratoria migratoriodes,* and *Anacrydium aegyptium* (Orthopterorida) intestine.
108. *G. geopetiti* Théodoridès and Ormières, 1959 in *Loboptera decipiens* (Orthopterorida) location in host not given.
109. *G. geusi* Levine, 1980 (syn., *G. pumila* Geus, 1968; [*non*] *G. pumila* H. Hoshide, 1958) in *Octotemnus glabriculus* and *Sulcacis affinis* (Coleopterorida) midgut.
110. *G. gibbii* (Kaschef and Roshdy, 1966) Geus, 1968 (syns., *G. gibbi* [Kaschef and Roshdy] Geus, 1968 of Amoji and Rodgi [1976] *lapsus calami; Pyxinia gibbii* Kaschef and Roshdy, 1966) in *Gibbium psylloides* (Coleopterorida) intestine.
111. *G. gibbsi* Harrison, 1955 in *Temnopteryx phalerata* (Orthopterorida) intestine.
112. *G. globosa* M. E. Watson, 1915 in *Coptotomus interrogatus* (Coleopterorida) intestine.
113. *G. golvani* Théodoridès and Jolivet, 1959 in *Erionomus planiceps, E. platypleura,* and *Didimus africanus* (Coleopterorida) intestine.
114. *G. gonimusi* H. Hoshide, 1952 in *Amara chalcites* (Coleopterorida) intestine.
115. *G. gonocephali* Obata, 1953 in *Gonocephalum pubens* (Coleopterorida) intestine.
116. *G. gorokae* Théodoridès, Desportes, and Jolivet, 1972 in *Conocephalus modestus* (Orthopterorida) intestine.
117. *G. gracilis* M. E. Watson, 1915 in Elateridae gen. sp. (Coleopterorida) intestine.
118. *G. grandicephala* H. Hoshide, 1958 in *Anisodactylus signatus* (Coleopterorida) intestine.
119. *G. grassei* Théodoridès, Ormières, and Jolivet, 1958 in *Enyaliopsis maculipes* and *Cosmoderus erinaceus* (Orthopterorida) intestine.
120. *G. grisea* Ellis, 1913 in *Tenebrio castaneus, Amarygmus morio,* and *A. laniger* (Coleopterorida) intestine.
121. *G. grylli* Corbel, 1964 in *Gryllus campestris* (Orthopterorida) intestine.
122. *G. gryllodis* Haldar and Sarkar, 1980 emend. (syn., *G. gryllodesii* Haldar and Sarkar, 1980) in *Gryllodes* sp. (Orthopterorida) intestinal ceca and midgut.
123. *G. gryllotalpae* Devdhar, 1962 in *Gryllotalpa africana* (Orthopterorida).
124. *G. guatemalensis* Ellis, 1912 in *Ninus interstitialis* (Orthopterorida) intestine.
125. *G. guttiventra* Haldar and Sarkar, 1980 in *Plebeogryllus guttiventris* (Orthopterorida) intestinal ceca and midgut.
126. *G. haasi* Geus, 1969 in *Nauphoeta cinerea* (Orthopterorida) intestine.
127. *G. hadenoeci* Semans, 1939 (syn., *G. hadenocei* Semans, 1939 of Amoji and Rodgi [1976] *lapsus calami)* in *Hadenoecus putaneus* (Orthopterorida) intestine.
128. *G. haranti* Théodoridès and Jolivet, 1959 in *Prioscelis serrata, P. fabricii, P. westwoodi, Chiroscelis digitata, Pristophilus passaloides, Opatrinus latipes, Uloma* sp., *Eutochia pulla,* and Tenebrionidae gen. sp. (Coleopterorida) intestine.
129. *G. heteroceri* Rauchalles in Geus, 1969 in *Heterocerus sericans* and *Heterocerus* sp. (Coleopterorida) location in host not stated.
130. *G. heterochirae* Théodoridès, 1959 in *Heterochira fryeri* (Coleopterorida) intestine.
131. *G. hirakawaenis* K. Hoshide and Shiranita, 1981 in *Hemidia sauteri* (Collembolorida) midgut.
132. *G. hoshidei* Levine, 1980 (syn., *G. ctenolepismae* H. Hoshide, 1954) in *Ctenolepisma villosa* (Thysanurorida) intestine.
133. *G. hyalocephala* Dufour, 1837 in *Tridactylus variegatus* (Orthopterorida) intestine.
134. *G. hydrophili* Hasselmann, 1919 in *Hydrophilus* sp. (Coleopterorida) intestine.
135. *G. hylastidis* Rauchalles in Geus, 1969 in *Hylaster ater, H. cunicularius,* and *H. opacus* (Coleopterorida) midgut.
136. *G. hylobii* (Fuchs, 1915) Kamm, 1922 (syn., *Clepsidrina hylobii* Fuchs, 1915) in *Hylobius abietis* (Coleopterorida) intestine.

137. *G. hypophloei* Lipa, 1967 in *Hypophloeus unicolor* (Coleopterorida) intestine.
138. *G. illinensis* M. E. Watson, 1915 in *Parcoblatta* (syn., *Ischnoptera) pennsylvanica* (Orthopterorida) intestine.
139. *G. impetuosa* Harrison, 1955 in *Melanosilpha capensis* (Orthopterorida) intestine.
140. *G. inago* H. Hoshide, 1958 in *Oxa velox, O. japonica*, and *Oxa gavisa* (Orthopterorida) intestine.
141. *G. inclinata* K. Hoshide, 1979 in *Hemicera zigzaga* (Coleopterorida) intestine.
142. *G. indianensis* Semans, 1939 in *Chorthippus curtipennis, Catantops melanostictus, Melanoplus differentialis, M. bivittatus, M. mexicanus mexicanus*, and *Thericles* sp. (Orthopterorida) intestine.
143. *G. indica* Devdhar, 1962 in *Heterogonia indica* (Insectasida).
144. *G. intestinalis* M. E. Watson, 1915 in *Pterostichus stygicus* (Coleopterorida) intestine.
145. *G. ipidiae* Foerster, 1938 in *Ipidia quadrimaculata* (Coleopterorida) intestine.
146. *G. ischnoptera* Datta and Haldar, 1984 in *Ischnoptera* sp. (Orthopterorida) midgut.
147. *G. isotomae* Geus, 1969 in *Isotoma viridis* (Collembolorida) location in host not stated.
148. *G. joliveti* Théodoridès, 1958 (syn., *G. jolivet* Théodoridès, 1958 of Amoji and Rodgi [1976] *lapsus calami*) in *Hegeter tristis, Pachychila punctata, P. trisulcata*, and *Pseudogena parvicollis* (Coleopterorida) intestine.
149. *G. katherina* M. E. Watson, 1915 in *Coccinella novemnotata, C. quatuordecimpustulata, C. septempunctata, C. bruckii, Pachychila punctata mamorensis, P. trisulcata*, and *Pseudogena parvicollis* (Coleopterorida) intestine.
150. *G. kazumii* Levine, 1985 (syn., *G. parva* K. Hoshide, 1978) in *Pteronemobius taprobanensis* (Orthopterorida) intestine and gastric ceca.
151. *G. kingi* Crawley, 1907 (syn., *Gigaductus kingi* [Crawley, 1907] Ellis, 1913) in *Gryllus abbreviatus* (Orthopterorida) intestine.
152. *G. kokunusuto* H. Hoshide, 1958 in *Tenebrionoides mauritanicus* (Orthopterorida) intestine.
153. *G. korogi* H. Hoshide, 1952 in *Gryllus mitratus* and *G. yemma* (Orthopterorida) intestine (gastric ceca).
154. *G. labidurae* Théodoridès, Ormières, and Jolivet, 1982 in *Labidura riparia* (Dermapterorida).
155. *G. lagena* Tsvetkov, 1929 in Tipulidae gen. sp. (Dipterorida) intestine.
156. *G. lagenoides* (Léger, 1892) Labbé, 1899 (syn., *Clepsidrina lagenoides* Léger, 1892) in *Lipisma saccharina* (Thysanurorida) intestine.
157. *G. larvarum* Filipponi, 1951 in *Blaps gibba* (Coleopterorida) intestine.
158. *G. latumerata* Geus, 1967 in *Elmis maguei* (Coleopterorida) intestine.
159. *G. laucournetensis* (Schneider, 1885) Labbé, 1899 (syn., *Clepsidrina laucournetensis* Schneider, 1885) in *Dryops* (syn., *Parnus) auriculatus* and *Dryops* sp. (Coleopterorida) intestine.
160. *G. legeri* Pinto, 1918 in *Periplaneta* (syn., *Stylopyga) americana* (Orthopterorida) intestine.
161. *G. leucophaeae* Geus, 1969 in *Leucophaea madeirae* (Orthopterorida) intestine.
162. *G. levinei* Haldar and Sarkar, 1980 emend. Levine, 1985 (syn., *G. levinii* Haldar and Sarkar, 1980) in *Pteronemobius fascipense* (Orthopterorida) midgut.
163. *G. limnadiae* Uttangi, 1954 in *Limnadia* sp. (Crustacea).
164. *G. limnophili* Tsvetkov, 1929 in *Limnophilus rhombicus, L. flavicornis, L. bipunctatus, L. nigriceps, L. stigma, Glyphotaelius punctatolineatus, G. pellucidus, Halesus interpunctatus*, and *Anabia nervosa* (Trichopterorida) intestine.
165. *G. lipai* Levine, 1980 (syns., *G. harpali* Lipa, 1967; [*non*] *G. harpali* Crawley, 1903) in *Harpalus aeneus* (Coleopterorida) intestine.
166. *G. litargi* Geus, 1969 in *Litargus connexus* (Coleopterorida) midgut.

167. *G. locustae* Lankester, 1863 (syns., *G. locustae carolinae* Leidy, 1853 in part; *G. fimbriata* Diesing, 1859 in part; *G. locustaecarolinae* [Leidy, 1853] Labbé, 1899; *Stephanophora locustaecarolinae* [Leidy, 1853] Crawley, 1903 in part) in *Dissosteria carolina* (Orthopterorida) intestine.

168. *G. longa (Léger, 1892) Labbé, 1899 (syn., Clepsidrina longa* Léger, 1892) in *Ctenophora* sp., *Pales crocata, P. pratensis, Tipula oleracea, T. paludosa,* and *Tipula* sp. (Dipterorida) intestine.

169. *G. longiducta* Ellis, 1913 in *Ceuthophilus latens* and *C. maculatus* (Orthopterorida) intestine.

170. *G. longirostris* (Léger, 1892) Labbé, 1899 (syn., *Clepsidrina longirostris* Léger, 1892) in *Thanasimus formicarius* (Coleopterorida) digestive tract.

171. *G. lunata* Rauchalles in Geus, 1969 in *Limnophilus lunatus* and *Neuronia reticulata* (Trichopterorida) intestine.

172. *G. lypropsi* H. Hoshide, 1951 in *Lyprops sinensis* (Coleopterorida) intestine.

173. *G. macrocephala* (Schneider, 1875) Labbé, 1899 (syn., *Clepsidrina macrocephala* Schneider, 1875 in part) in *Acheta domesticus, Anaepoda lamellata, Gryllus sylvestris, Mecopoda elongata,* and *Nemobius silvestris gata* (Orthopterorida) intestine.

174. *G. macrocephalia* Lipa, 1967 (syn., *G. macracephalia* Lipa, 1967 of Amoji and Rodgi [1976] *lapsus calami)* in *Aphodius depressus* (Coleopterorida) intestine and gastric ceca.

175. *G. macrotermitis* Kalavati, 1977 in *Macrotermes estherae* (Isopterorida) midgut.

176. *G. maculata* Léger, 1904 in *Phylan gibbus* (Coleopterorida) intestine.

177a. *G. maculata* Léger, 1904 var. *banyulensis* Théodoridès, 1955 in *Phylan abbreviatus* (Coleopterorida) location in host not given.

178. *G. malachiidarum* Foerster, 1938 in *Anthocomus coccineus, Axinotarsus pulicarius, A. ruficollis, A. marginalis, Malachius bipustulatus, Psilothrix nobilis,* and *Troglops albicans* (Coleopterorida) intestine.

179. *G. maliensis* Corbel, 1968 in *Phymateus cinctus* (Orthopterorida) midgut.

180. *G. marteli* Léger, 1904 in *Haploembia solieri* (Embiopterorida) midgut.

181. *G. megaspora* Amoji and Rodgi, 1976 in *Forficula ambigua* (Dermapterorida) midgut.

182. *G. melolonthae* Leidy, 1856 (syns., *G. melolonthae brunneae* Leidy, 1856; *G. melolonthaebrunneae* Crawley, 1903) in *Melolontha brunnea* (Coleopterorida) intestine.

183. *G. mesomorphi* Devdhar, 1962 in *Mesomorphus velliger* (Coleopterorida).

184. *G. microcephala* Leidy, 1889 (syn., *Echinomera microcephala* [Leidy, 1889] Crawley, 1903) in *Hoplocephala bicornis* (Coleopterorida) location in host not given.

185. *G. minor* H. Hoshide, 1952 in *Allecula fuliginosa* (Coleopterorida) intestine.

186. *G. minuta* Ishii, 1914 (syn., *Didymophyes minuta* [Ishii, 1914] Watson, 1916) in *Tribolium ferrugineum, T. confusum, T. castaneum,* and *T. destructor* (Coleopterorida) intestine.

187. *G. mirotermitis* Kirby, 1927 in *Mirotermes panamaensis* (Isopterorida) location in host not given.

188. *G. monarchia* M. E. Watson, 1915 in *Pterostichus stygicus* (Coleopterorida) intestine.

189. *G. monoducta* H. Hoshide, 1953 in *Diestrammena japonica* (Orthopterorida) intestine.

190. *G. morioni* Théodoridès and Jolivet, 1959 in *Morion guineensis* (Coleopterorida) intestine.

191. *G. mukundai* Haldar and Kundu, 1980 in Blattellidae gen. sp. (Orthopterorida) midgut.

192. *G. munieri* (Schneider, 1875) Labbé, 1899 (syns., *Clepsidrina munieri* Schneider, 1875; *C. chrysomelae* von Wasielewski, 1896; *G. diabrotica* Kamm, 1918) in *Timarcha tenebricosa, Agelastica alni, Altica nitens, A. pyritosa, A. rothschildii, Aspidomorpha isparetta, A. mutata, A. quadrimaculata, A. togata, Aulacophora* sp., *Aulomorphus variabilis, Bequertinia nodicornis, Blosyvus obliquatus, Candezea bas-*

alis, C. duvivieri, C. haematura, Cassida altiuscula, C. inaequalis, C. numerosa, C. testa, C. japana, Chrysolina affinis, C. aupichalcea, C. aurichalces, C. banksi, C. diluta, C. menthastri, C. opulenta, C. staphylea, Chrysomela coerulans, C. fuliginosa, C. graminis, C. haemoptera, C. lurida, C. menthastri, C. polita, C. serialis, C. varians, C. violacea, Cryptonychus macrorhinus, Cyrtonus almeriensis, C. plumbeus, Diabrotica vittata, D. trivittata, Entypotrachelus micans, E. sjoestedti, Galerucas lineola, G. nympheae, G. tanaceti, Gynandrophthalma apicalis, Hyperacantha bifasciata, H. bifossulata, H. diffusa, H. dubia, H. duplicata, H. flavodorsata, H. humilis, H. sinuosa, H. stuhlamanni, Idacantha hastata, Idacantha sp., *Laccoptera gydenstolpei, Laetiacantha distincta, Leptaulaca basalis, L. fissicollis, Megaleruca triloba, Megalognatha ruandana, Monolepta apicalis, Nisotra delecta, N. punctatosulcata, Nisotra* sp., *Oidosoma coccinella, Ootheca mutabilis, Otiorhynchus meridionalis, Paracantha multicolor, P. vicinia, Parasbecesta costalis, P. ruwenzorica, Plagiodera ferrugata, P. thoracica, Platycorinus (Corynodes) peregrinus, Platyxantha carinata, Rhaphidor palpa similis, Schematizella erythrocephala, Timarcha affinis, T. atlantica, T. balaerica, T. goettingensis, T. interstitialis, T. nicaensis,* and *Xenarthracella sylvatica* (Coleopterorida) intestine.

193. *G. mystacidarum* von Frantzius, 1848 (syn., *Clepsidrina mystacidarum* [von Frantzius, 1848] Schneider, 1875) in *Leptocerus* sp., *Limnophilus flavicornis, L. ignotus, L. politus, Mystacides quadrifasciatus,* and *Mystacides* sp. (Trichopterorida) coelom.

194. *G. nalae* Datta and Haldar, 1984 in *Nala lividipes* (Dermapterorida) midgut.

195. *G. neglecta* Watson, 1916 in *Ceuthophilus neglectus* (Orthopterorida) intestine.

196. *G. nematiforma* Rauchalles in Geus, 1969 in *Ennearthron affine* (Coleopterorida) intestine.

197. *G. neobrasiliensis* Pinto, 1918 in *Periplaneta americana* (Orthopterorida) intestine.

198. *G. nigra* M. E. Watson, 1915 in *Melanoplus femurrubrum, Abisares viridipennis, Acantharis ruficornis, Anacridium* sp., *Atractomorpha aberrans, A. gerstaeckeri, Atractomorpha* sp., *Aularches miliaris, Auloserpusia* sp., *Catantops quadratus, C. spissus, Chordacris* sp., *Cyrtacanthacris aeruginosa, C. ruficornis, C. tatanica, Encoptolophus sordidis, Epistaurus succineus, Eyprepocnemis ibandana, Heteropterus obscurella, Lentula* sp., *Maura* sp., *Melanoplus differentialis, Odontomelus* sp., *Orbillus coerulus, Oxycatantops congoensis, Paracinema luculenta, Paracoptaera cauta, Patanga succincta, Serpusia lemarineli, Taphronota calliparea, Thisoicetrus guineensis, Thisoicetrus* sp., *Valanga g. gohieri, V. papuasica, Zonoceros elegans,* and *Z. variegatus* (Orthopterorida) intestine.

199. *G. nipti* Rauchalles in Geus, 1969 in *Niptus hololeucus* (Coleopterorida) location in host not stated.

200. *G. noncylindrica* Levine, 1984 (syn., *G. cylindrica* Devdhar, 1962 and also Devdhar and Deshpande, 1976; [non] *G. cylindrica* Foerster, 1938) in *Mesomorphus velliger* (Coleopterorida).

201. *G. ohioensis* Semans, 1939 in *Parcoblatta virginica* (Orthopterorida) midgut.

202. *G. omalina* Foerster, 1938 in *Omalium rivulare, Heterothops dissimilis, Ditoma crenata, Onychophilonthus marginatus,* and Cucujidae gen. sp. (Coleopterorida) intestine.

203. *G. opatroidea* Patil, 1982 in tenebrionid beetle (Coleopterorida) gut (?).

204. *G. ophoni* Tuzet and Ormières, 1956 in *Ophonus pubescens, O. similis,* and *Notiobia bradytoides* (Coleopterorida) intestine.

205. *G. ormierei* Théodoridès, 1955 in *Gonocephalum rusticum, G. controversum, G. prolixum inornatum, G. simplex, Mesomorphus setosus, M. villiger, Taraxides laevigatus,* and Tentyria taurica (Coleopterorida) location in host not stated.

206. *G. ovata* Dufour, 1828 (syns., *Clepsidrina ovata* [Dufour, 1826] Schneider, 1873; *C. conoidea* Hammerschmidt, 1838) (TYPE SPECIES) in *Anisolobus maritima, A. an-*

nulipes, Anechura bipunctata, Euborellia moesta, Forficula auricularia, and *F. de-cipiens* (Dermapterorida) intestine.

207. *G. oviceps* Diesing, 1859 (syns., *G. achetae abbreviatae* Leidy, 1853; *G. achetaeab-breviatae* Labbé, 1899) in *Gryllodes melanocephalus, Gryllus abbreviatus, G. amer-icanus, G. assimilis, G. campestris, Gryllus* sp., Gryllidae gen. spp., *Metrioptera saussureina, Nemobius sylvestris, Phaeophilacris pilipennis,* and *Phaeophilacris* sp. (Orthopterorida) intestine.

208. *G. ovoidea* Wellmer, 1911 in *Crypticus quisquilius, Adesmia metallica,* and *Uloma* sp. (Coleopterorida) intestine.

209. *G. oxytelopsis* Geus, 1969 in *Oxytelus tetracarinatus* (Coleopterorida) midgut.

210. *G. panchlorae* Frenzel, 1892 (syn., *G. planchlorae* Frenzel, 1892 of Geus [1969] *lapsus calami*) in *Panchlora exoleta* and *Panchlora* sp. (Orthopterorida) midgut.

211. *G. paranensis* (Kunckel d'Herculais, 1899) Watson, 1916 (syn., *Clepsidrina para-nensis* Kunckel d'Herculais, 1899) in *Schistocerca paranensis* (Orthopterorida) intestine.

212. *G. parcoblattae* Semans, 1939 in *Parcoblatta uhleriana* and *P. pennsylvanica* (Or-thopterorida) midgut.

213. *G. passali* Lankester, 1863 (syns., *G. passali cornuti* Leidy, 1853; *G. passalicornuti* Leidy, 1853 of Crawley [1903]) in *Passalus cornutus* (Coleopterorida) intestine.

214. *G. pediciae* Rauchalles in Geus, 1969 in *Pedicia rivosa* (Dipterorida) intestine.

215. *G. peruviana* Corbel, 1968 (syn., *G. peruvinae* Corbel, 1967 of Amoji and Rodgi [1976] *lapsus calami*) in *Gryllus peruviensis* (Orthopterorida) intestine.

216. *G. phaleriae* Tuzet and Ormières, 1955 in *Phaleria cadaverina* and *P. prolixa* (Co-leopterorida) intestine.

217. *G. pharaxonothae* Rauchalles in Geus, 1969 in *Pharaxonotha hirschi* (Coleopterorida) location in host not stated.

218. *G. philonthi* Rauchalles in Geus, 1969 in *Philonthus concinnus* and *P. immundus* (Coleopterorida) intestine.

219. *G. phyllotretae* H. Hoshide, 1952 in *Phyllotreta vittata* (Coleopterorida) intestine.

220. *G. piktyoktenidis* Rauchalles in Geus, 1969 in *Piktyoktenes curvidens* (Coleopterorida) intestine.

221. *G. pityogenidis* Rauchalles in Geus, 1969 in *Pityogenes bidentatus* (Coleopterorida) intestine.

222. *G. platycephala* H. Hoshide, 1952 in *Neatus* (syn., *Tenebrio) picipes* (Coleopterorida) intestine.

223. *G. platydema* Kamm, 1918 in *Platydema excavatum* (Coleopterorida) intestine.

224. *G. platyni* Watson, 1916 in *Platynus ruficollis* and *Dromius longiceps* (Coleopterorida) intestine.

225. *G. plegaderi* Foerster, 1938 in *Plegaderus saucius* (Coleopterorida) midgut.

226. *G. plesiophthalmi* H. Hoshide, 1952 in *Plesiophthalmus nigrocyaneus* (Coleopterorida) intestine.

227. *G. pocadii* Raushalles in Geus, 1969 in *Pocadius ferrugineus* (Coleopterorida) location in host not stated.

228. *G. poecili* Rauchalles in Geus, 1969 in *Poecilus lipidus* (Coleopterorida) intestine.

229. *G. poeciloceri* Ganapati and Mrutyunjayedevi, 1954 emend. (syn., *G. poecilocerum* Ganapati and Mrutyunjayedevi, 1954) in *Poecilocerus pictus* (Orthopterorida) intestine.

230. *G. polaris* Massot and Ormières, 1979 in *Relor (Iberozabrus) obesus* (Coleopterorida) intestine.

231. *G. polyaulia* Wellmer, 1910 in *Harpalus aeneus, H. ruficornis, H. affinis, H. servus, Amara aulica, A. familiaris, A. fulva, A. similata,* and *Dromius longiceps* (Coleop-terorida) intestine.

232. *G. polymorpha* (Hammerschmidt, 1838) von Stein, 1848 (syn., *Clepsidrina polymorpha* Hammerschmidt, 1838 in part) in *Gonocephalum* sp., *Gonocnemis* sp., *Peltoides senegalensis, Pogonobasis* sp., *Taraxides punctatus, Tenebrio guineensis, T. molitor, T. obscurus* (Coleopterorida), and dubiously *Chortoglyphus arcuatus* (Acarinorida) intestine.

233. *G. polyphagae* Geus, 1969 in *Polyphaga aegyptiaca* (Orthopterorida) intestine.

234. *G. prima* Semans, 1939 (syn., *G. prisma* Semans, 1939 of Amoji and Rodgi [1976] *lapsus calami*) in *Ceuthophilus uhleri* (Orthopterorida) midgut.

235. *G. princisi* Peregrine, 1970 in *Blaberus boliviensis* (Orthopterorida) intestine.

236. *G. proteocephala* Semans, 1939 (syn., *G. protocephala* Semans, 1939 of Amoji and Rodgi [1976] *lapsus calami*) in *Ceuthophilus gracilipes* (Orthopterorida) midgut.

237. *G. ptini* Foerster, 1938 in *Ptinus fur* and *P. latro* (Coleopterorida) location in host not stated.

238. *G. pumila* H. Hoshide, 1958 in Tenebrionidae gen. sp. (Coleopterorida) intestine.

239. *G. pupa* Tsvetkov, 1929 in a trichopteran larva (Trichopterorida) intestine.

240. *G. pusilla* Baudoin, 1967 (syn., *G. posilla* Baudoin, 1967 of Amoji and Rodgi [1976] *lapsus calami*) in *Tinodes waeneri* (Trichopterorida) intestine.

241. *G. pycnoceri* Théodoridès, Desportes, and Mateu, 1976 in *Pycnocerus sulcatus* (Coleopterorida) intestine.

242. *G. quenui* Tuzet and Ormières, 1963 in *Halictus scabiosae* (Hymenopterorida) intestine.

243. *G. raphidiae* Achtelig, 1974 in *Raphidia major, R. notata,* and *R. ophiopsis* (Raphidiopterorida) intestine.

244. *G. raptoris* Geus, 1969 in *Cyphoderes raptor* (Coleopterorida) intestine.

245. *G. rhomborrhinae* H. Hoshide, 1952 in *Rhomborrhina japonica* (Coleopterorida) intestine.

246. *G. rhyparobiae* J. M. Watson, 1945 (syn., *G. rhyparobia* J. M. Watson, 1945 of Amoji and Rodgi, [1976] *lapsus calami*) in *Leucophaea* (syn., *Rhyparobia) maderae* (Orthopterorida) intestine.

247. *G. rigida* (Hall, 1907) Ellis, 1913 (syns., *Hirmocystis rigida* Hall, 1907; *G. melanopli* Crawley, 1907) in *Melanoplus femoratus, Acanthacris ruficornis, Arphia sulphurea, Auloserpusia* sp., *Brachystola magna, Calliptamus italicus, Catantops quadratus, Chortophaga viridifasciata, Dissosteira carolina, Encoptolophus sordidus, Ephippiger curciger, Ephippigerida nigromarginata, Hesperotettix pratensis, Homorocoryphus nitidulus, Melanoplus angustipennis, M. atlantis, M. bivitattus, M. coloradoensis, M. differentialis, M. femurrubrum, M. keeleri, M. luridus, M. mexicanus, M. obovatipennis, M. scudderi, Oedipoda germanica, O. miniata, Parapropacris rhodoptera, Parasphena ruandensis, Pardalophora apiculata, Phaneroptera* sp., *Saga pedo, Schistocerca americana, Serpusia lemarineli, Spharagemon bolli,* and *S. collare* (Orthopterorida) intestine.

248. *G. rivularis* Geus, 1969 in *Omalium rivulare* (Coleopterorida) intestine.

249. *G. rondoni* Théodoridès and Desportes, 1966 in *Cyrphaeus gazella* (Coleopterorida) intestine.

250. *G. rostrata* Wellmer, 1911 in *Lagria hirta, L. villosa, Chrysolagria arthritica, C. basicornis, C. ituriensis, C. metallica,* and *C. rothschildi* (Coleopterorida) intestine.

251. *G. rotundicephala* H. Hoshide, 1958 in *Mycetophagus* sp. (Coleopterorida) intestine.

252. *G. ruszkowskii* Lipa, 1967 in *Coccinella septempunctata* and *C. quinquepunctata* (Coleopterorida) intestine.

253. *G. sandoni* Harrison, 1955 in *Melanosilpha capensis* (Orthopterorida) intestine.

254. *G. scaphosomatis* Rauchalles in Geus, 1969 in *Scaphosoma agaricinum* (Coleopterorida) intestine.

255. *G. scapsipedi* H. Hoshide, 1958 emend. (syn., *G. scapsipedae* H. Hoshide, 1958) in *Scapsipedus asperno* (Orthopterorida) intestine.

256. *G. scarabaei* Leidy, 1851 (syns., *G. scarabaei relicti* Leidy, 1851; *G. scarabaeirelicti* Crawley, 1903) in *Scarabaeus relictus* (Coleopterorida) location in host not stated.

257. *G. segmentata* Vincent, 1924 ([*non*] *G. segmentata* Georgévitch, 1951) in *Cis bidentatus* (Coleopterorida) midgut.

258. *G. sericostomae* A. M. Baudoin, 1966 in *Sericostoma personatum* and *Notidobia ciliaris* (Trichopterorida) intestine.

259. *G. serpentula* de Magalhaes, 1900 in *Periplaneta orientalis, Epilampra kerandrenii,* and *Paranauphaeta discoidalis* (Orthopterorida) intestine.

260. *G. signata* Geus, 1969 in *Ecdyonorus forcipula* (Ephemeropterorida) midgut.

261. *G. socialis* Léger, 1906 in *Eryx ater* (Lepidopterorida) intestine.

262. *G. soroniae* Foerster, 1938 in *Soronia punctatissima, S. grisea,* and *Rhizophagus bipustulatus* (Coleopterorida) intestine.

263. *G. soror* Dufour, 1837 in *Phymata crassipes* (Insectasida) intestine.

264. *G. spinata* Geus, 1969 in *Silvanus unidentatus* (Coleopterorida) intestine.

265. *G. spraguei* Haldar and Chakraborty, 1978 in Cuculionidae gen. sp. (Coleopterorida) midgut.

266. *G. statirae* Frenzel, 1892 in *Statira unicolor* (Coleopterorida) intestine.

267. *G. steini* Berndt, 1902 in *Tenebrio molitor* (Coleopterorida) intestine.

268. *G. stenophylacis* Tsvetkov, 1929 in *Stenophylax* sp. (Trichopterorida) intestine.

269. *G. stigmae* Shtein, 1960 in *Limnophilus stigma* and *L. rhombicus* (Trichopterorida) intestine.

270. *G. straeleni* Théodoridès and Jolivet, 1959 in *Solanophila karisimbica* ab. *atrometra, Epilachna annulata, E. apicalis, E. bissexpustulata, E. lucifera, E. scutellaris, E. serva, E. connectens* ab. *enucleata, E. c.* ab. *joliveti, E. wittei* ab. *bentensis, E. w.* ab. *sine nomen, Solanophila apicornis, S. a.* ab. *fulvicollis, S. a.* ab. *separata, S. a.* ab. *sine nomen, S. gibbosa* ab. *negligens, S. kaffaensis* ab. *patherina, S. karisimbica, S. rubropustulata,* and *Solanophila* sp. (Coleopterorida) intestine.

271. *G. stygia* M. E. Watson, 1915 in *Ceuthophylus stygicus* (Orthopterorida) intestine.

272. *G. subtilis* Geus, 1967 in *Ennearthron affine* and *Ennearthron* sp. (Coleopterorida) midgut.

273. *G. tahitiensis* Corbel, 1968 in *Gryllodes sigillatus* (Orthopterorida) intestine.

274. *G. tenebrionella* M. E. Watson, 1915 in Tenebrionidae gen. sp. (Coleopterorida) intestine.

275. *G. tenuis* (Hammerschmidt, 1838) von Frantzius, 1848 (syns., *Clepsidrina tenuis* Hammerschmidt, 1838; *G. tenius* von Frantzius, 1848 of Geus [1969] *lapsus calami*) in *Allecula morio* (Coleopterorida) location in host not stated.

276. *G. tetricis* Corbel, 1968 in *Tetrix tenuicornis* (Orthopterorida) intestine.

277. *G. thomasi* Semans, 1939 in *Parcoblatta pennsylvanica* (Orthopterorida) intestine.

278. *G. thoracotermitis* Gisler, 1967 in *Thoracotermes macrothorax* and *Euchilotermes tensus* (Isopterorida) intestine.

279. *G. tokonoi* Obata, 1953 in *Uloma latimanus* (Coleopterorida) intestine.

280. *G. trichopterae* Levine, 1980 (syn., *G. segmentata* Georgévitch, 1951; [*non*] *G. segmentata* Vincent, 1924) in Trichopteroridae gen. sp. (Trichopterorida) intestine.

281. *G. trogophloei* Rauchalles in Geus, 1969 in *Trogophloeus bilineatus* and *T. rivularis* (Coleopterorida) midgut.

282. *G. tubuliformis* Geus, 1967 in *Sericostoma timidum* (Trichopterorida) intestine.

283. *G. tuerkorum* Geus, 1969 in *Ptinus latro* (Coleopterorida) location in host not stated.

284. *G. tuzetae* Théodoridès, Ormières, and Jolivet, 1958 in *Serpusia lemarineli* (Orthopterorida) location in host not stated.

285. *G. typographi* Fuchs, 1915 in *Ips typographicus, I. sexdentatus,* and *Dryocoetes autographus* (Coleopterorida) intestine.

286. *G. udeopsyllae* M. E. Watson, 1916 in *Udeopsylla nigra* (Orthopterorida) midgut.

287. *G. ulomae* H. Hoshide, 1951 in *Uloma latimanus* and *Taraxides laevigatus* (Coleopterorida) intestine.

288. *G. umbilicata* Geus, 1968 in *Cis boleti, C. hispidus, C. nitidus,* and *Sulcacis affinis* (Coleopterorida) intestine.

289. *G. uniducta* Devdhar, 1962 in *Tenebrionides mauritanica* (Coleopterorida).

290. *G. vellardi* (Corbel, 1968) Levine, 1979 (syn., *Gigaductus vellardi* Corbel, 1968) in *Gryllus* sp. *aff. assimilis* (Orthopterorida) intestine.

291. *G. verroni* Théodoridès, Desportes, and Jolivet, 1965 in *Pogonobasis raffrayi* (Coleopterorida) intestine.

292. *G. vittata* Rauchalles in Geus, 1969 in *Limnophilus bipunctatus, L. griseus,* and *L. vittatus* (Trichopterorida) intestine.

293. *G. vizri* Lipa, 1968 in *Zabrus tenebrioides* (Coleopterorida).

294. *G. vulgata* Foerster, 1938 in *Ephemera vulgata* (Ephemeropterorida) intestine.

295. *G. wahrmani* Théodoridès, 1955 (syn., *G. wharmani* Théodoridès, 1955 of Amoji and Rodgi [1976] *lapsus calami*) in *Scaurus puncticollis* var. *rugicollis* (Coleopterorida) location in host not stated.

296. *G. watsonae* Pinto, 1918 emend. (syn., *G. watsoni* Pinto, 1918) in *Omoplata normalis* (Coleopterorida) intestine.

297. *G. wellmeri* Tsvetkov, 1929 in *Sminthurus fuscus, S. marginatus, Dicyrtoma atra, D. ornata,* and *Tetrodontophora bielanensis* (Collembolorida) intestine.

298. *G. wolfi* Geus, 1969 in *Nauphoeta* sp. (Orthopterorida) intestine.

Genus *Erhardovina* Levine, 1985. Similar to *Gregarina,* but in mites. TYPE SPECIES *E. scutovertexi* (Erhardová, 1955) Levine, 1985.

1. *E. bisphaera* (Purrini and Ormières, 1981) nov. comb. (syn., *Gregarina bisphaera* Purrini and Ormières, 1981) in *Damaeus onustus, D. clavipes,* and *Eupelops hirtus* (Acarinorida) intestine.

2. *E. carabodis* (Purrini and Ormières, 1981) emend., nov. comb. (syn., *Gregarina carabodesi* Purrini and Ormières, 1981) in *Carabodes coriaceus* (Acarinorida) intestine. (*Remarks.* The genitive of words ending in es is is, not esi, so the name must be emended.)

3. *E. euzeti* (Lipa, 1982) Levine, 1985 (syn., *Gregarina euzeti* Lipa, 1982) in *Euzetes seminulum* (Acarinorida) intestine.

4. *E. fuscozetis* (Purrini, Bukva, and Bäumler, 1979) nov. comb. (syns., *Gregarina fuscozetis* Purrini, Bukva, and Bäumler, 1979; *G. fuscocetis* Purrini, Bukva, and Bäumler, 1979 *lapsus calami; G. fusczetis* Purrini, Bukva, and Bäumler, 1979 *lapsus calami*) in *Fuscozetes setosus* (Acarinorida) location in host not given.

5. *E. phtiracari* (Purrini and Ormières, 1981) nov. comb. (syn., *Gregarina phtiracari* Purrini and Ormières, 1981) in *Phtiracarus globosus* and *P. piger* (Acarinorida) intestine.

6. *E. platynothri* (Purrini and Ormières, 1981) nov. comb. (syn., *Gregarina platynothri* Purrini and Ormières, 1981) in *Platynothrus peltifer* (Acarinorida) intestine.

7. *E. postneri* (Purrini, Bukva, and Bäumler, 1979 (nov. comb. (syn., *Gregarina postneri* Purrini, Bukva, and Bäumler, 1979) in *Hermannia gibba* (Acarinorida) ceca.

8. *E. scutovertexi* (Erhardová, 1955) Levine, 1985 (TYPE SPECIES) (syns., *Gregarina scutovertexi* [Erhardová, 1955; *Arachnocystis scutovertexi* [Erhardová, 1955]) in *Scutovertex minutus* (Acarinorida) intestine.

Genus *Gymnospora* Moniez, 1886. Gametocyst with up to eight sporoducts; with spherical oocysts. TYPE SPECIES *G. nigra* Moniez, 1886.

1. *G. nigra* Moniez, 1886 (TYPE SPECIES) in *Vanessa urticae* (Lepidopterorida) presumably intestine.

Genus *Triseptata* H. Hoshide, 1958. Gamonts biassociative, cylindrical, with body composed of three segments separated by septa; epimerite a simple ovoid knob; oocysts ellipsoidal, extruded in chains. TYPE SPECIES *T. fungicola* H. Hoshide, 1958.

1. *T. fungicola* H. Hoshide, 1958 (TYPE SPECIES) in Elotylidae gen. sp. (Coleopterorida) intestine.

Genus *Gamocystis* Schneider, 1875. Protomerite transitory (i.e., only early gamonts septate, a feature which differentiates this genus from *Gregarina*); syzygy very early; gamont body cylindrical, round, or tongue-shaped; oocysts cylindrical; in intestine of insects. TYPE SPECIES *G. tenax* Schneider, 1875.

1. *G. cloenonis* Geus, 1969 in larva of *Cloeon* sp. (Ephemeropterorida) location in host not stated.
2. *G. fimetarii* Cordua, 1953 in *Aphodius fimetarius* (Coleopterorida) intestine.
3. *G. humilis* Geus, 1967 in *Ephemera vulgata* (Ephemeropterorida) intestine.
4. *G. obliterati* Cordua, 1953 in *Aphodius obliteratus* and *A. contaminatus* (Coleopterorida) intestine.
5. *G. salganeae* Théodoridès, Desportes, and Jolivet, 1972 in *Salganea morio* (Orthopterorida) intestine.
6. *G. tachytae* Geus, 1969 in *Tachyta nana* (Coleopterorida) location in host not stated.
7. *G. tenax* Schneider, 1875 (TYPE SPECIES) in *Ectobius* (syn., *Blatta) lapponicus, E. livens,* and *E. silvestris* (Orthopterorida) intestine.

Genus *Anisolobus* Vincent, 1924. Mucron apparently absent, even in the youngest stages; protomerite forms a strong fixation sucker; syzygy very early; gametocysts ellipsoidal, with thick wall; six to eight sporoducts present; oocysts doliform, emitted in chains; in insects. TYPE SPECIES *A. dacnecola* Vincent, 1924.

1. *A. aleocharae* Rauchalles in Geus, 1969 in *Aleochara intricata* (Coleopterorida) intestine.
2. *A. bulliardi* Théodoridès and Jolivet, 1959 in *Autispyris planicollis* (Coleopterorida) intestine.
3. *A. dacnecola* Vincent, 1924 (TYPE SPECIES) in *Dacne rufifrons* (Coleopterorida) intestine.
4. *A. desportis* Théodoridès and Jolivet, 1981 in *Cratopus frappieri* (Coleopterorida) intestine.
5. *A. gymnopholi* Théodoridès, Desportes, and Jolivet, 1972 in *Gymnopholus marquarti* (Coleopterorida) intestine.
6. *A. theodoridesi* Levine, 1984 (syn., *Anisolobus* sp. Théodoridès and Desportes, 1966) in *Cryphaeus gazella* (Coleopterorida) intestine.

Genus *Garnhamia* Crusz, 1957. Mucron papillate to acicular; without septum between protomerite and deutomerite; syzygy occurs while primite is still attached to midgut epithelium of host; oocysts ovoid, extruded in chains; in silverfish. TYPE SPECIES *G. aciculata* (Bhatia, 1938) Crusz, 1957.

1. *G. aciculata* (Bhatia, 1938) Crusz, 1957 (TYPE SPECIES) (syns., *Gregarina aciculata* Bhatia, 1938; *G. ctenolepismae* Lindsay, 1939) in *Lepisma saccharina, Ctenolepisma longicaudata, Peliolepisma calva,* and other Lepismatidae (Thysanurorida) intestine.

Genus *Torogregarina* Geus, 1969. Protomerite of young gamont with a broad swelling at its base; gametocysts round; oocysts ovoid. TYPE SPECIES *T. stammeri* Rauchalles in Geus, 1969.

1.   *T. stammeri* Rauchalles in Geus, 1969 (TYPE SPECIES) in *Neodendron fasciculare* (Coleopterorida) location in host not stated.

Genus *Faucispora* Baudoin, 1967. Oocysts with ellipsoidal endospore and with epispore distended at both poles to form spouts closed by a small clap-valve. TYPE SPECIES *F. phryganeae* (Berg-von-Emme, 1912) Baudoin, 1967.

1.   *F. phryganeae* (Berg-von-Emme, 1912) Baudoin, 1967 (syn., *Diplocystis phryganeae* Berg-von-Emme, 1912) (TYPE SPECIES) in *Phryganea striata, Phryganea* sp., *Agrypnia pagetana,* and *A. obsoleta* (Trichopterorida) gamonts on coelomic surface of larval intestine; gametocysts in coelom of adult.

Genus *Spinispora* Baudoin, 1967. Oocysts fusiform, covered with spines over their whole surface. TYPE SPECIES *S. dytisci* Baudoin, 1967.

1.   *S. dytisci* Baudoin, 1967 (TYPE SPECIES) in *Dytiscus marginalis* (Coleopterorida) coelom.

Genus *Bolivia* Corbel, 1968. Gamonts filiform; protomerite with an anterior depression and often with bulb at base; mucron (epimerite) not seen; deutomerite cylindrical, with fine longitudinal striations; gametocysts spherical, surrounded by a mucilaginous sheath, with a single short, broad sporoduct and a large residuum; oocysts doliform, not emitted in chains but in an agglomerated mass. TYPE SPECIES *B. vellardi* Corbel, 1968.

1.   *B. vellardi* Corbel, 1968 (TYPE SPECIES) in *Gryllus* sp. (Orthopterorida) intestine.

Genus *Degiustia* Levine, 1979. Gamont broadly oval, very small; mucron (epimerite) simple; microgametes with four pairs of flagella; in amphipods. TYPE SPECIES *D. hyalella* (Batten and DeGiusti, 1949) Levine, 1979.

1.   *D. hyalella* (Batten and DeGiusti, 1949) Levine, 1979 (syn., *Gregarina hyalella* Batten and DeGiusti, 1949) (TYPE SPECIES) in *Hyalella azteca* (Amphipodorida) intestine.

Genus *Cirrigregarina* Levine, 1979. Similar to *Gregarina,* but in barnacles. TYPE SPECIES *C. spissa* (Henry, 1938) Levine, 1979.

1.   *C. kamenote* (H. Hoshide, 1951) Levine, 1979 (syn., *Gregarina kamenote* H. Hoshide, 1951) in *Mitella mitella* (Cirripedasina) intestine.
2.   *C. spissa* (Henry, 1938) Levine, 1979 (TYPE SPECIES) (syn., *Gregarina spissa* Henry, 1938) in *Balanus crenatus, B. cariosus,* and *B. glandula* (Cirripedasina) intestine.

Genus *Molluskocystis* Levine, 1979. Similar to *Gregarina,* but in mollusks. TYPE SPECIES *M. pterotracheae* (Stuart, 1871) Levine, 1979.

1.   *M. pterotracheae* (Stuart, 1871) Levine, 1979 (syns., *Zygocystis pterotracheae* Stuart, 1871; *Gregarina pterotracheae* [Stuart, 1871] Labbé, 1899) in *Pterotrachea* sp. (Gastropodasida) coelom.

Family METAMERIDAE *Levine, 1979*
   Mucron simple; gamont with secondary segmentation of protomerite and deutomerite; in annelids.

Genus *Metamera* Duke, 1910. Mucron (epimerite) subconical, eccentric, with many branched digitiform processes; gametocysts open by simple rupture; oocysts biconical, navicular; in leeches. TYPE SPECIES *M. schubergi* Duke, 1910.

1.   *M. reynoldsi* Jones, 1943 in *Glossosiphonia complanata* (Hirudinasida) intestine.
2.   *M. schubergi* Duke, 1910 (TYPE SPECIES) in *Glossosiphonia complanata* and *Hemiclepsis marginata* (Hirudinasida) intestine.

Genus *Gopaliella* Ganapati, Kalavati, and Sundaram, 1974. Mucron (epimerite) umbrella-like, with a central, deeply stained rod; body with many segments (usually 8, occasionally 11); gametocysts dehisce by simple rupture; gametes similar; oocysts biconical, with 8 sporozoites; in polychetes. TYPE SPECIES *G. marphysae* Ganapati, Kalavati, and Sundaram, 1974.

1.   *G. marphysae* Ganapati, Kalavati, and Sundaram, 1974 (TYPE SPECIES) in *Marphysa gravelyi* (Polychaetasida) intestine.

Genus *Deuteromera* Bhatia and Setna, 1938. Mucron subconical, with a cup-shaped apex; protomerite and deutomerite of gamont with incomplete secondary segmentation; syzygy, gametocysts, and oocysts unknown; in polychetes. TYPE SPECIES *D. cleava* Bhatia and Setna, 1938.

1.   *D. cleava* Bhatia and Setna, 1938 (TYPE SPECIES) in *Eunice siciliensis* (Polychaetasida) intestine.

Genus *Cognettiella* Pižl, Chalupský, and Levine, 1983. Mucron simple; protomerite and deutomerite of gamont with complete secondary segmentation; syzygy, gametocysts, and oocysts unknown; in oligochetes. TYPE SPECIES *C. legeri* (Cognetti de Martilis, 1911) Pižl, Chalupský, and Levine, 1983.

1.   *C. legeri* (Cognetti de Martiis, 1911) Pižl, Chalupský, and Levine, 1983 (syn., *Taeniocystis legeri* Cognetti de Martiis, 1911; *Cognettia legeri* [Cognetti de Martiis, 1911] Levine, 1979) (TYPE SPECIES) in *Kynotus pitarellii* (Oligochaetasida) coelom.

Family DIDYMOPHYIDAE *Léger, 1892*
   Septum of satellite resorbed more or less slowly during syzygy; gametocysts spherical or somewhat elongate; oocysts with a loose epispore and an ellipsoidal endospore; oocysts emitted in packets enveloped by a thin membrane (secondary cysts) by means of a single, very long sporoduct with a thin wall.

Genus *Didymophyes* von Stein, 1848. Mucron (epimerite) cylindroconical, very degenerate (a small, pointed papilla); deutomerite with needle-like crystalloids of unknown chemical nature in endoplasm; gametocyst usually spherical; oocysts grouped in spherical packets in gametocysts. TYPE SPECIES *D. gigantea* von Stein, 1848.

1.   *D. africanus* Théodoridès and Jolivet, 1959 in *Heliocopris colossus* and *Onitis fabricii* (Coleopterorida) intestine.

2.   *D. aphodii* Foerster, 1938 in *Aphodius distinctus, A. rufus, A. scrutator, A. prodromus, Aphodius* sp., *Onthophagus fracticornis,* and *O. nuchicornis* (Coleopterorida) intestine.
3.   *D. bessei* Ormières, 1968 in *Aphodius haemorhoidalis* (Coleopterorida) intestine.
4.   *D. caudata* Cordua, 1953 in *Aphodius fossor* and *A. rufipes* (Coleopterorida) intestine.
5.   *D. cercyonis* Cordua, 1953 in *Cercyon impressus* and *C. lateralis* (Coleopterorida) intestine.
6.   *D. cerylonis* Geus, 1969 in *Cerylon ferrugineum* and *C. histeroides* (Coleopterorida) intestine.
7.   *D. chaudefouri* Ormières, 1968 in *Aphodius depressus* and *Aphodius* sp. (Coleopterorida) intestine.
8.   *D. crassa* Cordua, 1953 in *Aphodius distinctus, A. prodromus, A. sphacelatus,* and *A. sordidus* (Coleopterorida) intestine.
9.   *D. cryptopleuri* Geus, 1969 in *Cryptopleurum minutum* (Coleopterorida) location in host not stated.
10.   *D. cuneicystis* Ormières, 1968 in *Aphodius rufipes* (Coleopterorida) intestine.
11.   *D. diminuta* Obata, 1953 in *Aphodius rectus biformia* (Coleopterorida) intestine.
12.   *D. fimetarii* Cordua, 1953 in *Aphodius fimetarius* (Coleopterorida) intestine.
13.   *D. gigantea* von Stein, 1848 (syns., *Gregarina gigantea* Diesing, 1859; *G. gigantea* Lankester, 1863) (TYPE SPECIES) in *Oryctes nasicornis, Oryctes* sp., *Phyllognathus* sp., and *Xylotrupes dichotomus* (Coleopterorida) intestine.
14.   *D. gracilis* Cordua, 1953 in *Aphodius distinctus, A. prodromus,* and *A. sphacelatus* (Coleopterorida) intestine.
15.   *D. guttiformis* Cordua, 1953 in *Aphodius fimetarius, A. foetens, A. alpinus,* and *Aphodius* sp. (Coleopterorida) intestine.
16.   *D. hydrobiina* Foerster, 1938 in *Hydrobius fuscipes, Enochrus frontalis,* and *Anacaena limbata* (Coleopterorida) intestine.
17.   *D. joliveti* Théodoridés and Desportes, 1965 in *Scarabaeus* sp. (Coleopterorida) intestine.
18.   *D. lanceolata* Cordua, 1953 in *Cercyon lateralis, C. quisquilius,* and *C. haemorhoidalis* (Coleopterorida) intestine.
19.   *D. lata* Cordua, 1953 in *Aphodius scyabalarius* (Coleopterorida) intestine.
20.   *D. leuckarti* Marshall, 1893 in *Aphodius prodromus, A. nitidulus, A. coenosus, A. contaminatus, A. distinctus, A. eraticus, A. fossor, A. granarius, A. haemorhoidalis, A. luridus, A. niger, A. rufipes, A. rufus, A. sordidus, A. sphacelatus, A. sticticus, A. subterraneus, A. tessulatus,* and *Aphodius* sp. (Coleopterorida) intestine.
21.   *D. longa* Cordua, 1953 in *Onthophagus fracticornis* and *O. nuchicornis* (Coleopterorida) intestine.
22.   *D. macrocystis* Cordua, 1953 in *Cercyon impressus* and *C. lateralis* (Coleopterorida) intestine.
23.   *D. microcephala* Cordua, 1953 in *Cercyon impressus* and *C. lateralis* (Coleopterorida) intestine.
24.   *D. minima* Geus, 1969 in *Cercyon pygmaeus* (Coleopterorida) midgut.
25.   *D. onthophagi* Foerster, 1938 in *Onthophagus fracticornis* and *O. coenobita* (Coleopterorida) intestine.
26.   *D. ovalocephala* Cordua, 1953 in *Onthophagus vacca* (Coleopterorida) intestine.
27.   *D. ovalocystis* Cordua, 1953 in *Aphodius merdarius, A. nitidulus,* and *A. granarius* (Coleopterorida) midgut.
28.   *D. ovati* Cordua, 1953 in *Onthophagus fracticornis* and *O. ovatus* (Coleopterorida) intestine.
29.   *D. paradoxa* von Stein, 1848 (syns., *Didymophyes rara* Léger, 1892; *Gregarina paradoxa* [von Stein, 1848] Lankester, 1863) in *Geotrupes stercorarius, G. mutator, G. niger, G. pyrenaeus, G. spiniger, G. stercorosus, G. vernalis,* and *Geotrupes* sp. (Coleopterorida) intestine.

30. *D. parva* Cordua, 1953 in *Aphodius distinctus, A. prodromus,* and *A. sphacelatus* (Coleopterorida) intestine.
31. *D. rotunda* Foerster, 1938 in *Onthophagus ovatus* (Coleopterorida) intestine.
32. *D. scarabaei* Théodoridès, 1955 in *Scarabaeus laticollis* (Coleopterorida) location in host not stated.
33. *D. sisyphi* Théodoridès, 1955 in *Sisyphus schaefferi* (Coleopterorida) gamonts in intestine, gametocysts in coelom.
34. *D. sphaeridii* Cordua, 1953 in *Sphaeridium bipustulatum* and *S. scarabaeoides* (Coleopterorida) intestine.
35. *D. sphaerocephala* Cordua, 1953 in *Sphaeridium bipustulatum* (Coleopterorida) intestine.
36. *D. tuzetae* Théodoridès, 1955 in *Copris lunaris* (Coleopterorida) intestine.

Genus *Liposcelisus* Sarkar and Haldar, 1980. Epimerite spherical, hyaline, with a corona of 14 to 16 ridges; gametocysts spherical, dehiscing by simple rupture; oocysts spindle-shaped. TYPE SPECIES *L. coronatus* Sarkar and Haldar, 1980.

1.   *L. coronatus* Sarkar and Haldar, 1980 in *Liposcelis* sp. (Psocopterorida) midgut.

Genus *Quadruhyalodiscus* Kundu and Haldar, 1984. Epimerite spherical, with a corona of four sucker-like hyaline disks at the anterior end provided with a very short stalk; gamonts biassociative; gametocysts dehisce by simple rupture; oocysts spindle-shaped, with polar thickenings; extracellular development unknown. TYPE SPECIES *Q. gallerucidae* Kundu and Haldar, 1984.

1.   *Q. gallerucidae* Kundu and Haldar, 1984 in *Gallerucida bicolor* (Coleopterorida) midgut.

Family HIRMOCYSTIDAE *Grassé, 1953*
  Mucron (epimerite) ordinarily papilla-like or simple knob-like; gametocysts dehisce by simple rupture; oocysts ellipsoidal, prismatic, fusiform, ovoid, or even spherical.

Genus *Hirmocystis* Labbé, 1899. Mucron (epimerite) very labile, a conical button; gamonts most often associated in chains of up to 12 individuals; oocysts ovoid, ellipsoidal, or cylindroid; in insects. TYPE SPECIES *H. ventricosa* (Léger, 1892) Labbé, 1899.

1.   *H. asidae* (Léger, 1896) Labbé, 1899 (syn., *Eirmocystis asidae* Léger 1896) in *Asida servillei* (Coleopterorida) digestive tract.
2.   *H. bengalensis* Haldar and Chakraborty, 1979 in *Myllocerus* sp. (Coleopterorida) intestine.
3.   *H. capritermes* Uttangi and Desai, 1961 in *Capritermes incola* (Isopterorida) esophagus and foregut.
4.   *H. completa* Hasselmann, 1926 in Locustidae gen. sp. (Orthopterorida) intestine.
5.   *H. ctenocephali* (Kamm, 1922) Levine, 1980 (syns., *Gregarina ctenocephalus canis* Ross, 1909; *G. ctenocephali* Kamm, 1922) in *Ctenophalides canis* (Siphonapterorida) intestine.
6.   *H. dharwarensis* Uttangi and Desai, 1962 in *Speculitermes cyclops sinhalensis* (Isopterorida) foregut.
7.   *H. gryllotalpae* (Léger, 1892) Labbé, 1899 (syns., *Eirmocystis gryllotalpae* Léger, 1892; *Gregarina sphaerulosa* Dufour, 1837) in *Gryllotalpa gryllotalpa, Gryllotalpa* sp., and *Oedipoda coerulescens* (Orthopterorida) intestine.

8.   *H. harpali* M. Watson, 1916 in *Harpalus pennsylvanicus erythropus (Coleopterorida)* intestine.
9.   *H. hoplasomae* Kundu and Haldar, 1981 in *Hoplasoma unicolor* (Coleopterorida) midgut.
10. *H. inaequalis* Tuzet and Théodoridès, 1951 in *Asida sericea* (Coleopterorida) intestine.
11. *H. incola* Uttangi and Desai, 1961 in *Capritermes incola* (Isopterorida) gut.
12. *H. indica* Uttangi and Desai, 1961 in *Capritermes incola* (Isopterorida) gut.
13. *H. laemophloei* (Foerster, 1938) Levine, 1980 (syn., *Gregarina laemophloei* Foerster, 1938) in *Laemophloeus ferrugineus* (Coleopterorida) intestine.
14. *H. lepropi* Haldar and Chakraborty, 1981 in *Lepropus* sp. (Coleopterorida) gut.
15. *H. locustanae* (Fantham, 1925) Corbel, 1968 (syn., *Gregarinoides locustanae* Fantham, 1925) in *Locustana pardalina, Gastrimargus verticalis, Lentula* sp., and other "common species of grasshoppers" in Natal (Orthopterorida) intestine.
16. *H. minima* Geus, 1969 in *Omalium caesum* (Coleopterorida) location in host not stated.
17. *H. mirabilis* H. Hoshide, 1951 in *Lyprops sinensis* (Coleopterorida) intestine.
18. *H. mycetocharae* Rauchalles in Geus, 1969 in *Mycetochara linearis* (Coleopterorida) intestine.
19. *H. ovalis* Crawley, 1903 (syns., *Gregarina ovalis* [Crawley, 1903] Watson, 1916; *G. elaterae* Ellis, 1913) in Cucujidae gen. sp. (Coleopterorida) intestine.
20. *H. pierrei* (Théodoridès, 1959) Théodoridès, 1980 (syn., *Gregarina pierrei* Théodoridès, 1959) in *Coxelinus pauliana* (Coleopterorida) intestine.
21. *H. pitcharis* Haldar and Chakraborty, 1979 in *Xanthoprochilus* sp. (Coleopterorida) midgut.
22. *H. polymorpha* (Léger, 1892) Labbé, 1899 (syn., *Eirmocystis polymorpha* Léger, 1892) in *Limnobia* sp., *Erioptera* sp., *Symplecta* sp. (Dipterorida), and *Systenocerus caraboides* (Coleopterorida) digestive tract.
23. *H. pseudoductis* Haldar and Chakraborty, 1979 in *Mylloceros* sp. (Coleopterorida) midgut.
24. *H. pterygospora* Crumpton, 1974 in *Pericoptus truncatus* (Coleopterorida) midgut.
25. *H. socialis* Geus, 1969 in *Scaphosoma agaricinum* (Coleopterorida) location in host not stated.
26. *H. speculitermitis* Uttangi and Desai, 1962 emend. (syn., *H. speculitermis* Uttangi and Desai, 1962) in *Speculitermes cyclops sinhalensis* (Isopterorida) foregut.
27. *H. termitis* (Leidy, 1881) Henry, 1933 (syn., *Gregarina termitis* Leidy, 1881) in *Zootermopsis angusticollis, Z. nevadensis,* and Isopterorida gen. sp. (Isopterorida) intestine.
28. *H. theodoridesi* Kundu and Haldar, 1981 in *Gryllotalpa africana* (Orthopterorida) midgut.
29. *H. trichopteri* Tsvetkov, 1929 in *Limnophilus nigriceps* (Trichopterorida) intestine.
30. *H. ventricosa* (Léger, 1892) Labbé, 1899 (syn., *Eirmocystis ventricosa* Léger, 1892) (TYPE SPECIES) in *Tipula oleracea, T. paludosa, Tipula* sp., and *Pales pratensis* (Dipterorida) digestive tract.

Genus *Hyalospora* Schneider, 1875. Mucron (epimerite) a globular button; young gamont endoplasm orange-yellow; oocysts ellipsoidal or fusiform. TYPE SPECIES *H. roscoviana* Schneider, 1875.

1.   *H. affinis* Schneider, 1882 in *Lepisma saccharina* (syn., *Machilus cylindrica* (Thysanurorida) intestine.
2.   *H. hemerobii* Geus, 1969 in *Hemerobium pini* (Neuropterorida) intestine.

3. *H. psocorum* (von Siebold, 1839) Wellmer, 1911 (syns., *Gregarina psocorum* von Siebold, 1839; *G. ovata* Diesing, 1851; *? psocorum* [von Siebold, 1839] Kamm, 1922) in *Psocus quadripunctatus, P. longicornis, Psocus* sp., *Amphigerontia bifasciata, Caecilius flavitus, Craphopsocus cruciatus, Lachesilla quercus, Mesopsocus unipunctatus,* and *Stenopsocus maculatus* (Psocopterorida) intestine.

4. *H. roscoviana* Schneider, 1875 (TYPE SPECIES) in *Holomachilis* (syn., *Petrobius) maritimus, Machilis tenuis,* and *Machilis* sp. (Thysanurorida) intestine.

5. *H. stenopsoci* Geus, 1969 in *Stenopsocus* sp. (Psocopterorida) intestine.

6. *H. volsella* Tuzet and Ormières, 1956 in *Atelura pseudolepisma* (Thysanurorida) intestine.

Genus *Tettigonospora* L. M. Smith, 1930. Poorly studied; with characters of *Hirmocystis;* mucron (epimerite) spherical; gamont cytoplasm opaque white; oocysts hyaline and spherical. TYPE SPECIES *T. stenopelmati* (L. M. Smith, 1929) L. M. Smith, 1930.

1. *T. stenopelmati* (L. M. Smith, 1929) L. M. Smith, 1930 (syns., *Coccospora stenopelmati* L. M. Smith, 1929; *Hirmocystis stenopelmati* [L. M. Smith, 1929] Corbel, 1968) in *Stenopelmatus fuscus* and *S. pictus* (Orthopterorida) intestine.

Genus *Euspora* Schneider, 1875. Poorly defined genus; gametocysts without sporoducts; oocysts prismatic. TYPE SPECIES *E. fallax* Schneider, 1875.

1. *E. fallax* Schneider, 1875 (TYPE SPECIES) in *Rhizotrogus aestivus, Rhizotrogus* sp., *Melolontha melolontha, M. vulgaris, Melolontha,* sp., *Allacula letestoi, Allacula* sp., *Asida servillei,* and *Monomma giganteum* (Coleopterorida) intestine.

2. *E. lucani* Crawley, 1903 (?) in *Lucanus dama* (Coleopterorida) location in host not stated.

3. *E. mesomorphi* Devdhar and Deshpande, 1977 in *Mesomorphus velliger* (Coleopterorida) gut.

4. *E. zealandica* Allison, 1969 in *Costelytra zealandica* and *Oryctes* sp. (Coleopterorida) intestine.

Genus *Tintinospora* H. Hoshide, 1959. Gamonts in associations of two or three; mucron (epimerite) a simple globular papilla; gametocysts dehisce by simple rupture; oocysts extruded in chains. TYPE SPECIES *T. soroniae* H. Hoshide, 1959.

1. *T. soroniae* H. Hoshide, 1959 (TYPE SPECIES) in *Soronia japonica* (Coleopterorida) intestine.

Genus *Arachnocystis* Levine, 1979. Mucron (epimerite) simple, spherical; gametocysts spherical, dehiscing by simple rupture, without residuum; oocysts biconical, with truncate ends; in arachnids. TYPE SPECIES *A. arachnoidea* (Devdhar and Gourishankar, 1971) Levine, 1979.

1. *A. arachnoidea* (Devdhar and Gourishankar, 1971) Levine, 1979 (syn., *Sycia arachnoidea* Devdhar and Gourishankar, 1971) (TYPE SPECIES) in *Opalnia* sp. (Arachnidasida) intestine.

2. *A. nitida* (Geus, 1969) Levine, 1979 (syn., *Gregarina nitida* Geus, 1969) in *Damaeus* (syn., *Oribata) clavipes* (Acarinorida) intestine.

3. *A. oribatae* (Geus, 1969) Levine, 1979 (syn., *Gregarina oribatae* Geus, 1969) in *Damaeus onustus* (syn., *Oribata geniculata*) (Acarinorida) intestine.

Genus *Protomagalhaensia* Pinto, 1918. Development intracellular; young gamonts always elongate; mucron (epimerite) unknown; syzygy head to tail, tongue-like; gametocysts dehisce by simple rupture; oocysts keg-shaped, with one spine at each corner. TYPE SPECIES *P. serpentula* (de Magalhaes, 1900) Pinto, 1918.

1. *P. blaberi* Peregrine, 1970 emend. (syn., *P. blaberae* Peregrine, 1970) in *Blaberus boliviensis* (Orthopterorida) intestine.
2. *P. granulosae* Peregrine, 1970 in *Blaberus discoidalis* (Orthopterorida) intestine.
3. *P. serpentula* (de Magalhaes, 1900) Pinto, 1918 (syn., *Gregarina serpentula* de Magalhaes, 1900) (TYPE SPECIES) in *Periplaneta orientalis* and *Stylopyga americana* (Orthopterorida) intestine and coelom.

Genus *Pintospora* Carini, 1944. Gamonts at first intra- and then extracellular; gametocysts with smooth wall; oocysts united in pairs, each one with thickened, rather pointed ends, with smooth walls, frequently free in coelomic cavity of host. TYPE SPECIES *P. bigemina* Carini, 1944.

1. *P. bigemina* Carini, 1944 (TYPE SPECIES) in *Astylus atromaculatus* (Coleopterorida) intestine.

Genus *Endomycola* Théodoridès, Desportes, and Jolivet, 1972. Gamonts solitary and globular, with button-like epimerite (mucron?) and spherical nucleus; entocyte and ectocyte very chromophilic; gametocysts and oocysts unknown. TYPE SPECIES *E. baiyeri* Théodoridès, Desportes, and Jolivet, 1972.

1. *E. baiyeri* Théodoridès, Desportes, and Jolivet, 1972 (TYPE SPECIES) in *Encymon ruficollis* (Coleopterorida) intestine.

Genus *Retractocephalus* Haldar and Chakraborty, 1976. Epimerite globular, retractile into protomerite; initial development intracellular; syzygy head to tail; dehiscence of gametocyst by simple rupture; oocysts barrel-shaped, extruded in chains. TYPE SPECIES *R. raphidopalpae* Haldar and Chakraborty, 1976.

1. *R. aulacophorae* Haldar, Chakraborty, and Kundu, 1982 in *Aulacophora intermedia* (Coleopterorida) midgut.
2. *R. halticae* Haldar, Chakraborty, and Kundu, 1982 emend. Levine, 1985 (syn., *Retractocephalus halticus* Haldar, Chakraborty, and Kundu, 1982) in *Haltica* sp. (Coleopterorida) midgut.
3. *R. raphidopalpae* Haldar and Chakraborty, 1976 emend. Haldar, Chakraborty, and Kundu, 1982 (syn., *R. raphidopalpii* Haldar and Chakraborty, 1976) in *Raphidopalpa* (syn., *Aulacophora) foveicollis* (Coleopterorida) midgut.
4. *R. spatulatus* Haldar, Chakraborty, and Kundu, 1982 in *Lema* sp. (Coleopterorida) midgut.
5. *R. spinosus* Haldar, Chakraborty, and Kundu, 1982 in *Monolepta signata* (Coleopterorida) midgut.

Superfamily STENOPHORICAE *Levine, 1984 (syn.,* SOLITARICAE *Chakravarty, 1960)*
    Gamonts solitary; development intra- or extracellular; mucron (epimerite) simple or complex; gametocysts without sporoducts; homoxenous.

Family STENOPHORIDAE *Léger and Duboscq, 1904*
    Early development intracellular; epimerite absent or reduced to an invaginable mucron;

syzygy head to tail, late (just before encystment); anisogamy accentuated; gametocysts open by simple rupture, eliminating oocysts singly; oocysts ovoid or ellipsoidal, with very ample epispore, with or without equatorial suture; in diplopods.

Genus *Stenophora* Labbé, 1899. Development intracellular; oocysts ovoid, with prominent equatorial suture. TYPE SPECIES *S. juli* (von Frantzius, 1846) Labbé, 1899.

1. *S. aculeata* Léger and Duboscq, 1904 in *Craspedosoma rawlinsii simile* (*C. slavum*) (Diplopodasida) intestine.
2. *S. adjanohouni* Gasc and Bouix, 1976 in *Habrodesmus falx* (Diplopodasida) stomodeum.
3. *S. akiyoshiensis* H. Hoshide, Wakagi, and K. Hoshide, 1970 in *Skleroprotopus ikedai* (Diplopodasida) intestine.
4. *S. beroni* Golemansky, 1973 in *Balkanopetalum armatum* (Diplopodasida) intestine.
5. *S. broelemanni* Léger and Duboscq, 1903 in *Blaniulus hirsutus, Brachydesmus superus,* and *B. pusillus lusitanus* (Diplopodasida) intestine.
6. *S. bulgarosomae* Golemansky, 1973 in *Bulgarosoma bureschi* (Diplopodasida) intestine.
7. *S. cassidiformis* Rodgi and Ball, 1961 in *Strongylosoma* sp. (Diplopodasida) intestine.
8. *S. caudata* (Ishii, 1915) Kamm, 1922 (syn., *Spirosoma caudata* Ishii, 1915) in *Fontaria coarctata* (Diplopodasida) intestine.
9. *S. chordeumae* Léger and Duboscq, 1904 (syn., *S. chordueme* Léger and Duboscq, 1904 of Watson [1916] *lapsus calami*) in *Chordeuma sylvestre* and *Chordeuma* sp. (Diplopodasida) intestine.
10. *S. cockerellae* Ellis, 1912 in *Paraiulus* sp., *Orthomorpha coarctata,* and *Orthomorpha* sp. (Diplopodasida) intestine.
11. *S. congoi* Tuzet, Manier, and Jolivet, 1957 in *Spirostreptus ibanda* and provisionally *Brachyspirobolus cyclopygus* and *B. caudatus* (Diplopodasida) midgut.
12. *S. conjugata* Rodgi and Ball, 1961 in *Thyropygus nigralabiatus* (Diplopodasida) intestine.
13. *S. cordylopori* Gasc and Bouix, 1976 in *Cordyloporus ornatus* (Diplopodasida) stomodeum.
14. *S. cruzi* Pinto, 1918 in *Rhinocricus* sp. (Diplopodasida) intestine.
15. *S. cunhai* Pinto, 1918 in *Rhinocricus puglo* (Diplopodasida) intestine.
16. *S. dahomeensis* Gasc, Ormières, and Bouix, 1975 in *Habrodesmus falx* (Diplopodasida) intestine, stomodeum.
17. *S. dauphinia* M. Watson, 1916 (syn., *S. juli* Léger, and Duboscq, 1904) in *Schizophyllum* (syn., *Julus) mediterraneum, Cylindroiulus boleti* (syns., *Julus boleti, J. londinensis), Ophyiulus fallax* (syn., *Julus fallax), Choneilus palmatus,* and *Tachypodoiulus* (syn., *Julus) albipes* (Diplopodasida) intestine.
18. *S. diplocorpa* M. E. Watson, 1915 in *Euryurus erythropygus* (Diplopodasida) intestine.
19. *S. ellipsoidi* Chakravarty, 1934 in *Diplopoda* sp. (Diplopodasida) midgut.
20. *S. elongata* Ellis, 1912 in *Orthomorpha coarctata* (Diplopodasida) intestine.
21. *S. falciformis* Rodgi and Ball, 1961 in *Thyropygus* sp. (Diplopodasida) intestine.
22. *S. flexuosa* H. Hoshide, 1958 in *Orthomorpha* sp. (Diplopodasida) intestine.
23. *S. fontariae* (Crawley, 1903) M. Watson, 1916 (syn., *Amphoroides fontariae* Crawley, 1903) in *Polydesmus denticulatus, Polydesmus* sp., and *Fontaria* sp. (Diplopodasida) intestine.
24. *S. gongylorrhi* Amoji and Rodgi, 1970 emend. (syn., *S. gongylorrha* Amoji and Rodgi, 1970) in *Gongylorrhus* sp. (Diplopodasida) intestine.
25. *S. habrodesmi* Gasc and Bouix, 1976 in *Habrodesmus falx* (Diplopodasida) stomodeum.
26. *S. hagiensis* H. Hoshide, 1958 in *Rhysodesmus* sp. (Diplopodasida) intestine.
27. *S. haplothysani* Tuzet, Manier, and Jolivet, 1957 in *Haplothysanus emini* (Diplopodasida) intestine.

28. *S. hoshidei* Théodoridès, Desportes, and Jolivet, 1975 in *Orthomorpha uncinata* (Diplopodasida) intestine.
29. *S. impressi* M. E. Watson, 1915 emend. (syn., *S. impressa* M. E. Watson, 1915) in *Parajulus impressus* (Diplopodasida) intestine.
30. *S. jeekeli* Tuzet, Manier, and Jolivet, 1957 in *Spirostreptus virgator* (Diplopodasida) intestine.
31. *S. juli* (von Frantzius, 1846) Labbé, 1899 (syns., *Gregarina juli* von Frantzius, 1846; *G. paradoxa* Gabriel, 1880; *Sporodina juli* von Frantzius, 1848; *Stenocephalus juli* [von Frantzius, 1846] Schneider, 1875) (TYPE SPECIES) in *Julus terrestris, Craspedosoma alemannicum, Cylindroiulus londinensis, C. silvarum, Cylindroiulus* sp., *Ophyiulus fallax, Rhinocricus padbergi, Rhinocricus* sp., *Scaphiostreptus* sp., *Schizophyllum rutilans, S. mediterraneum, S. sabulosum,* and *Tachypodoiulus albipes* (Diplopodasida) intestine.
32. *S. julipusilli* Crawley, 1903 (syns., *Gregarina juli pusilli* Leidy, 1853; *G. juli* [von Frantzius, 1846] of Lankester [1863] in part) in *Julus pusillus, Julus* sp., *Craspedosoma alemannicum, Cylindroiulus* sp., *Hypsoiulus alpivagus, Leptophyllum nanum, Leptoiulus proximus, Paraiulus* sp., *Schizophyllum sabulosum, Tachypodoiulus albipes,* and *Unciger foetidus* (Diplopodasida) intestine.
33. *S. kampinosi* Lipa and Stachurska, 1968 in *Polydesmus complanatus* (Diplopodasida) intestine.
34. *S. karnataki* Karandikar and Rodgi, 1956 in *Thyropygus* sp. and *Isoporostreptos bouixi* (Diplopodasida) intestine, stomodeum.
35. *S. khagendrae* Ray, 1933 in *Zikadesmus* sp. and Diplopodasida gen. sp. (Diplopodasida) intestine.
36. *S. kojiroensis* H. Hoshide, 1958 in *Orthomorpha* sp. (Diplopodasida) intestine.
37. *S. lactaria* M. E. Watson, 1915 in *Callipus lactarius* (Diplopodasida) intestine.
38. *S. lagenata* Geus, 1969 in *Glomeris* sp. (Diplopodasida) midgut.
39. *S. larvata* (Leidy, 1849) Ellis, 1913 (syns., *Gregarina larvata* Leidy, 1849; *G. juli marginati* Leidy, 1853; *G. juli* von Frantzius, 1846 of Lankester [1863] in part; *S. juli* [von Frantzius, 1846] Labbé, 1899 in part) in *Spirobolus spinigerus* (syn., *Julus marginatus*) and *Schizophyllum sabulosum* (Diplopodasida) intestine.
40. *S. lipai* Levine, 1980 (syn., *S. caudata* Lipa, 1967) in *Chromatoiulus projectus* (Diplopodasida) intestine.
41. *S. lutzi* Pinto, 1918 in *Rhinocricus* sp. (Diplopodasida) intestine.
42. *S. mahabaleshwari* Amoji and Rodgi, 1972 in *Strongylosoma* sp. (Diplopodasida) intestine.
43. *S. mandrakae* Tuzet and Manier, 1955 in Spiropteridae gen. sp. (Diplopodasida) intestine.
43a. *S. mandrakae* forma *minima* Gasc and Bouix, 1976 in *Peridontopyge gasci, Isoporostreptus bouixi,* and *Plethocrossus acutiformis* (Diplopodasida) stomodeum.
44. *S. mruthunjayi* Rodgi and Amoji, 1970 in *Condromorpha kelaarti* (Diplopodasida) intestine.
45. *S. murozumiensis* H. Hoshide, 1952 in *Fusiulus* sp. (Diplopodasida) intestine.
46. *S. negotiosus* Ramachandran, 1974 in *Phyllogonostreptus negotiosus* (Diplopodasida) midgut.
47. *S. nematoides* Léger and Duboscq, 1903 in *Entothalassinum* (syn., *Strongylosoma) italicum, Orthomorpha gracilis,* and *Strongylosoma pallipes* (Diplopodasida) intestine.
48. *S. orobainosomatis* Geus, 1968 in *Orobainosoma* sp. (Diplopodasida) intestine.
49. *S. orthomorphae* H. Hoshide, 1952 in *Orthomorpha* sp. (Diplopodasida) intestine.
50. *S. ovoidalis* Karandikar and Rodgi, 1956 in *Chondromorpha kelaarti* (Diplopodasida) intestine.

51. *S. oxydesmi* Gasc and Ormières, 1974 in *Oxydesmus granulosus* and *O. granulatus* (Diplopodasida) intestine, stomodeum.
52. *S. ozakii* Hukui, 1955 in *Orthomorpha gracilis, Chondromorpha kelaarti,* and *C. severini* (Diplopodasida) intestine.
53. *S. pachyboli* Gasc and Bouix, 1976 in *Pachybolus ligulatus* (Diplopodasida) stomodeum.
54. *S. papillata* Karandikar and Rodgi, 1956 in *Thyropygus nigrolabiatus* (Diplopodasida) intestine.
55. *S. papua* Théodoridès, Desportes, and Jolivet, 1972 in *Polyconoceras* sp. (Diplopodasida) intestine.
56. *S. pintoi* Hasselmann, 1926 in *Rhinocricus* sp. (Diplopodasida) hindgut.
57. *S. plagiodesmi* Jeekel et al., 1958 in *Plagiodesmus occidentialis tuberosus* (Diplopodasida) intestine.
58. *S. polydesmi* (Lankester, 1863) M. Watson, 1916 (syns., *Gregarina polydesmi virginiensis* Leidy, 1853 in part; *G. polydesmi* Lankester, 1863; *G. polydesmivirginiensis* [Leidy, 1853] Crawley, 1903; *Amphoroides polydesmi* [Léger, 1892] Labbé, 1899; *A. polydesmivirginiensis* [Leidy, 1863] Ellis, 1913) in *Fontaria* (syn., *Polydesmus) virginiensis, Plagiodesmus oatypus, Rhinocricus* sp., and probably *Habrodesmus cagnii* (Diplopodasida) intestine.
59. *S. polyxeni* Léger and Duboscq, 1900 in *Polyxenus lagurus* (Diplopodasida) intestine.
60. *S. portonovoensis* Gasc and Bouix, 1976 in *Plethocrossus acutiformis* (Diplopodasida) stomodeum.
61. *S. poznanensis* Lipa, 1967 (syn., *S. orthomorphae* Rauchalles in Geus, 1969) in *Orthomorpha gracilis* (Diplopodasida) intestine.
62. *S. prionopetali* Tuzet, Manier, and Jolivet, 1957 in *Prionopetalum megalacanthum* (Diplopodasida) intestine.
63. *S. procidua* Rodgi and Ball, 1961 in Diplopodasida gen. sp. (Diplopodasida) intestine.
64. *S. producta* Léger and Duboscq, 1904 in *Cylindroiulus teutonicus* (syns., *Pachyiulus varius, Julus varius, Parajulus varius*) (Diplopodasida) intestine.
65. *S. rauchallesi* Levine, 1980 (syn., *S. orthomorphae* Lipa, 1967) in *Orthomorpha gracilis* and *O. coarctata* (Diplopodasida) intestine, stomodeum.
66. *S. robusta* Ellis, 1912 in *Paraiulus venustus, Orthomorpha gracilis,* and *Orthomorpha* sp. (Diplopodasida) alimentary canal.
67. *S. sarmatiuli* Lipa, 1967 in *Sarmatiulus vilnensis* (Diplopodasida) intestine.
68. *S. schizophylli* Lipa, 1967 in *Schizophyllum sabulosum* (Diplopodasida) intestine.
69. *S. shyamaprasadi* Chakravarty, 1939 in *Carmocephalus dentipes* (Chilopodasida) intestine.
70. *S. silene* Léger and Duboscq, 1904 in *Lysiopetalum foetidissimum* (Diplopodasida) intestine.
71. *S. spiroboli* Crawley, 1903 (syn., *Cnemidospora spiroboli* [Crawley, 1903] Crawley, 1903) in *Spirobolus* sp. (Diplopodasida) intestine (?).
72. *S. strongylosomae* Lipa, 1967 in *Strongylosoma pallipes* (Diplopodasida) intestine.
73. *S. suhoensis* H. Hoshide, 1952 in *Fusiulus simodanus* (Diplopodasida) intestine.
74. *S. tenuicollis* Pinto, 1918 in *Rhinocricus* sp. (Diplopodasida) intestine.
75. *S. thyrogluti* Ganapti and Narasimhamurti, 1962 (syn., *S. tynoghiti* Ganapti and Narasimhamurti, 1959) in *Thyroglutus* sp. (Diplopodasida) intestine.
76. *S. thyropygi* (Rodgi and Ball, 1961) Théodoridès and Desportes, 1963 (syn., *Actinocephalus thyropygi* Rodgi and Ball, 1961) in *Thyropygus minusculus* (Diplopodasida) intestine.
77. *S. triangula* Hukui, 1951 in *Nedyopus patrioticus patrioticus* (Diplopodasida) esophagus.
78. *S. tubulosus* Karandikar and Rodgi, 1956 in *Chondromorpha kelaarti* (Diplopodasida) intestine.

79.   *S. typhloiuli* Golemansky and Taschev, 1973 in *Typhloiulus bureschi* (Diplopodasida) intestine.
80.   *S. umbilicata* Pinto, 1918 in *Rhinocricus* sp. (Diplopodasida) intestine.
81.   *S. varians* Léger and Duboscq, 1903 in *Schizophyllum corsicum* (Diplopodasida) intestine.
82.   *S. vermiformis* Rauchalles in Geus, 1969 in *Orthomorpha gracilis* (Diplopodasida) intestine.
83.   *S. viannai* Pinto, 1918 in *Rhinocricus* sp. (Diplopodasida) intestine.
84.   *S. xenoboli* Ganapati and Narasimhamurti, 1957 in *Xenobolus* sp. (Diplopodasida) intestine.

Genus *Fonsecaia* Pinto, 1918. Development intracellular; oocysts elongate ovoid, without equatorial ridge. TYPE SPECIES *F. polymorpha* Pinto, 1918.

1.   *F. polymorpha* Pinto, 1918 in *Orthomorpha gracilis* (Diplopodasida) intestine.

Genus *Hyalosporina* Chakravarty, 1935. Epimerite diskoid with a central mucron (and with fine roots on the edges of the disk in aged individuals); nucleus attached to posterior end of body by two bundles of myonemes; anisogamous; oocysts ovoid, with a rather large epispore at one end. TYPE SPECIES *H. cambolopsis* Chakravarty, 1935.

1.   *H. cambolopsis* (Chakravarty, 1935) emend. (syn., *H. cambolopsisae* Chakravarty, 1935) (TYPE SPECIES) in *Cambolopsis* sp. (Diplopodasida) intestine.
2.   *H. froilanoi* Karandikar and Rodgi, 1956 in *Chondromorpha kelaarti* and *C. severini* (Diplopodasida) intestine.
3.   *H. rayi* Chakravarty, 1936 in *Polydesmus* sp., *Strongylosoma contortipes*, *Chondromorpha kelaarti*, and *C. severini* (Diplopodasida) intestine.
4.   *H. zebriaca* Karandikar and Rodgi, 1956 in *Arthrosphaera zebriaca* (Diplopodasida) intestine.

Family LEIDYANIDAE *Kudo, 1954*
   Similar to Gregarinidae, but solitary; mucron (epimerite?) simple, knob-like, gametocysts with several sporoducts; oocysts doliform.

Genus *Leidyana* M. E. Watson, 1915. Syzygy late; mucron (epimerite?) a simple, globular, sessile knob; oocysts emerge from sporoducts in long chains. TYPE SPECIES *L. erratica* (Crawley, 1907) Watson, 1916.

1.   *L. aglossae* H. Hoshide, 1958 in *Aglossa dimidiata* (Lepidopterorida) intestine.
2.   *L. berkeleyi* Lipa and Martignoni, 1984 in *Phryganidia californica* (Lepidopterorida) intestine.
3.   *L. erratica* (Crawley, 1907) Watson, 1916 (syns., *Gregarina achetaeabbreviatae* Leidy of Crawley [1903]; *Stenophora erratica* Crawley, 1907; *L. solitaria* M. E. Watson, 1915) (TYPE SPECIES) in *Gryllus abbreviatus*, *G. pennsylvanicus*, *G. yemma*, and *Gryllodes sigillatus* (Orthopterorida) intestine, gastric ceca.
4.   *L. gnyanagangai* Patil and Amoji, 1979 in *Rhytinota tristis* (Coleopterorida) gut.
5.   *L. gryllorum* (Cuénot, 1897) M. Watson, 1916 (syns., *Clepsidrina gryllorum* Cuénot, 1897; *C. macrocephala* Schneider, 1875 in part; *Gregarina gryllorum* [Cuénot, 1897] Cuénot, 1900; *Stenophora gryllodes-sigillatae* Narain, 1961; *L. gryllodes-sigillatae* [Narain, 1961] Théodoridés and Echard, 1962) in *Acheta domesticus*, *Gryllus assimilis*, *G. campestris*, *G. yemma*, *Gryllodes sigillatus*, *Gymnogryllus novaeguineae*, *Nemo-*

bius fasciatus, N. silvestris, Phaeophilacris pilipennis, Phaeophilacris, sp., and *Teleogryllus mitratus* (Orthopterorida) intestine.

6. *L. lancea* H. Hoshide, 1958 in *Aphonia gularis* (Lepidopterorida) intestine.
7. *L. latifolia* (Braune, 1930) Geus, 1969 (syn., *Gregarina latifolia* Braune, 1930) in *Niptus hololeucus* (Coleopterorida) intestine.
8. *L. latiformis* H. Hoshide, 1958 in *Tinea granella* (Lepidopterorida) intestine.
9. *L. limnophili* Baudoin, 1966 in *Limnophilus stigma, L. flavicornis,* and *L. lunatus* (Trichopterorida) intestine.
10. *L. linguata* Haldar and Sarkar, 1979 in *Pteronemobius concolor* (Orthopterorida) midgut.
11. *L. oblonga* (Dufour, 1837) Geus, 1969 (syn., *Gregarina oblonga* Dufour, 1837) in *Gryllus campestris, Nemobius silverstris, Locusta migratoria,* and *Psophus stridulus* (Orthopterorida) intestine.
12. *L. obtusa* Geus, 1969 in *Limnophilus flavicornis* (Trichopterorida) intestine.
13. *L. oviformis* K. Hoshide, 1978 in *Pteronemobius fascipes* (Orthopterorida) midgut, gastric ceca.
14. *L. peregrinata* Ormières, 1967 in *Aglossa pinguinalis* and *Asopia farinalis* (Lepidopterorida) intestine.
15. *L. phryganidiae* Lipa and Martignoni, 1984 in *Phryganidia californica* (Lepidopterorida) intestine.
16. *L. (?) reduvii* (Ramdohr, 1811) Levine, 1980 (syns., *Vibrio reduvii* Ramdohr, 1811; *Sporadina reduvii [Ramdohr, 1811] von stein, 1848; Gregarina reduvii [Ramdohr, 1811] Diesing, 1859; Hyalospora reduvii* [Ramdohr, 1811] Labbé, 1899; *? reduvii* [Ramdohr, 1811] Kamm, 1922) in *Reduvius personatus* (Hemipterorida) intestine.
17. *L. saigonensis* Corbel, 1968 in *Gryllus bimaculatus* and *Gryllodes sigillatus* (Orthopterorida) midgut.
18. *L. stejskali* Geus, 1967 in *Achroea grisella* (Lepidopterorida) intestine.
19. *L. suzumushi* K. Hoshide, 1973 in *Homoeogryllus japonicus* (Orthopterorida) intestine, gastric ceca.
20. *L. tinei* Keilin, 1918 (syns., *Gregarina mrazeki* Milojevic, 1921; *G. mirazeki* Milojevic, 1921 of Geus [1969] *lapsus calami; L. ephestiae* Daviault, 1929) in *Endrosis fenestrella, Pachytella* sp., *Canephora unicolor, Ephestia kuehniella,* and *Fumea casta* (Lepidopterorida) midgut.
21. *L. vierlingi* Geus, 1969 in *Stenophylax rotundipennis* (Trichopterorida) intestine.
22. *L. xylocopae* Bhatia and Setna, 1924 in *Xylocopa aestuans* (Hymenopterorida) alimentary canal.

Family CNEMIDOSPORIDAE *Grassé, 1953*

Young gamont at first fixed to intestinal cell by a voluminous mucron which, in the course of development, regresses and becomes reduced to a small cap or spherule on the protomerite; sporoducts absent (?); oocysts ellipsoidal.

Genus *Cnemidospora* Schneider, 1882. With the characters of the family. TYPE SPECIES *C. lutea* Schneider, 1882.

1. *C. lutea* Schneider, 1882 (TYPE SPECIES) in *Glomeris marginata, G. conspersa, G. hexasticha, G. pustulata, G. connexa,* and *Glomeris* sp. (Diplopodasida) intestine.
2. *C. rhysodesmi* (H. Hoshide, 1952) H. Hoshide, 1958 (syn., *Stenophora rhysodesmi* H. Hoshide, 1952) in *Rhysodesmus semicircularis hosidei* (Diplopodasida) intestine.
3. *C. schizophylli* Tuzet and Guerin, 1947 emend. (syn., *C. schizophyllae* Tuzet and Guerin, 1947) in *Schizophyllum mediterraneum* and *S. sabulosum* (syn., *S. rutilans [?]) (Diplopodasida)* intestine.

4.   *C. takaneensis* H. Hoshide, 1958 in *Rhysodesmus* sp. (Diplopodasida) intestine.

Family MONODUCTIDAE *Ray and Chakravarty, 1933*
    Initial development intracellular; epimerite present or absent, if present, with prongs; gametocyst with a single sporoduct; oocysts ellipsoidal or ovoid, with hyaline epispore; oocyst with eight sporozoites.

Genus *Monoductus* Ray and Chakravarty, 1933. Syzygy late; anisogamous; epimerite present, with a deep myocyte composed of two axial bundles, one retracting the epimerite; oocysts extruded in chains; in millipedes. TYPE SPECIES *M. lunatus* Ray and Chakravarty, 1933.

1.   *M. kelaarti* Karandikar and Rodgi, 1956 in *Chondromorpha kelaarti* (Diplopodasida) intestine.
2.   *M. longicollis* Amoji, 1978 in *Strongylosoma* sp. (Diplopodasida) proventriculus.
3.   *M. lunatus* Ray and Chakravarty, 1933 (TYPE SPECIES) in unnamed millipede, *Chondromorpha kelaarti* and *C. severini* (Diplopodasida) intestine.
4.   *M. spinicephalus* Amoji, 1978 in *Chondromorpha kelaarti* (Diplopodasida) proventriculus.
5.   *M. tubulosus* Karandikar and Rodgi, 1956 in *Chondromorpha kelaarti* and *C. severini* (Diplopodasida) intestine.

Genus *Stenoductus* Ramachandran, 1976. Epimerite absent in mature gamont; isogamous; oocysts extruded in chains; in millipedes. TYPE SPECIES *S. penneri* Ramachandran, 1976.

1.   *S. carlogoni* Janardanan, 1978 (syn., *S. carlogoni* Janardanan and Ramachandran, 1979) in *Carlogonus palmatus* (Diplopodasida) intestine.
2.   *S. chondromorphi* Janardanan, 1978 in *Chondromorpha kelaarti* (Diplopodasida) intestine.
3.   *S. disticta* Janardana, 1978 (syn., *S. disticta* Janardanan and Ramachandran, 1983) in *Arthrosphaera disticta*, *A. dalyi*, and *Arthrosphaera* sp. (Diplopodasida) intestine.
4.   *S. gordanus* Ramachandran, 1976 in *Narceus gordanus* (Diplopodasida) intestine.
5.   *S. ktenostrepti* Janardanan, 1978 in *Ktenostreptus calcaratus* (Diplopodasida) intestine.
6.   *S. nitida* Janardanan, 1978 (syn., *S. nitida* Janardanan and Ramachandran, 1983) in *Arthrosphaera nitida* (Diplopodasida) intestine.
7.   *S. organognathi* Janardanan, 1978 (syn., *S. organognathi* Janardanan and Ramachandran, 1982) in *Organognathus janardhanani* and *Fageostreptus hyatti* (Diplopodasida) intestine.
8.   *S. penneri* Ramachandran, 1976 (TYPE SPECIES) in *Floridobolus penneri* (Diplopodasida) intestine.
9.   *S. polydesmi* Janardanan, 1978 (syn., *S. polydesmi* Janardanan and Ramachandran, 1983) in *Polydesmus* sp. (Diplopodasida) intestine.
10.   *S. trigoniuli* Janardanan, 1978 (syn., *S. trigoniuli* Janardanan and Ramachandran, 1982) in *Trigoniulus goesi* and *Xenobolus acuticonus* (Diplopodasida) intestine.
11.   *S. xenoboli* Janardanan, 1978 (syn., *S. xenoboli* Janardanan and Ramachandran, 1982) in *Xenobolus acuticonus* and *Aulacobolus graveleyi* (Diplopodasida) intestine.

Genus *Phleobum* Haldar and Chakraborty, 1974. Mucron (epimerite?) absent; gametocyst with ectocyst, endocyst, and single sporoduct; oocysts "oval", not extruded in chains; in insects. TYPE SPECIES *P. gigantinum* Haldar and Chakraborty, 1974.

1.   *P. gigantinum* Haldar and Chakraborty, 1974 (TYPE SPECIES) in *Phleoba antennata* (Orthopterorida) presumably intestine.

Family SPHAEROCYSTIDAE *Chakravarty, 1960*
   Epimerite (mucron?) sessile, a thick horizontal disk with milled border, or absent; protomerite in young stage only, atrophied in mature gamont; gametocysts dehisce by simple rupture; oocysts ovoid or biconical.

Genus *Sphaerocystis* Léger, 1892. Mucron (epimerite?) very labile; gamonts apparently with a single segment, solitary, probably parthenogenetic; oocysts ellipsoidal, with adherent endospore and epispore; in insects. TYPE SPECIES *S. simplex* Léger, 1892.

1.   *S. odontotermitis* Kalavati and Narasimhamurti, 1980 emend. (syn., *S. odontotermi* Kalavati and Narasimhamurti, 1980) in *Odontotermes* sp. (Isopterorida).
2.   *S. simplex* Léger, 1892 (TYPE SPECIES) in *Cyphon pallidus* (Coleopterorida) location in host not stated.
3.   *S. tentyriae* Tuzet and Théodoridès, 1951 in *Tentyria mucronata* (Coleopterorida) intestine.
4.   *S. termitis* Desai and Uttangi, 1962 in *Capritermes incola* (Isopterorida) foregut.

Genus *Schneideria* Léger, 1892. Mucron (epimerite?) sessile, with a thick horizontal disk with milled border and a style arising from its center; septum between protomerite and deutomerite resorbed so that the organism appears aseptate; oocysts biconical, smooth, swollen; in mesenteron and gut of insects. TYPE SPECIES *S. mucronata* Léger, 1892.

1.   *S. caudata* (von Siebold, 1839) Labbé, 1899 (syns., *Gregarina caudata* von Siebold, 1839; *Actinocephalus caudatus* [von Siebold, 1839] von Frantzius, 1848; *S. coronata* Léger, 1892) in *Sciara nitidicollis* and *Sciara* sp. (Dipterorida) intestine.
2.   *S. mucronata* Léger, 1892 (TYPE SPECIES) in *Bibio marci* (Dipterorida) intestine.
3.   *S. pusilla* H. Hoshide, 1959 in *Penthetria japonica* (Dipterorida) intestine.
4.   *S. quadrinotati* Amoji and Rodgi, 1973 emend. (syn., *S. quadrinotota* Amoji and Rodgi, 1983) in *Pirates quadrinotatus* (Hemipterorida) gut.
5.   *S. schneideri* da Cunha et al., 1975 (syn., *S. schneiderae* da Cunha et al., 1975 *lapsus calami*) in *Trichosia pubescens* and (experimentally with great difficulty) *Plastosciara* sp. (Dipterorida) intestinal ceca.

Genus *Paraschneideria* Nieschulz, 1924. Mucron (epimerite?) simple, button-like; development intracellular; mature gamont not septate; oocysts biconical; in insects. TYPE SPECIES *P. metamorphosa* (Nowlin, 1922) Nieschulz, 1924.

1.   *P. metamorphosa* (Nowlin, 1922) Nieschulz, 1924 (syn., *Schneideria metamorphosa* Nowlin, 1922) (TYPE SPECIES) in *Sciara coprophila* (Dipterorida) intestine.

Genus *Neoschneideria* Ormières, Manier, and Mathiez, 1965. Epimerite composed of a lobed pad in the center of which is a sharp mucron; epicyte longitudinally striated; gamont divided into segments, generally six; gametocyst with an external wall forming two long prolongations opposite each other; oocysts biconical, with very loose epispore, emitted in chains; in insects. TYPE SPECIES *N. douxi* (Hesse, 1926) Ormières, Manier, and Mathiez, 1965.

1.   *N. aprilinae* (Hesse, 1926) Ormières, Manier, and Mathiez, 1965 (syn., *Asterophora aprilinae* Hesse, 1926) in *Dixa aprilina* (= *D. aestivalis*) (Dipterorida) intestine.

2. *N. douxi* (Hesse, 1926) Ormières, Manier, and Mathiez, 1965 (syn., *Asterophora douxi* Hesse, 1926) in *Dixa autumnalis* and *Dixa* sp. (Dipterorida) intestine.

Family TRICHORHYNCHIDAE *Ormières, Marquès, and Puisségur, 1977*
   Epimerite slightly lobed, carried on a long neck, limited by three unit membranes; epicyte ornamented by numerous digitations forming a dense "beard"; epicytic folds absent but arches and longitudinal "apical" filaments regularly disposed in the membranes; gametocyst studded with mamelons; dehiscence by means of a lateral residuum ("pseudocyst"), gametocysts often with an equatorial suture; oocysts cylindrical, emitted in chains; in chilopods.

Genus *Trichorhynchus* Schneider, 1882. With the characters of the family. TYPE SPECIES *T. pulcher* Schneider, 1882.

1. *T. lithobii* Crawley, 1903 in *Lithobius* sp. (Chilopodasida) presumably intestine.
2. *T. pulcher* Schneider, 1882 (syns., *T. insignis* Schneider, 1882; *Gregarina megacephala* Leidy, 1889) in *Scutigera* sp., *S.* (syn., *Cermatia) coleoptrata*, *S. forceps*, *Theleuonema tuberculata*, and *Thereuopoda clunifera* (Chilopodasida) intestine.

Family DACTYLOPHORIDAE *Léger, 1892*
   Epimerite (mucron?) complex, digitate, without a long neck; nucleus in protomerite; syzygy late, frontal; anisogamy very accentuated; gametocyst generally with a single residuum ("pseudocyst") coming from the male and serving for expulsion of the oocysts; oocysts cylindrical, with rounded ends; in intestine of chilopods and sometimes millipedes and insects.

Genus *Dactylophorus* Balbiani, 1889. Protomerite dilated, with a large number of baguettes; gametocyst spherical; oocysts in more or less long chains. TYPE SPECIES *D. robustus* (Léger, 1892) Labbé, 1899.

1. *D. macedonicus* Geus, 1967 in *Scolopendra cingulata* and *Lithobius* sp. (Chilopodasida) midgut.
2. *D. robustus* (Léger, 1892) Labbé, 1899 (syn., *Dactylophora robusta* Léger, 1892) in *Cryptops hortensis*, *C. anomalus lusitanus*, and *C. triculatus* (Chilopodasida) intestine.

Genus *Echinomera* Labbé, 1899. Mucron (epimerite?) forming a cone with an eccentric summit, bearing a more or less large number of short, digitiform "rhizoids"; gametocysts without sporoducts; residuum formed by male gamont; oocysts cylindrical, in chains. TYPE SPECIES *E. hispida* (Schneider, 1875) Labbé, 1899.

1. *E. caudata* Théodoridès and Ormières, 1959 in *Lithobius inermis pyrenaicus, L. lapidicola,* and *L. melanops* (Chilopodasida) intestine.
2. *E. cryptopsis* Rauchalles in Geus, 1969 in *Cryptops hortensis* (Chilopodasida) presumably intestine.
3. *E. erythrocephali* Rauchalles in Geus, 1969 emend. (syn., *E. erythrocephala)* Rauchalles in Geus, 1969) in *Lithobius erythrocephalus* (Chilopodasida) intestine.
4. *E. hispidi* (Schneider, 1875) Labbé, 1899 emend. (syns., *Echinocephalus hispidus* Schneider, 1875; *Echinomera hispida* [Schneider, 1875] Labbé, 1899) (TYPE SPECIES) in *Lithobius forficatus, L. coloradensis, L. calcaratus, L. muticus,* and *L. piceus* (Chilopodasida) intestine.
5. *E. horridi* (Léger, 1902) M. Watson, 1916 (syns., *E. horrida* [Léger, 1902] M. Watson, 1916; *Echinocephalus horridus* Léger, 1902) in *Lithobius calcaratus* and *L. mutabilis* (Chilopodasida) intestine.

6. *E. leptoiuli* Lipa, 1967 in *Leptoiulus proximus* (Diplopodasida) intestine.
7. *E. lithobii* (Tuzet and Ormières, 1951) Ormières, 1966 (syn., *Capillicephalus lithobii* Tuzet and Ormières, 1951) in *Lithobius piceus* (Chilopdasida) intestine.
8. *E. magalhaesi* Pinto, 1918 in *Scolopendra* sp. (Chilopodasida) intestine.

Genus *Grebnickiella* Bhatia, 1938 emend. *auctores*. Protomerite spread out transversely, with numerous delicate rhizoids, made up of two long, narrow horizontal lobes fused and turned up spirally at one end, peripheral portion with many teeth from which long filaments project; oocysts long ovoid, mostly in chains. TYPE SPECIES *G. gracilis* (Grebnitskii, 1873) Bhatia, 1938.

1. *G. giardi* (Léger, 1902) Geus, 1969 (syns., *Pterocephalus giardi* Léger, 1902; *P. giardi* var. *corsicum* Léger and Duboscq, 1903; *Nina giardi* [Léger, 1899] Sokolow, 1911; *N. g. corsicum* [Leger and Duboscq, 1903] Sokolow, 1911) in *Scolopendra africana* and *S. oranensis* (Chilopodasida) intestine.
2. *G. gracilis* (Grebnitskii, 1873) Bhatia, 1938 emend. *auctores* (syns., *Nina gracilis* Grebnitskii, 1873; *Pterocephalus nobilis* Schneider, 1887) (TYPE SPECIES) in *Scolopendra cingulata* and *S. morsitans* (Chilopodasida) intestine.
3. *G. indica* (Merton, 1911) M. Watson, 1916; Levine, 1985 (syns., *Nina indica* Merton, 1911; *Pterocephalus indica* [Merton, 1911] *Zool. Rec., 1911; N. indicia* [Merton, 1911] M. Watson, 1916 *lapsus calami; G. indicia* [Merton, 1911] M. Watson, 1916) in *Scolopendra subspinipes* (Chilopodasida) intestine.
4. *G. japonica* (H. Hoshide, 1952) Levine, 1980 (syn., *Nina japonica* H. Hoshide, 1952) in *Scolopendra subspinipes mutilans* (Chilopodasida) intestine.
5. *G. leitaodacunhai* (Hasselmann, 1918) Levine, 1980 (syn., *Pterocephalus leitaodacunhai* Hasselmann, 1918) in *Scolopendra* sp. (Chilopodasida) intestine.
6. *G. navillae* (Mitra and Chakravarty, 1937) Bhatia, 1938 (syn., *Nina navillae* Mitra and Chakravarty, 1937) in *Scolopendra* sp. (Chilopodasida) presumably intestine.
7. *G. pixellae* Misra, 1942 in *Scolopendra morsitans* (Chilopodasida) intestine.
8. *G. scolopendrae* (von Kölliker, 1848) Levine, 1980 (syns., *Gregarina scolopendrae* von Kölliker, 1848; *Pterocephalus scolopendrae* [von Kölliker, 1848] Ellis, 1913) in *Scolopendra morsitans* (Chilopodasida) intestine.

Genus *Rhopalonia* Léger, 1894. Epimerite a subspherical button, with ten or more digitiform processes; gametocyst with residuum (''pseudocyst''); oocysts cylindrical. TYPE SPECIES *R. geophili* Léger, 1894.

1. *R. geophili* Léger, 1894 (syn., *R. hispido* [Schneider] of Kudo [1966] *lapsus calami)* in *Geophilus* sp., *G. gabrielis, Haplophilus* (syn., *Stigmogaster) gracilis,* and *H. subterraneus* (Chilopodasida) intestine.
2. *R. lithobii* Lipa, 1967 in *Lithobius calcaratus* (Chilopodasida) intestine.
3. *R. longicornis* Rauchalles in Geus, 1969 in *Geophilus longicornis* (Chilopodasida) intestine.

Genus *Acutispora* Crawley, 1903. Mucron (epimerite?) a button at the end of a long neck; gametocyst with or without residuum; oocysts biconical, with a thick, blunt endocystic rod at each end. TYPE SPECIES *A. macrocephala* Crawley, 1903.

1. *A. macrocephala* Crawley, 1903 (TYPE SPECIES) in *Lithobius forficatus* (Chilopodasida) presumably intestine.
2. *A. procera* Geus, 1969 in *Lithobius dentatus* (Chilopodasida) presumably intestine.
3. *A. pulchra* Ormières, 1969 in *Lithobius dentatus* (Chilopodasida) intestine.

Genus *Seticephalus* Kamm, 1922. Epimerite forming a thick tuft of short, anteriorly directed, brush-like bristles; protomerite broad and flat. TYPE SPECIES *S. elegans* (Pinto, 1918) Kamm, 1922.

1.   *S. elegans* (Pinto, 1918) Kamm, 1922 (syn., *Gregarina elegans* Pinto, 1918) (TYPE SPECIES) in *Scolopendra* sp. (Chilopodasida) intestine.

Genus *Dendrorhynchus* Keilin, 1920. Epimerite a sucker with border of ramified lobes (papillae); transverse fibrils conspicuous; septate or almost always so; protodeutomerite divided into several compartments; gametocysts ellipsoidal; oocysts biconical (fusiform). TYPE SPECIES *D. systeni* Keilin, 1920.

1.   *D. keilini* Ahamed and Narasimhamurti, 1979 in *Ceriagrion coromandelianum* (Odonatorida) midgut.
2.   *D. systeni* Keilin, 1920 (TYPE SPECIES) in *Systenus* sp. (Dipterorida) midgut.

Genus *Mecistophora* Ganapati and Narasimhamurti, 1960. Mucron (epimerite?) cup-shaped, with numerous intracellular filaments; gamont with septum between protomerite and deutomerite; gametes dissimilar; gametocyst with residuum ("pseudocyst"); oocysts simple, octozoic. TYPE SPECIES *M. legeri* Ganapati and Narasimhamurti, 1960.

1.   *M. legeri* Ganapati and Narasimhamurti, 1960 (TYPE SPECIES) in *Mecistocephalus punctiformis* (Chilopodasida) intestine.

Family STYLOCEPHALIDAE *Ellis, 1912*
    Mucron (epimerite?) elongated, with or without appendages; development extracellular; syzygy frontal, late; gametocyst residuum present; oocysts purse- or hat-shaped, emitted in chains, with a single brown or blackish wall, with a dehiscence line on the convex border; in arthropods.

Genus *Stylocephalus* Ellis, 1912. Mucron (epimerite?) elongated into a neck, swollen posteriorly at its free end; gametocysts covered by papillae, with residuum ("pseudocyst"); oocysts purse-like, usually emerging in chains; in insects. TYPE SPECIES *S. oblongatus* (Hammerschmidt, 1838) M. Watson, 1916.

1.   *S. africanus* Théodoridès and Jolivet, 1965 in *Zophosis lineata* (Coleopterorida) intestine.
2.   *S. apapillatus* Haldar and Chakraborty, 1979 in *Gonocephalum* sp. (Coleopterorida) midgut.
3.   *S. bahli*, Misra, 1941 (syn., *S. japonicus* H. Hoshide, 1951) in *Gonocephalum helopioides, G. coriaceum, G. pubens, G. japanum, Gonocephalum* sp., and *Mesomorphus villiger* (Coleopterorida) alimentary canal.
4.   *S. brevirostris* (von Kölliker, 1848) M. Watson, 1916 (syns., *Gregarina brevirostra* von Kölliker, 1848; *Stylorhynchus brevirostris* [von Kölliker, 1848] von Frantzius, 1848; *Phialoides ornata* [Léger, 1892] Labbé, 1899 in part) in *Hydrophilus* sp., *Hydrous aterrimus*, and *Hydrous* sp. (Coleopterorida) intestine.
5.   *S. carabi* Lipa, 1967 in *Carabus glabratus* (Coleopterorida) intestine.
6.   *S. coniontis* Nelson, 1970 in *Coniontis* sp. (Coleopterorida) intestine.
7.   *S. conoides* Desai, 1965 in *Gonocephalum hoffmanseggi* (Coleopterorida) intestine.
8.   *S. convexus* Nelson, 1970 in *Eleodes* sp. (Coleopterorida) intestine.
9.   *S. depressicus* Patil, 1982 in *Gonocephalum depressum* (Coleopterorida) gut.
10.   *S. devdhari* Corbel, 1971 (syn., *S. elongatus* Desai, 1965) in *Gonocephalum hoffmanseggi* (Coleopterorida) intestine.

11. *S. eastoni* Théodoridès, 1955 in *Blaps inflata* and *B. polychresta* (Coleopterorida) intestine.
12. *S. eledonae* Foerster, 1938 in *Eledona agaricola, Mycetophagus piceus, M. atomarius, M. quadripustulatus,* and *Pentaphyllus testaceus* (Coleopterorida) intestine.
13. *S. elongatus* (von Frantzius, 1846) Geus, 1969 (syns., *Gregarina elongata* von Frantzius, 1846; *Stylorhynchus elongatus* [von Frantzius, 1846] Wellmer, 1911) in *Adesmia metallica, Crypticus quisquilius,* and *Crypticus* sp. (Coleopterorida) intestine.
14. *S. ensiferus* Ellis, 1912 in *Leptochirus edax* (Coleopterorida) intestine.
15. *S. filiformis* Théodoridès, 1959 (syn., *S. phalloides* Théodoridès, 1955 in part) in *Gonocephalum simplex, G. prolixum inornatum, G. controversum, G. planatum,* and *Gonocephalum* sp. (Coleopterorida) intestine.
16. *S. giganteus* Ellis, 1912 in *Asida* sp., *A. opaca, Blaps* sp., *Eleodes* sp., *E. hispilabris,* and *Eusattrus* sp. (Coleopterorida) intestine.
17. *S. gigas* Filipponi, 1949 in *Blaps gigas, B. lethifera, B. mucrorata, B. similis,* and *Eleodes tricostata* (Coleopterorida) intestine.
18. *S. gonocephali* Bhushana Rao, 1966 in *Gonocephalum brachyelytra* (Coleopterorida) intestine.
19. *S. heterotarsus* Patil and Amoji, 1984 in *Heterotarsus indicus* (Coleopterorida) gut.
20. *S. hoffmannseggi* Devdhar and Amoji, 1977 in *Gonocephalum hoffmannseggi* (Coleopterorida) intestine.
21. *S. indicus* Misra, 1942 in *Opatroides (Penthicus) vicinis* (Coleopterorida) midgut.
22. *S. ingeri* Gibbs, 1946 in *Gonocephalum arenarium* and *G. simplex* (Coleopterorida) intestine.
23. *S. japonicus* H. Hoshide, 1951 in *Gonocephalum pubens* and *G. japanum* (Coleopterorida) intestine.
24. *S. lanatus* Nelson, 1970 in *Eleodes* sp. (Coleopterorida) intestine.
25. *S. lingulus* Nelson, 1970 in *Eleodes extricata* (Coleopterorida) intestine.
26. *S. longicollis* (von Stein, 1848) Watson, 1916 (syns., *Stylorhynchus longicollis* von Stein, 1848; *Gregarina longicollis* [von Stein, 1848] Lankester, 1863; *G. mortisagae* Diesing, 1851; *Stylocephalus mucronatus* Filipponi, 1951) in *Adasmia metallica, Blaps* sp., *B. mortisaga, B. gibba, B. gigas, B. lethifera, B. lusitanica, B. mucronata, B. persica, B. pterophaga, B. bifurcata, B. sulcata, Crypticus quisquilius, Hoplarion (Atlasion) bedeli, Morica planata, Pimelia rugosa* var. *laeyisulcata, Prionychus ater,* and *Scaurus tristis* (Coleopterorida) intestine.
27. *S. mesomorphi* Devdhar, 1962 in *Gonocephalum hoffmannseggi* (Coleopterorida) intestine.
28. *S. micranterei* Théodoridès, Desportes, and Jolivet, 1964 in *Micrantereus gerstaeckeri laevior* (Coleopterorida) intestine.
29. *S. oblongatus* (Hammerschmidt, 1838) M. Watson, 1916 (syns., *Rhizinia oblongata* Hammerschmidt, 1838; *Sporadina oblongata* [Hammerschmidt, 1838] von Frantzius, 1848; *Gregarina oblongata* [Hammerschmidt, 1838] Diesing, 1851; *Stylorhynchus oblongatus* [Hammerschmidt, 1838] Schneider, 1875) (TYPE SPECIES) in *Asida grisea, A. jurinei, A. sericea, A. dejeani, Dendarus piceus, D. messenius, Heliopathes littoralis, H. montivagus, Himatisonus villosus, Macellocerus* sp., *Micrositus semicostatus, Opatrum sabulosum, O. alternatum, Pedinus* sp., *P. quadratus, Phylan abbreviatus, P. gibbus, Pimelia interjecta, Pseudoblaps javana, Scleron orientale, Setenis semisulcata,* and *S. foveicollis* (Coleopterorida), and also *Acanthaclisis baeticus* (Neuropterorida) intestine.
30. *S. pauliani* Théodoridès, 1959 in *Phrynocolus tenuesculptus, P. ovipennis,* and *Selinus abacoides* (Coleopterorida) intestine.
31. *S. phaleriae* Tuzet and Ormières, 1955 in *Phaleria cadaverina* (Coleopterorida) intestine.

32. *S. phalloides* Théodoridès, 1955 in *Blaps requieni, Blaps* sp., *Erodius puncticollis ecostatus, Gonocephalum* sp., *Gonocnemis* sp., *Mesomorphus longulus,* and *Quadrideres interioris* (Coleopterorida) intestine.
33. *S. pseudoblapsis* Patil, 1982 in *Pseudoblaps* sp. (Coleopterorida) gut.
34. *S. punctulatus* Patil, 1982 in *Opatroides punctulatus* (Coleopterorida) gut.
35. *S. reticulatus* Filipponi, 1949 in *Blaps gigas* (Coleopterorida) intestine.
36. *S. variabilis* Tuzet and Théodoridès, 1951 in *Asida sericea, Psammophanes raffrayi, P. rubrolineatus, Stenosis angustata, Tentyria mucronata, T. tessulata,* and *Zophosis sulcatus* (Coleopterorida) intestine.

Genus *Stylocephaloides* Théodoridès, Desportes, and Jolivet, 1965. Mucron (epimerite?) a button; older gamonts with epicytic ribs; protomerite often with thin axis; oocysts purselike; in insects. TYPE SPECIES *S. monodi* Théodoridès, Desportes, and Jolivet, 1965.

1. *S. monodi* Théodoridès, Desportes, and Jolivet, 1965 (TYPE SPECIES) in *Adesmia antiqua, Pimelia grandis,* and *Prionotheca coronata* (Coleopterorida) intestine.
2. *S. setenis* H. Hoshide, 1952 emend. K. Hoshide, 1979 (syns., *Sphaerorhynchus sedenis* H. Hoshide, 1952; *Stylocephalus sedenis* [H. Hoshide, 1953] Corbel, 1971) in *Setenis valgipes* (Coleopterorida) intestine.

Genus *Cystocephalus* Schneider, 1886. Mucron (epimerite?) broad at the base, swollen into a bulb, and surmounted with a small, olive-shaped prolongation; oocysts hat-shaped; in insects. TYPE SPECIES *C. algerianus* Schneider, 1886.

1. *C. aethiopicus* Théodoridès, Desportes, and Jolivet, 1964 in *Phanerotoma* sp. and *Rhytinota praelonga* (Coleopterorida) intestine.
2. *C. albrechti* Théodoridès and Jolivet, 1963 in *Trachyderma hispida* (Coleopterorida) midgut.
3. *C. algerianus* Schneider, 1886 (TYPE SPECIES) in *Pimelia* sp. (Coleopterorida) intestine.
3a. *C. algerianus* var. *mauritanicus* Tuzet and Théodoridès, 1951 in *Dailognatha quadricollis, Adesmia dilatata* var. *drakei, A. procera, Mesomorphus longulus, Pimelia angulata angulosa, P. angulata* var. *syriaca, P. derasa, P. bipunctata, Tenthyrina orbiculata subsulcata, Tentyria herculanea, Thriptera asphaltidis,* and *Zophosis punctata* group (Coleopterorida) intestine.
3b. *C. algerianus* var. *persica* Théodoridès, 1961 in *Adesmia karelina* (Coleopterorida) intestine.
4. *C. devdharii* Patil and Amoji, 1979 in *Rhytinota impolita* and *Himatismus fasciculatus* (Coleopterorida) gut.
5. *C. gabei* Théodoridès, Desportes, and Jolivet, 1964 in *Trachyderma hispida* (Coleopterorida) intestine.
6. *C. harpalus* Nelson, 1970 in *Harpalus* sp. (Coleopterorida) intestine.
7. *C. hollandei* Théodoridès, Desportes, and Jolivet, 1964 in *Oxycara* sp. and *Rhytinota praelonga* (Coleopterorida) intestine.
8. *C. leidyi* (Kamm, 1917) Théodoridès, 1954 (syn., *Leidyana leidyi* Kamm, 1917) in *Nyctobates pennsylvanica* (Coleopterorida) intestine.
9. *C. rhytinotus* Patil and Amoji, 1980 in *Rhytinota tristis* (Coleopterorida) midgut.
10. *C. trapezus* Nelson, 1970 in *Cypherotylus californicus* (Coleopterorida) intestine.

Genus *Bulbocephalus* M. Watson, 1916. Mucron (epimerite?) with a long neck with a bulbous swelling in the middle; oocysts unknown; in insects. TYPE SPECIES *B. wardi* M. Watson, 1916.

1.   *B. elongatus* M. Watson, 1916 in *Cucujus* sp. (Coleopterorida) intestine.
2.   *B. indicus* Narain, 1961 in *Gryllus bimaculatus* (Orthopterorida) midgut.
3.   *B. wardi* M. Watson, 1916 (TYPE SPECIES) in Cleridae gen. sp. (Coleopterorida) intestine.

Genus *Xiphocephalus* Théodoridès, 1964 emend. Corbel, 1971. Mucron (epimerite?) very long, with a sharp or blunt point, on a filiform neck; in insects. TYPE SPECIES *X. gladiator* (Blanchard, 1905) Corbel, 1971.

1.   *X. africanus* (Théodoridès, Desportes, and Jolivet, 1965) Corbel, 1971 (syn., *Stylocephalus africanus* Théodoridès, Desportes, and Jolivet, 1965) in *Erodius granipennis* var. *maximus, Pimelia grandis, P. platynota, P. rotundipennis, Pogonobasis raffrayi, Prionotheca coronata,* and *Thriptera crinita* (Coleopterorida) intestine.
2.   *X. gladiator* (Blanchard, 1905) Corbel, 1971 (syns., *Stylorhynchus gladiator* Blanchard, 1905; *Stylocephalus gladiator* [Blanchard, 1905] Watson, 1916) in *Helenophorus collaris* (Coleopterorida) intestine.
3.   *X. gonocephali* Patil, 1982 (syn., *X. gonocephali* Patil and Amoji, 1985) in tenebrionid beetle (Coleopterorida) intestine.
4.   *X. karnatakaensis* (Devdhar and Amoji, 1977) Levine, 1984 (syn., *Stylocephalus (X.) karnatakaensis* Devdhar and Amoji, 1977) in *Gonocephalum hoffmannseggi* (Coleopterorida) intestine.
5.   *X. latipes* Patil and Amoji, 1985 in *Scleron latipes* (Coleopterorida) intestine.
6.   *X. phaleriae* (Tuzet and Ormières, 1955) Corbel, 1971 (syn., *Stylocephalus phaleriae* Tuzet and Ormières, 1955) in *Phaleria cadaverina* (Coleopterorida) intestine.
7.   *X. reitteri* Patil, 1982 (syn., *X. reitterae* Patil and Amoji, 1985) in tenebrionid beetle (Coleopterorida) intestine.
8.   *X. serpentula* Devdhar and Amoji, 1977 (Devdhar and Amoji, 1977) Levine, 1984 (syn., *Stylocephalus* [X.] *Serpentula* Devdhar and Amoji, 1977) in *Gonocephalum hoffmannseggi* (Coleopterorida) intestine.

Genus *Lophocephalus* Labbé, 1899. Mucron (epimerite?) forming a large cupule surrounded by a crenulate border at its base and with a crown of small processes which seem to penetrate the endoplasm; nucleolus a long, twisted sausage; in insects. TYPE SPECIES *L. insignis* (Schneider, 1882) Labbé, 1899.

1.   *L. fernandezi* Théodoridès, 1958 in *Hegeter amaroides* and *H. tristis* (Coleopterorida) intestine.
2.   *L. insignis* (Schneider, 1882) Labbé, 1899 (syn., *Lophorhynchus insignis* Schneider, 1882) (TYPE SPECIES) in *Helops striatus* and *H. quisquilius* (Coleopterorida) intestine.

Genus *Lophocephaloides* Théodoridès, Desportes, and Jolivet, 1965. Young gamonts with a long, button-shaped mucron (epimerite?) on a short neck; mucron of older gamonts with a more or less flaccid floral corolla somewhat resembling that of *Lophocephalus;* gametocysts of the usual cystocephalid type; oocysts in strings like beads; in insects. TYPE SPECIES *L. thripterae* Théodoridès, Desportes, and Jolivet, 1965.

1.   *L. ardoini* Théodoridès, Desportes, and Jolivet, 1965 in *Pogonobasis raffrayi* (Coleopterorida) intestine.
2.   *L. thripterae* Théodoridès, Desportes, and Jolivet, 1965 (TYPE SPECIES) in *Thriptera crinita* (Coleopterorida) intestine.

Genus *Sphaerorhynchus* Labbé, 1899. Mucron (epimerite?) a small sphere at the end of a long or sometimes short neck; in insects. TYPE SPECIES *S. ophioides* (Schneider, 1886) Labbé, 1899.

1. *S. chabaudi* Tuzet and Théodoridès, 1951 in *Akis elegans* (Coleopterorida) intestine.
2. *S. hamoni* Tuzet and Théodoridès, 1951 in *Akis punctata* (Coleopterorida) intestine.
3. *S. ophioides* (Schneider, 1886) Labbé, 1899 (syns., *Sphaerocephalus ophioides* Schneider, 1886; *Stylocephalus ophioides* [Schneider, 1886] Corbel, 1971) (TYPE SPECIES) in *Akis* sp., *A. acuminata*, *A. elegans*, *A. elevata*, *A. punctata* (syn., *A. baccarozzo*), *Cyphogenia lucifuga*, *Morica favieri*, *M. hybrida*, *Pimelia atlantis*, and *P. brisouti* (Coleopterorida) intestine.
4. *S. stenosiae* Tuzet and Ormières, 1955 in *Stenosis angustata* (Coleopterorida) intestine.

Genus *Oocephalus* Schneider, 1886. Mucron (epimerite?) a rounded button borne on a short, conoidal neck hardly higher than wide; in insects. TYPE SPECIES *O. hispanus* Schneider, 1886.

1. *O. dromii* Geus, 1969 (syn., *Cystocephalus dromii* [Geus, 1969] Corbel, 1971) in *Dromius quadrinotatus* (Coleopterorida) intestine.
2. *O. hispanus* Schneider, 1886 (syn., *Stylocephalus hispanus* [Schneider, 1886] Corbel, 1971) in *Megaladacne abnormalis*, *Morica* sp., and *M. planata* (Coleopterorida) intestine.

Genus *Campanacephalus* Théodoridès, 1955. Mucron (epimerite?) without differentiated neck, in the form of a button sometimes quadrangular or truncated; oocysts emitted in a chain; in intestine of beetles. TYPE SPECIES *C. villiersi* Théodoridès, 1955.

1. *C. constrictus* Nelson, 1970 in *Coniontis* sp. (Coleopterorida) intestine.
2. *C. villiersi* Théodoridès, 1955 (TYPE SPECIES) in *Gargilius* sp., *Macropodia variolaris*, and *Taraxides punctatus* (Coleopterorida) intestine.

Genus *Clavicephalus* Théodoridès, 1959. Mucron (epimerite?) a quadrangular knob which prolongs the protomerite; mucron (epimerite?) finely punctate at its apex and having a punctate "islet" in its interior; syzygy head to tail; gametocyst and oocyst unknown; in insects. TYPE SPECIES *C. madagascariensis* Théodoridès, 1959.

1. *C. laosensis* Théodoridès and Desportes, 1967 in *Setenis semisulcatus* (Coleopterorida) intestine.
2. *C. madagascariensis* Théodoridès, 1959 (syn., *Cystocephalus madagascariensis* [Théodoridès, 1959] Corbel, 1971) (TYPE SPECIES) in *Melanocratus* sp. (Coleopterorida) intestine.
3. *C. papuensis* Théodoridès, Desportes, and Jolivet, 1972 in *Setenis subfoveata* (Coleopterorida) intestine.

Genus *Cystocephaloides* Théodoridès and Jolivet, 1963. Mucron (epimerite?) in the form of a button on a short neck and surrounded by an epicytic collarette; later the collarette disappears and the mucron (epimerite?) takes the form of a quadrangular button carried on a short neck; late gamonts hat-shaped; in insects. TYPE SPECIES *C. prionothecae* Théodoridès and Jolivet, 1963.

1. *C. prionothecae* Théodoridès and Jolivet, 1963 (TYPE SPECIES) in *Prionotheca coronata* (Coleopterorida) midgut.

Genus *Orbocephalus* Nelson, 1970. Mucron (epimerite?) a dilated papilla surrounded by a flat disk; diamerite present; gametocysts and oocysts unknown; in insects. TYPE SPECIES *O. bulbus* Nelson, 1970.

1. *O. bulbus* Nelson, 1970 (TYPE SPECIES) in *Eleodes* sp. (Coleopterorida) intestine.
2. *O. eleodis* Nelson, 1970 in *Eleodes* sp. (Coleopterorida) intestine.

Genus *Lepismatophila* Adams and Travis, 1935. Mucron (epimerite?) at first elongated into a neck but then turning into a simple knob; gamonts with a broad protomerite; gametocyst with a residuum; gametocyst walls ornamented with channels delimited by small granular areas; oocysts ovoid or ellipsoidal, emitted in chains; in diploplurids and thrips. TYPE SPECIES *L. thermobiae* Adams and Travis, 1935.

1. *L. campodea* Tuzet, Manier, and Ormières, 1953 in *Campodea augens* and *C. monspessulana* (Diploplurorida) intestine.
2. *L. cornwalli* (Bhatia, 1938) Levine, 1980 (syn., *Gregarina cornwalli* Bhatia, 1938) in *Lepisma saccharina* (?) (Thysanurorida) midgut.
3. *L. ctenolepismae* Lindsay, 1939 in *Ctenolepisma longicaudata* (Thysanurorida) intestine.
4. *L. karnataki* Nimbargi and Rodgi, 1974 in *Ctenolepisma* sp. (Thysanurorida) intestine.
5. *L. orientalis* Crusz, 1960 in *Ctenolepisma nigrum* and *Peliolepisma calvum* (Thysanurorida) midgut.
6. *L. parva* Tuzet, Manier, and Ormières, 1953 in *Ctenolepisma lineata* (Thysanurorida) intestine.
7. *L. piliformis* Rauchalles in Geus, 1969 in *Campodes staphilinus* (Diploplurorida) intestine.
8. *L. plusiocampae* Taschev and Golemansky, 1973 in *Plusiocampa bureschi* (Diploplurorida) intestine.
9. *L. rhombocephala* Haldar and Chakraborty, 1977 in *Ctenolepisma nigra* and *C. rhombocephala* (Thysanurorida) midgut and ventricular ceca.
10. *L. thermobiae* Adams and Travis, 1935 (TYPE SPECIES) in *Thermobia domestica* (Thysanurorida) alimentary canal.

Genus *Colepismatophila* Adams and Travis, 1935. Mucron (epimerite?) globular; oocysts hat-shaped, with two curved, filamentous processes at opposite ends, in wavy chains; in silverfish. TYPE SPECIES *C. watsonae* Adams and Travis, 1935.

1. *C. buckleyi* Crusz, 1960 in *Ctenolepisma nigrum* (Thysanurorida) midgut.
2. *C. burti* Crusz, 1960 in *Peliolepisma calvum* (Thysanurorida) midgut.
3. *C. watsonae* Adams and Travis, 1935 (TYPE SPECIES) in *Thermobia domestica* (Thysanurorida) intestine.

Family ACTINOCEPHALIDAE *Léger, 1892*
Mucron (epimerite?) symmetrical, with or without appendage, syzygy late, frontal; gametocysts without sporoducts, generally dehiscing by simple rupture; oocysts biconical, cylindrobiconical, or irregular — simple or spiny; in intestine or arthropods and chordates.

Subfamily ACTINOCEPHALINAE *Léger, 1899*
Oocysts without spines or thickenings at their poles; in insects, diplopods, chilopods, opilionids, and chordates.

Genus *Actinocephalus* von Stein, 1848. Mucron (epimerite?) sessile, with a short neck with eight to ten simple digitiform processes at its apex; neck persists more or less in gamont, but digitiform processes (tentacles) disappear; gametocysts dehisce by formation of hole in wall through which oocysts are extruded in a single thread; oocysts biconical or lemon-shaped; in insects. TYPE SPECIES *A. conicus* (Dufour, 1837) von Frantzius, 1848.

1. *A. acanthaclisis* Marquès and Ormières, 1978 in *Acanthaclisis baetica* and *A. occitanica* (Neuropterorida) intestine.
2. *A. acutispora* Léger, 1892 in *Ablatteria* (syn., *Silpha) laevigata* and *Silpha* sp. (Coleopterorida) intestine.
3. *A. americanus* Crawley, 1903 in *Galerita bicolor* (Coleopterorida) presumably intestine.
4. *A. blaniuli* Triffitt, 1927 (syn., *A. blanuli* Triffitt, 1927 *lapsus calami*) in *Blaniulus guttulatus* (Diplopodasida) intestine.
5. *A. brachydactylus* Ellis, 1913 in *Aeshna* sp. (Neuropterorida) intestine.
6. *A. bradinopygi* Narasimhamurti and Ahamed, 1980 in *Bradinopyga geminata* (Odonatorida) gut.
7. *A. capritermes* Desai and Uttangi, 1966 in *Capritermes incola* (Isopterorida) esophagus.
8. *A. ceriagrionae* Sarkar and Chakravarty, 1969 in *Ceriagrion coromandelianum* (Coleopterorida) intestine.
9. *A. ceropriae* Théodoridès, and Desportes, 1967 in *Ceropria subocellata* (Coleopterorida) intestine.
10. *A. conicus* (Dufour, 1837) von Frantzius, 1848 (syns., *Gregarina conica* Dufour, 1837; *G. lucani* [von Stein, 1848] Diesing, 1851; *A. lucani* von Stein, 1848; *A. lucanus* [von Stein, 1848] von Frantzius, 1848; *Stephanophora radiosa* Léger, 1892; *S. lucani* [von Stein, 1848] Labbé, 1899; *A. conicus* var. *magna* Théodoridès, 1955) (TYPE SPECIES) in *Dorcus* (syn., *Lucanus) parallelopipedus, Ludius ferrugineus,* and *Lucanus cervus* (Coleopterorida) intestine.
11. *A. crassus* (Ellis, 1912) Ellis, 1913 (syn., *Stephanophora crassa* Ellis, 1912) in *Leptochirus edax* (Coleopterorida) intestine.
12. *A. dermestica* Uttangi and Desai, 1964 in *Dermestes vulpinus* (Coleopterorida) intestine.
13. *A. dicaeli* (Crawley, 1903) M. Watson, 1916 (syns., *Gregarina discaeli* Crawley, 1903 *lapsus calami; G. dicaeli* Crawley, 1903; *A. discaeli* [Crawley, 1903] Ellis, 1913 *lapsus calami*) in *Dicaelus ovalis* (Coleopterorida) intestine.
14. *A. digitatus* Schneider, 1875 in *Chlaenius vestitus, Carabus cancellatus, C. clathratus, C. glabratus, C. nemoralis, Pterostichus vulgaris,* and *Pterostichus* sp. (Coleopterorida) intestine.
15. *A. dytiscorum* (von Frantzius, 1848) M. Watson, 1916 (syns., *Gregarina dytiscorum* von Frantzius, 1848; *Ancyrophora uncinata* of Labbé, 1899) in *Dytiscus* sp., *D. marginalis,* and *Acilius sulcatus* (Coleopterorida) intestine.
16. *A. echinatus* Wellmer, 1910 in *Pterostichus niger, P.* (syn., *Omaseus) vulgaris, P. gracilis, P. diligens,* and *Harpalus rufipes* (Coleopterorida) intestine.
17. *A. ellipsoidus* Sarkar and Haldar, 1981 in *Ischnura delicata* (Odonatorida) midgut.
18. *A. elongatus* Semans, 1939 in *Dichromorpha viridis, Arphia sulphurea, Chortophaga viridifasciata, Schistocerca americana,* and *Melanoplus mexicanus* (Orthopterorida) intestine.
19. *A. enigmaticus* Théodoridès and Jolivet, 1959 in *Megacantha dentata, Taraxides punctatus,* and *Prioscelis serrata* (Coleopterorida) intestine.
20. *A. fimbriatus* (Diesing, 1859) M. Watson, 1916 (syns., *Gregarina fimbriata* Diesing, 1859 in part; *G. locustae carolinae* Leidy, 1853 in part; *Stephanophora locustaecarolinae* [Leidy, 1853] Crawley, 1903 in part; *S. pachyderma* Crawley, 1907; *A. pachydermus* [Crawley, 1907] Ellis, 1913) in *Dissosteira carolina* (Orthopterorida) intestine.
21. *A. giganteus* Bush, 1928 in *Cataloipus cognatus, Pnorisa capensis, Gastrimargus verticalis,* and *Acrida sulphuripennis* (Orthopterorida) intestine.

22. *A. gimbeli* (Ellis, 1913) M. Watson, 1916 (syn., *Stenophora gimbeli* Ellis, 1913) in *Harpalus pennsylvanicus* (Coleopterorida) intestine.
23. *A. grassei* Théodoridès and Jolivet, 1959 in *Gargilius* sp., *Monomma giganteum*, and *M. triplacinum* (Coleopterorida) intestine.
24. *A. harpali* (Crawley, 1903) Crawley, 1903 (syn., *Gregarina harpali* Crawley, 1903) in *Harpalus caliginosus* (Coleopterorida) intestine.
25. *A. histoliticus* Filipponi, 1949 in *Laemostenus* (= *Pristonychus*) *algerinus* (Coleopterorida) intestine.
26. *A. kintaikyoensis* H. Hoshide, 1953 in Perlidae gen. sp. (Plecopterorida) intestine.
27. *A. laosensis* Théodoridès and Desportes, 1966 in *Alidria parallela* (Coleopterorida) intestine.
28. *A. laticaudatus* K. Hoshide, 1982 in *Parastenopsyche sauter* (Trichopterorida) intestine.
29. *A. licini* Tuzet and Théodoridès, 1951 in *Licinus punctatulus* (Coleopterorida) intestine.
30. *A. megabuni* Ormières and Baudoin, 1973 in *Megabunus diademi* (Opilionorida) intestine.
31. *A. nitiophili* Foerster, 1938 in *Notiophilus biguttatus* and *N. aquaticus* (Coleopterorida) intestine.
32. *A. parvus* Wellmer, 1910 in *Ctenocephalides canis*, *Ceratophyllus fringillae*, and *C. gallinae* (Siphonapterorida) intestine.
33. *A. permagnus* Wellmer, 1910 in *Carabus* (syn., *Procrustes*) *coriaceus*, *C. cancellatus*, *C. catenulatus*, *Cathoplius asperatus*, and *Ceutosphodrus oblongus* (Coleopterorida) intestine.
34. *A. puisseguri* Tuzet and Tarroux, 1959 in *Chrysocarabus punctatoauratus barthei* (Coleopterorida) intestine.
35. *A. teffli* Théodoridès and Jolivet, 1959 in *Tefflus gracilentus* (Coleopterorida) intestine.
36. *A. theodoridis* Corbel, 1968 in *Locusta migratoria capito* (Orthopterorida) intestine.
37. *A. tipulae* (Hammerschmidt, 1838) Léger, 1892 (syns., *Bululina tipulae* Hammerschmidt, 1838; *Gregarina tipulae* [Hammerschmidt, 1838] von Frantzius, 1846) in *Tipula* sp., *T. abdominalis*, *T. oleracea*, *T. paludosa*, and *Pales pratensis* (Dipterorida) intestine.
38. *A. zophus* (Ellis, 1913) Ellis, 1913 (syn., *Stephanophora zopha* Ellis, 1913) in *Nyctobates barbata* and *N. pennsylvanica* (Coleopterorida) intestine.

Genus *Caulocephalus* Bhatia and Setna, 1924. Mucron (epimerite?) dilated into a cauliflower at the top and narrow at the base; protomerite with a specialized cytoplasmic zone; mucron (epimerite?) of primite persists; oocysts ovoid or spherical; in beetles. TYPE SPECIES *C. bhatiasetnai* Théodoridès, Desportes, and Jolivet, 1964.

1. *C. bhatiasetnai* Théodoridès, Desportes, and Jolivet, 1964 (syn., *C. crenatus* Bhatia and Setna, 1924 in part) (TYPE SPECIES) in *Aulacophora foveicollis*, *A. africana*, *Leptaulaca vinula*, and *Monolepta pauperata* (Coleopterorida) alimentary canal.
2. *C. japonicus* H. Hoshide, 1958 in *Chrysomela aurichalces* and *Aulacophora foveicollis* (Coleopterorida) intestine.
3. *C. sastrae* Théodoridès, Desportes, and Jolivet, 1972 in *Sastra* sp. (Coleopterorida) intestine.

Genus *Cornimeritus* H. Hoshide, 1959. Gamonts solitary; mucron (epimerite?) an acute claw with a long, large, flexible stalk; gametocyst spherical, dehiscing by simple rupture; oocysts biconical, extruded in lateral chains. TYPE SPECIES *C. ovalis* H. Hoshide, 1959.

1. *C. ovalis* H. Hoshide, 1959 (TYPE SPECIES) in Nitidulidae gen. sp. (Coleopterorida) intestine.

Genus *Umbracephalus* H. Hoshide, 1959. Gamonts solitary, elongate cylindrical, with elongate ellipsoidal nucleus; mucron (epimerite?) with a very long neck bearing an anterior crown with about 20 recurved hooks; gametocyst spherical; oocysts unknown. TYPE SPECIES *U. longicollis* H. Hoshide, 1959.

1.   *U. longicollis* H. Hoshide, 1959 (TYPE SPECIES) in *Lithobius* sp. (Chilopodasida) intestine.

Genus *Urnaepimeritus* H. Hoshide, 1959. Gamonts solitary, elongate, with spherical nucleus; mucron (epimerite?) with a short neck and a bowl-shaped crown with 30 or more recurved hooks; gametocyst spherical, dehiscing by simple rupture; oocysts spindle-shaped. TYPE SPECIES *U. spathiformis* H. Hoshide, 1959.

1.   *U. spathiformis* H. Hoshide, 1959 (TYPE SPECIES) in *Prolamnonyx holstii* (Chilopodasida) intestine.

Genus *Asterophora* Léger, 1892. Mucron (epimerite?) a thick, horizontal disk with a milled border and a stout style projecting from the center; oocysts cylindrobiconical; in intestine of arthropods. TYPE SPECIES *A. mucronata* Léger, 1892.

1.   *A. caliciformis* Baudoin, 1967 in *Agrypnia obsoleta* (Trichopterorida) intestine.
2.   *A. caloglyphi* Rauchalles in Geus, 1969 in *Caloglyphus moniezi* (Acarinorida) presumably intestine.
3.   *A. capitata* Baudoin, 1966 in *Rhyacophila praemorsa* and *R. obliterata* (Trichopterorida) intestine.
4.   *A. cordiformis* Baudoin, 1967 in *Mystacides longicornis*, *Triaenodes bicolor*, and *Oecetis furva* (Trichopterorida) intestine.
5.   *A. cupressiformis* Baudoin, 1967 in *Oligoplectrum maculatum* (Trichopterorida) intestine.
6.   *A. elegans* Léger, 1892 in *Phryganea grandis*, *Limnophilus flavicornis*, *L. griseus*, *L. rhombicus*, *L. sparsus*, *Sericostoma* sp., and *Stenophylax* sp. (Trichopterorida) intestine.
7.   *A. hemicerae* K. Hoshide, 1979 in tenebrionid beetle *Hemicera zigzaga* (Coleopterorida) intestine.
8.   *A. hydropsyches* (H. Hoshide, 1953) Baudoin, 1967 (syn., *Pileocephalus hydropsyches* (H. Hoshide, 1953) in *Hydropsyches* sp. (Trichopterorida) intestine.
9.   *A. mucronata* Léger, 1892 (TYPE SPECIES) in *Rhyacophila* sp., *Agryphia obsoleta*, *Molanna angustata*, *M. septentrionis*, *Phryganea* sp., and *Neuronia ruficornis* (Trichopterorida) intestine.
10.   *A. orientalis* H. Hoshide, 1959 in *Holostrophus orientalis* (Coleopterorida) intestine.
11.   *A. pachydera* Baudoin, 1966 in *Rhyacophila evoluta* (Trichopterorida) intestine.
12.   *A. philica* (Leidy, 1889) Crawley, 1903 (syns., *Gregarina philica* Leidy, 1889; *Anthorhynchus philicus* [Leidy, 1889] Ellis, 1913) in *Nyctobates pennsylvanica* (Coleopterorida) intestine.
13.   *A. pygmaea* H. Hoshide, 1959 in *Mycetophagus* sp. (Coleopterorida) intestine.
14.   *A. tiaroides* Baudoin, 1967 in *Tricostegia minor* and *Ecnomus tenellus* (Trichopterorida) intestine.

Genus *Pileocephalus* Schneider, 1875. Mucron (epimerite?) lance-shaped, with a short neck; oocysts biconical; in insect larvae. TYPE SPECIES *P. chinensis* Schneider, 1875.

1. *P. agilis* Geus, 1969 in *Limnophilus auricula* (Trichopterorida) intestine.
2. *P. astaurovi* Lipa, 1967 in *Baicalina spinosa* (Trichopterorida) intestine.
3. *P. benli* Geus, 1969 in Lepismatidae gen. sp. (Thysanurorida) intestine.
4. *P. chinensis* Schneider, 1875 (syn., *Cardiocephalus sororculae* Tsvetkov, 1929) (TYPE SPECIES) in *Anabolia nervosa, A. sororcula, Brachycentrus montanus, Chaetopteryx villosa, Mesophylax asperus, Mystacides* sp., *M. quadrifasciatus* (Trichopterorida), *Isoperla oxylepis* (Plecopterorida) and also accidentally ephemeropterorid larvae (Ephemeropterorida) intestine.
5. *P. dinarthrodes* K. Hoshide, 1972 in *Dinarthrodes japonica* (Trichopterorida) intestine.
6. *P. glyphotaelii* Shtein, 1960 (syn., *P. hovassei* Baudoin, 1965) in *Nemotaulius* (syn., Glyphotaelius) *punctatolineatus, Limnophilus* sp., *L. flavicornis, L. igma, Glyphotaelius pellucidus,* and *Grammotaulius atomarius* (Trichopterorida) intestine.
7. *P. heeri* (von Kölliker, 1845) Schneider, 1887 (syns., *Gregarina heeri* von Kölliker, 1845; *Stylorhynchus heeri* [von Kölliker, 1845] von Frantzius, 1848; *G. frantziusiana* Diesing, 1851; *Asterophora heerii* [von Kölliker, 1848] Baudoin, 1967) in *Phryganea* sp., *P. grandis, P. varia, P. striata, Agrypnia obsoleta, Limnophilus vittatus, Rhyacophila* sp., *Athripsodes aterrimus,* and *Leptocerus aterrimus* (Trichopterorida) intestine.
8. *P. hovassei* Baudoin, 1965 in *Limnophilus* sp. (Trichopterorida) intestine.
9. *P. hydropsychus* H. Hoshide, 1953 in *Hydropsyche* sp. (Trichopterorida) intestine.
10. *P. lanceatus* Baudoin, 1967 in *Drusus discolor* (Trichopterorida) intestine.
11. *P. nemurae* (Foerster, 1938) Shtein, 1960 (syns., *Gregarina nemurae* Foerster, 1938; *P. nemouri* Sorcetti and di Giovanni, 1984 *lapsus calami*) in *Nemura* sp., *N. cinerea, Leuctra* sp., and *Perlodes* sp. (Plecopterorida) intestine.
12. *P. orientalis* Devdhar and Deshpande, 1977 in *Haliplus* sp. (Coleopterorida) gut.
13. *P. rhytinotae* Amoji and Rodgi, 1976 emend. (syn., *P. rhythinctus* Amoji and Rodgi, 1976) in *Rhytinota impolita* (Coleopterorida) intestine.
14. *P. sapporoensis* K. Hoshide, 1982 in Limnophilinae gen. sp. (Trichopterorida) intestine.
15. *P. scyphoides* Baudoin, 1967 in *Potamophylax nigricornis* (Trichopterorida) intestine.
16. *P. striatus* Léger and Duboscq, 1909 in *Ptychoptera* sp. (syn., *Liriope contaminata* [?]) (Dipterorida) intestine.
17. *P. suhoensis* H. Hoshide, 1952 in *Allecula fuliginosa* (Coleopterorida) intestine.
18. *P. tachycines* Semans, 1939 (syn., *Leidyana tachycines* [Semans, 1939] Corbel, 1968) in *Tachycines asinamorus* (Orthopterorida) midgut.

Genus *Gemmicephalus* Baudoin, 1967. Mucron (epimerite?) in the form of an oval bud; oocysts slightly biconical. TYPE SPECIES *G. mutabilis* Baudoin, 1967.

1. *G. japonicus* H. Hoshide, 1972 in *Allophylax* sp. (Trichopterorida) intestine.
2. *G. mutabilis* Baudoin, 1967 (TYPE SPECIES) in *Athripsodes cinereus* (Trichopterorida) intestine.

Genus *Pilidiophora* Baudoin, 1967. Mucron (epimerite?) without a differentiated neck and having the form of a small bonnet consisting of a dome prolonged toward the base by little tongues surrounding the diamerite. TYPE SPECIES *P. fragilis* Baudoin, 1967.

1. *P. fragilis* Baudoin, 1967 (TYPE SPECIES) in *Agrypnia pagetana* (Trichopterorida) intestine.

Genus *Geneiorhynchus* Schneider, 1875. Mucron (epimerite?) a long neck with a tuft of short bristles at its end; oocysts cylindrical or cylindroconical; in odonate larvae. TYPE SPECIES *G. monnieri* Schneider, 1875.

1.   *G. aeshnae* Crawley, 1907 (syns., *Geniorhynchus aschnae* Crawley, 1907 *lapsus calami; Actinocephalus brachydactylus* Ellis, 1913) (TYPE SPECIES) in *Aeshna constricta, A. cyanea, A. grandis,* and *Aeshna* sp. (Odonatorida) intestine.
2.   *G. monnieri* Schneider, 1875 (syn., *G. monieri* Schneider, 1875 of Geus [1969] *lapsus calami*) in *Libellula* sp., *L. depressa,* and *Aeshna cyanea* (Odonatorida) intestine.

Genus *Acanthoepimeritus* H. Hoshide, 1959. Gamonts solitary, with spherical nucleus; mucron (epimerite?) a swollen club, with nine or ten rows of hooks around it and with numerous recurved hooks covering its anterior surface. TYPE SPECIES *A. jimukade* H. Hoshide, 1959.

1.   *A. jimukade* H. Hoshide, 1959 (TYPE SPECIES) in *Mecistocephalus marmoratus* (Chilopodasida) intestine.

Genus *Phialoides* Labbé, 1899. Mucron (epimerite?) consisting of a long neck with a retractile papilla at its summit bordered by a cushion set peripherally with stout teeth and surrounded by a wider collarette with folds winding in a point; gametocysts spherical, without sporoducts; oocysts biconical, ventricose; in beetle larvae. TYPE SPECIES *P. ornata* (Léger, 1892) Labbé, 1899.

1.   *P. ornata* (Léger, 1892) Labbé, 1899 (syn., *Phialis ornata* Léger, 1892) (TYPE SPECIES) in *Hydrous piceus* (Coleopterorida) intestine.

Genus *Legeria* Labbé, 1899. Mucron (epimerite?) with a stalked, irregularly lobed, and folded plasma portion; protomerite surrounded by a collar; gametocysts without sporoducts; oocysts cylindrobiconical or subnavicular, with a thick wall; in beetle larvae. TYPE SPECIES *L. agilis* (Schneider, 1875) Labbé, 1899.

1.   *L. agilis* (Schneider, 1875) Labbé, 1899 (syns., *Sporadina dytiscorum* von Frantzius, 1848; *Dufouria agilis* Schneider, 1875) (TYPE SPECIES) in *Colymbetes* sp., *C. fuscus, Acilius sulcatus, Agatus* sp., *Dytiscus marginalis, D. semisulcatus, Ilybius* sp. (?), *Rhantus sulurellus,* and *R. punctatus* (Coleopterorida) intestine.

Genus *Pyxinia* Hammerschmidt, 1838. Mucron (epimerite?) a crenulate crateriform disk with or without hooks at its periphery, with a style in the center; gametocysts dehisce by formation of a hole in the wall through which oocysts are extruded in a single thread; oocysts biconical; in beetles. TYPE SPECIES *P. rubecula* Hammerschmidt, 1838.

1.   *P. anobii* Vincent, 1922 in *Sitodrepa* (syn., *Anobium) panicem* and *Dendrobium pertinax* (Coleopterorida) intestine.
2.   *P. bulbifera* Watson, 1916 in *Dermestes lardarius* (Coleopterorida) intestine.
3.   *P. crystalligera* Frenzel, 1892 in *Dermestes vulpinus* and *D. peruvianus* (Coleopterorida) midgut.
4.   *P. firma* (Léger, 1892) Wellmer, 1911 (syns., *Xiphorhynchus firmus* Léger, 1892; *Beloides firmus* [Léger, 1892] Labbé, 1899) in *Dermestes lardarius, D. atomarius, D. lanarius, D. murinus, D. vulpinus,* and *D. frischi* (Coleopterorida) intestine.
5.   *P. foliacea* Tuzet and Théodoridès, 1951 in *Dermestes frischi* (Coleopterorida) intestine.
6.   *P. frenzeli* Laveran and Mesnil, 1900 in *Attagenus pellio, A. megatoma, Attagenus* sp., *Anthrenus* sp., and *Dermestes* sp. (Coleopterorida) intestine.
7.   *P. japonica* H. Hoshide, 1952 in *Dermestes tesselatocollis* and *D. vulpinus* (Coleopterorida) intestine.

8. *P. major* H. Hoshide, 1959 in *Anthrenus vervaci* (Coleopterorida) intestine.
9. *P. moebuszi* Léger and Duboscq, 1900 in *Anthrenus moseorum, A. fuscus, A. scro-phulariae, A. verbasci, Anthrenus* sp., and *Attagenus* sp. (Coleopterorida) intestine.
10. *P. myelophili* H. Hoshide, 1952 emend. (syn., *P. myelophilia* H. Hoshide, 1952) in *Myelophilus piniperda* (Coleopterorida) intestine.
11. *P. rubecula* Hammerschmidt, 1838 (TYPE SPECIES) in *Dermestes lardarius* and *D. vulpinus* (Coleopterorida) intestine.
12. *P. tenuis* (Léger, 1892) Grassé, 1953 (syns., *Xiphorhynchus tenuis* Léger, 1892; *Beloides tenuis* [Léger, 1892] Labbé, 1899; *B. tenius* [Léger, 1892] Labbé, 1899 of Geus [1969] *lapsus calami*) in *Dermestes undulatus* (Coleopterorida) intestine.
13. *P. trogodermi* Levine, 1984 (syn., *Pyxinia* sp. Hall et al., 1971) in *Trogoderma simplex, T. sternale, T. variabile, T. inclusum, T. glabrum,* and *T. grassmani* (Coleopterorida) intestine.

Genus *Discorhynchus* Labbé, 1899. Mucron (epimerite?) a large spheroidal papilla with collar and short neck, dropping off early; oocysts biconical, slightly curved, a little swollen (ventricose); in insect larvae. TYPE SPECIES *D. truncatus* (Léger, 1892) Labbé, 1899.

1. *D. truncatus* (Léger, 1892) Labbé, 1899 (syn., *Discocephalus truncatus Léger, 1892) (TYPE SPECIES) in Sericostoma* sp. and *Plectocnemia conspersa* (Trichopterorida) intestine.

Genus *Steinina* Léger and Duboscq, 1904. Mucron (epimerite?) a short, motile, digitiform process which later changes into a flattened structure; oocyst biconical, swollen; in insects. TYPE SPECIES *S. ovalis* (von Stein, 1848) Léger and Duboscq, 1904.

1. *S. alphitobii* Sarkar and Chakravarty, 1969 emend. (syn., *S. alphitobiusae* Sarkar and Chakravarty, 1969) in *Alphitobius piceus* (Coleopterorida) intestine.
2. *S. amarygmi* Théodoridès, Desportes, and Jolivet, 1972 in *Amarygmus morio* (Coleopterorida) intestine.
3. *S. coptotermitis* Kalavati, 1977 emend. (syn., *S. coptotermi* Kalavati, 1977) in *Coptotermes heimi* (Isopterorida) mid- and hindgut.
4. *S. diaperis* Foerster, 1938 in *Diaperis boleti* (Coleopterorida) presumably intestine.
5. *S. dollfusi* Théodoridès and Desportes, 1967 in *Ceropria subocellata* and *Eucyrtus anthracinus* (Coleopterorida) intestine.
6. *S. ellipsoidalis* Tsvetkov, 1929 in *Phyllotreta undulata* (Coleopterorida) intestine.
7. *S. harpali* M. Watson, 1916 in *Harpalus pennsylvanicus longior* (Coleopterorida) coelom.
8. *S. maxima* Théodoridès, Desportes, and Jolivet, 1972 in *Gymnopholus gressitti* (Coleopterorida) intestine.
9. *S. microgoni* Sarkar and Chakravarty, 1969 emend. (syn., *S. microgoansae* Sarkar and Chakravarty, 1969) in *Anoplogonius microgonus* (Coleopterorida) intestine.
10. *S. minor* Obata, 1953 in Tenebrionidae gen. sp. (Coleopterorida) intestine.
11. *S. obconica* Ishii, 1914 in *Tribolium castaneum, T. ferruginium,* and *Lyprops sinensis* (Coleopterorida) intestine.
12. *S. ovalis* (von Stein, 1848) Léger and Duboscq, 1904 (syns., *Stylorhynchus ovalis* von Stein, 1848; *Gregarina ovalis* [von Stein, 1848] Diesing, 1851; *Clepsidrina polymorpha* Hammerschmidt, 1838 in part; *G. polymorpha* [Hammerschmidt, 1838] Lankester, 1863 in part) (TYPE SPECIES) in *Tenebrio molitor, T. obscurus, Tenebrio* sp., and Tenebrionidae gen. sp. (Coleopterorida) intestine.
13. *S. rotunda* M. E. Watson, 1915 in *Amara angustata* (Coleopterorida) intestine.

13a. *S. rotunda* var. *cryphaei* Théodoridès and Jolivet, 1959 in *Cryphaeus taurus, C. gazella,* and *Apomecyna kivuensis* (Coleopterorida) intestine.

14. *S. rotundata* Ashworth and Rettie, 1912 in *Ceratophyllus farreni, C. galligallinae,* and *C. styx* (Siphonapterorida) midgut.

15. *S. sphaerosphora* H. Hoshide, 1952 in *Neatus* (syn., *Tenebrio*) *picipes* (Coleopterorida) intestine.

16. *S. termitis* Uttangi and Desai, 1962 in *Speculitermes cyclops sinhalensis* (Isopterorida) foregut.

17. *S. trycheri* Théodoridès and Jolivet, 1959 in *Trycherus imperator* and *T. appendiculatus* (Coleopterorida) intestine.

Genus *Bothriopsides* Strand, 1928. Mucron (epimerite?) lost very early, lenticular, sessile, small, oval, with six or more filamentous processes directed upward; protomerite very large, very mobile, with septum convex anteriorly; gametocysts spherical; oocysts biconical. TYPE SPECIES *B. histrio* (Schneider, 1875) Strand, 1928.

1. *B. acilii* Baudoin, 1961 in *Acilius sulcatus* (Coleopterorida) intestine.

2. *B. agrypniae* Baudoin, 1967 in *Agrypnia pagetana* and *A. obsoleta* (Trichopterorida) intestine.

3. *B. claviformis* (Pinto, 1918) Foerster, 1938 (syn., *Bothriopsis claviformis* Pinto, 1918) in Aeshnidae gen. sp. (Odonatorida) and *Hydroporus palustris* and *Hyphydrus ovatus* (Coleopterorida) intestine.

4. *B. graphoderi* Baudoin, 1961 in *Graphoderes cinereus* (Coleopterorida) intestine.

5. *B. histrio* (Schneider, 1875) Strand, 1928 (syns., *Bothriopsis histrio* Schneider, 1875; *Iorella wegorecki* Lipa, 1967) (TYPE SPECIES) in *Agabus bipustulatus, Graphoderes* (syn., *Hydatiscus*) *cinereus, G. bilineatus, G. zonatus, Colymbetes fuscus, C. paykulli, Acilius sulcatus, Acilius* sp., *Dytiscus dimidiatus, D. marginalis,* Dytiscidae gen. sp., *Hydatiscus seminiger, Hydrophilus* sp., *Hyphydrus ovatus, Ilybius ater, I. fenestratus, Rhantus exoletus, R. notatus,* and *R. punctatus* (Coleopterorida) intestine.

6. *B. oswaldocruzi* (Hasselmann, 1918) Levine, 1980 (syn., *Bothriopsis oswaldocruzi* Hasselmann, 1918) in unnamed aquatic beetle (Coleopterorida) posterior intestine.

7. *B. pirajai* (Hasselmann, 1918) Levine, 1980 (syn., *Bothriopsis pirajai* Hasselmann, 1918) in unnamed aquatic beetle larva (Coleopterorida) hindgut.

8. *B. terpsichorella* (Ellis, 1913) Foerster, 1938 (syn., *Legeria terpsichorella* Ellis, 1913; *Bothriopsis terpsichorella* [Ellis, 1913] Watson, 1916) in *Hydrophilus* sp., *Ilybius* sp., and *I. ater* (Coleopterorida) intestine.

Genus *Pomania* Baudoin, 1967. Mucron (epimerite?) composed of a point which may be transformed into a sucker and a cap covering the protomerite. TYPE SPECIES *P. hovassei* Baudoin, 1967.

1. *P. hovassei* Baudoin, 1967 (TYPE SPECIES) in *Potamophylax nigricornis* (Trichopterorida) intestine.

Genus *Stictospora* Léger, 1893. Mucron (epimerite?) with a short neck, a spherical crateriform ball with 9 to 12 posteriorly directed laminations set close to the neck, shed very early; gametocysts with a gelatinous envelope and no sporoducts; oocysts biconical, slightly curved; in insects. TYPE SPECIES *S. provincialis* Léger, 1893.

1. *S. anomalae* H. Hoshide, 1952 in *Anomala* sp. (Coleopterorida) intestine.

2. *S. coelocystis* Obata, 1955 in Scarabaeidae gen. sp. (Coleopterorida) intestine (gametocyst always in hemocoele).

3. *S. costelytrae* Allison, 1969 in *Costelytra zealandica* (Coleopterorida) midgut.
4. *S. kabutomusi* H. Hoshide, 1952 in *Xylotrupes dichotomus* (Coleopterorida) intestine.
5. *S. kurdistana* Théodoridès, 1961 in *Oryctes* sp. (Coleopterorida) intestine.
6. *S. provincialis* Léger, 1893 (syn., *S. provincialis* var. *anomalae* Théodoridès, 1955) (TYPE SPECIES) in *Amphimallon solstitialis, Rhizotrogus* sp., *Melolontha* sp., *M. vulgaris,* and Melolonthinae gen. sp. (Coleopterorida), and *Lestes sponsa* and *Enallagma cyathigerum* (Odonatorida) intestine.

Genus *Coleorhynchus* Labbé, 1899. Mucron (epimerite?) diskoid, poorly developed; protomerite large, forming a sort of collarette or sucker with a muscular sucker serving for attachment; gamonts solitary; development parthenogenetic; oocysts biconical or navicular, octozoic; in intestine of insects. TYPE SPECIES *C. heros* (Schneider, 1885) Labbé, 1899.

1. *C. heros* (Schneider, 1885) Labbé, 1899 (syn., *Coleophora heros* Schneider, 1885) (TYPE SPECIES) in *Nepa cinerea* (Hemipterorida) intestine.

Genus *Amphoroides* Labbé, 1899. Mucron (epimerite?) a globular sessile papilla, shed very early; protomerite cup-shaped or globular; oocysts biconical, navicular, or curved, without epispore; in millipedes. TYPE SPECIES *A. polydesmi* (Léger, 1892) Labbé, 1899.

1. *A. calverti* (Crawley, 1903) M. E. Watson, 1915 (syn., *Gregarina calverti* Crawley, 1903) in *Lysiopetalum lactarium* and *Callipus lactarius* (Diplopodasida) intestine.
2. *A. circi* Triffitt, 1927 (syn., *A. ventosa* Tuzet and Guerin, 1947) in *Polydesmus complanatus* (Diplopodasida) intestine.
3. *A. polydesmi* (Léger, 1892) Labbé, 1899 (syn., *Amphorella polydesmi* Léger, 1892) (TYPE SPECIES) in *Polydesmus complanatus* and *P. denticulatus* (Diplopodasida) intestine.
4. *A. polydesmivirginiensis* Crawley, 1903 (syn., *Gregarina polydesmi virginiensis* Leidy, 1853 in part) in *Polydesmus virginiensis* (Diplopodasida) intestine.

Genus *Stylocystis* Léger, 1899. Mucron (epimerite?) a sharply pointed process, ordinarily recurved in a long, hyaline point, very sharp at its end; protomerite absent; gametocysts ripen entirely in host and do not contain a residuum; oocysts biconical; in intestine of insects. TYPE SPECIES *S. praecox* Léger, 1899.

1. *S. chowdhurya* Sarkar and Mazumder, 1983 in *Cryptophagus* sp. (Coleopterorida) midgut.
2. *S. ensifera* (Ellis, 1912) M. Watson, 1916 (syn., *Stylocephalus ensiferus* Ellis, 1912) in *Leptochirus edax, Oxytelus piceus, O. tetracarinatus,* and *Epomolytus sculptus* (Coleopterorida) intestine.
3. *S. praecox* Léger, 1899 (TYPE SPECIES) in *Tanypus* sp. (Dipterorida) intestine.
4. *S. riouxi* Tuzet and Ormières, 1964 in *Dasyhelea lithotelmatica* (Dipterorida) intestine.

Genus *Taeniocystis* Léger, 1905. Mucron (epimerite?) sessile or with a short neck, with six to ten recurved hooks at its apex; deutomerite divided by septa into many chambers; gametocysts spherical, opening by simple rupture; oocysts biconical; in insects. TYPE SPECIES *T. mira* Léger, 1905.

1. *T. mira* Léger, 1905 (TYPE SPECIES) in *Ceratopogon solstitialis* (Dipeterorida) intestine.
2. *T. parva* Foerster, 1938 in *Forcipomyia* sp. (Dipterorida) intestine.
3. *T. truncata* M. Watson, 1916 in *Sericostoma* sp. (Neuropterorida) intestine.

Genus *Sciadiophora* Labbé, 1899. Mucron (epimerite?) a large sessile disk with crenulate border, lost early; protomerite with numerous vertical laminations, broadening to an umbrella in the mature gamont; each rib (costule) of umbrella curved to form a spine pointing backward; oocysts biconical or ovoid, united into a string of beads; in opilionids. TYPE SPECIES *S. phalangii* (Léger, 1896) Labbé, 1899.

1. *S. caudata* (Rössler, 1882) Kamm, 1922 (syns., *Stylorhynchus caudatus* Rössler, 1882; *Stylocephalus caudatus* [Rössler, 1882] Ellis, 1913) in *Mitopus morio, Odius spinosus, Phalangium opilio,* and Phalangiidae gen. sp. (Opilionorida) intestine.
2. *S. claviformis* Ormières and Baudoin, 1973 in *Mitopus* sp. (Opilionorida) intestine.
3. *S. fissidens* (Rössler, 1882) Labbé, 1899 (syns., *Actinocephalus fissidens* Rössler, 1882; *Lycosella phalangii* Léger, 1896 in part) in Phalangiidae gen. sp. *Odiellus palpinalis* and *Phalangium opilio* (Opilionorida) intestine.
4. *S. gagrellula* Devdhar and Amoji, 1978 in *Gargrellula caddlana* (Opilionorida) intestine and ceca.
5. *S. goronowitschi* (Johansen, 1894) Labbé, 1899 (syns., *Actinocephalus goronowitschi* Johansen, 1894; *Lycosella phalangii* Léger, 1896 in part) in *Phalangium opilio* (Opilionorida) intestine.
6. *S. phalangii* (Léger, 1896) Labbé, 1899 (syn., *Lycosella phalangii* Léger, 1896 in part) (TYPE SPECIES) in *Phalangium crassum, P. opilio, P. cornutum, Platybunus bucephalus, P. pinetorum, P. triangularis, Opilio parietinus, Leiobunum hassiae, L. rotundum,* and *Mitopus morio* (Opilionorida) intestine.

Genus *Anthorhynchus* Labbé, 1899. Mucron (epimerite?) a large, flattened, fluted disk; oocysts biconical or ovoid, in lateral chains. TYPE SPECIES *A. sophiae* (Schneider, 1887) Labbé, 1899.

1. *A. hanumanthi* Kalavati and Narasimhamurti, 1978 in *Odonototermes* sp. (Isopterorida) midgut.
2. *A. longispora* Ormières and Baudoin, 1973 in *Liobunum rotundum, Mitopus morio, Opilio parietinus,* and *Platybunus bucephalus* (Opilionorida) intestine.
3. *A. sophiae* (Schneider, 1887) Labbé, 1899 (syn., *Antocephalus sophiae* Schneider, 1887) (TYPE SPECIES) in *Phalangium opilio, Lacinius ephippiatus, Mitopus morio, Platybunus triangularis, Leiobunum rotundum,* and *Oligolophus tridens* (Opilionorida) intestine.

Genus *Agrippina* Strickland, 1912. Mucron (epimerite?) digitate, with peripheral processes; oocysts symmetrical, unarmed, ellipsoidal; in flea larvae. TYPE SPECIES *A. bona* Strickland, 1912.

1. *A. bona* Strickland, 1912 (TYPE SPECIES) in *Nosopsyllus fasciatus* (Siphonapterorida) intestine.

Genus *Globulocephalus* Baudoin, 1965. Young gamont with globular mucron (epimerite?) with a permanent septum; syzygy late and ephemeral, intermediate between frontal and lateral; gametocysts dehisce by simple rupture; oocysts biconical; in intestine of trichopteran larvae. TYPE SPECIES *G. hydropsyches* Baudoin, 1965.

1. *G. hydropsyches* Baudoin, 1965 (TYPE SPECIES) in *Hydropsyche* sp. (Trichopterorida) intestine.

Genus *Alaspora* Obata, 1955. Mucron (epimerite?) sessile in a jar-shaped sucker, with large deep depression at anterior end; gamont solitary, elongate; gametocyst opens by simple rupture; oocysts cylindroconical, with three longitudinal-triangular thin plates extending radially from their trunk; in intestine of beetles. TYPE SPECIES *A. depressa* Obata, 1955.

1. *A. depressa* Obata, 1955 (TYPE SPECIES) in *Anoplogenius cyanescens* (Coleopterorida) intestine.

Genus *Ascocephalus* Obata, 1955. Mucron (epimerite?) sessile, sucker-like, with a thick collar at periphery, depressed very deeply at anterior end, with cavity widened at bottom and with a solid, rugged ring of tooth-like projections around the cavity; gamonts solitary, elongate ovoid; gametocyst spherical; gametocyst dehiscence and oocysts unknown; in intestine of beetles. TYPE SPECIES *A. armatus* Obata, 1955.

1. *A. armatus* Obata, 1955 (TYPE SPECIES) in *Chlaenius nigricans* (Coleopterorida) intestine.

Genus *Amphorocephalus* Ellis, 1913. Mucron (epimerite?) a sessile button fluted on its sides, on a short, dilated neck; protomerite with a transverse superficial constriction; in chilopods. TYPE SPECIES *A. amphorellus* Ellis, 1913.

1. *A. actinotus* (Leidy, 1889) Ellis, 1913 (syns., *Gregarina actinota* Leidy, 1889; *Hoplorhynchus actinotus* [Leidy, 1889] Crawley, 1903) in *Scolopocryptops* sp., *S. sexspinosus*, and *Scolopendra woodi* (Chilopodasida) intestine.
2. *A. amphorellus* Ellis, 1913 (TYPE SPECIES) in *Scolopendra heros* (Chilopodasida) intestine.
3. *A. aratoensis* (Hukui, 1952) Levine, 1980 (syn., *Hoplorhynchus aratoensis* Hukui, 1952) in *Cryptops japonicus* (Chilopodasida) intestine.
4. *A. bouruiensis* (Hukui, 1951) Levine, 1980 (syn., *Hoplorhynchus bouruiensis* Hukui, 1951) in *Otocryptops rubiginosus* (Chilopodasida) intestine.
5. *A. ozakii* (Hukui, 1952) Levine, 1980 (syn., *Hoplorhynchus ozakii* Hukui, 1952) in *Otocryptops rubiginosus* (Chilopodasida) intestine.
6. *A. scolopendrae* (Crawley, 1903) Levine, 1980 (syn., *Hoplorhynchus scolopendrae* Crawley, 1903) in *Scolopendra woodi* (Chilopodasida) presumably intestine.

Genus *Tricystis* Hamon, 1951. Only gamont known; intracellular or between cells of digestive epithelium; mucron (epimerite?), protomerite, and deutomerite present; in chaetognaths. TYPE SPECIES *T. planctonis* Hamon, 1951.

1. *T. leuckarti* (Mingazzini, 1893) Ormières, 1965 (syns., *Lecudina leuckarti* Mingazzini, 1893; *Lankesteria leuckarti* [Mingazzini, 1893] Labbé, 1899) in *Sagitta* sp. (Chaetognatha) intestine.
2. *T. planctonis* Hamon, 1951 (TYPE SPECIES) in *Sagitta lyra* and *S. bipunctata* (Chaetognatha) intestinal epithelial cells.

Genus *Thalicola* Ormières, 1965. Gamonts with or without longitudinal striations; mucron (epimerite?) simple; syzygy head to tail; gametocysts spherical; oocysts and mode of dehiscence unknown; in salps. TYPE SPECIES *T. salpae* (Frenzel, 1885) Ormières, 1965.

1. *T. ensiformis* (Bargoni, 1894) Ormières, 1965 (syn., *Gregarina ensiformis* Bargoni, 1894) in *Salpa (Thalia) democratica* (syn., *S. mucronata*) (Thallaceasida) intestine.

2. *T. filiformis* Théodoridés and Desportes, 1975 in *Cyclosalpa virgula* (Thaliaceasida) intestine.
3. *T. flava* (Roboz, 1886) Ormières, 1965 (syn., *Gregarina flava* Roboz, 1886) in *Salpa (Pegea) confoederata* (Thaliaceasida) intestine.
4. *T. salpae* (Frenzel, 1885) Ormières, 1965 (syn., *Gregarina salpae* Frenzel, 1885) (TYPE SPECIES) in *Salpa maxima* (syn., *S. africana), S. fusiformis,* and *Ilhea punctata* (Thaliaceasida) intestine.

Genus *Epicavus* Ormières and Daumal, 1970. Mucron (epimerite?) in the form of a deep cup with thick walls slightly striated longitudinally on their inner surface, carried on a neck; nucleus with a single large nucleolus; gametocysts spherical, dehiscing by simple rupture; oocysts subspherical, with rounded episporal polar plugs; in insects. TYPE SPECIES *E. araeoceri* Ormières and Daumal, 1970.

1. *E. araeoceri* Ormières and Daumal, 1970 (TYPE SPECIES) in *Araeocerus fasciculatus* (Coleopterorida) intestine.
2. *E. cratoparis* (Crawley, 1903) Ormières and Daumal, 1970 (syns., *Asterophora cratoparis* Crawley, 1903; *Anthorhynchus cratoparis* [Crawley, 1903] Ellis, 1913) in *Euparius* (syn., *Cratoparis) lunatus* (Coleopterorida) intestine.

Genus *Gryllotalpia* Hasselmann, 1926. Mucron (epimerite?) a large knob set on a long stalk; gametocysts dehisce by simple rupture; oocysts biconical; in insects. TYPE SPECIES *G. magalhaesi* Hasselmann, 1926.

1. *G. magalhaesi* Hasselmann, 1926 (TYPE SPECIES) in *Gryllotalpa* sp. (Orthopterorida) intestine.

Genus *Chilogregarina* Levine, 1979. Mucron (epimerite?) simple or with nonpersistent digitiform processes at apex; gametocysts and oocysts unknown; in chilopods. TYPE SPECIES *C. striata* (Léger and Duboscq, 1903) Levine, 1979.

1. *C. brasiliensis* (Pinto, 1918) Levine, 1979 (syn., *Gregarina brasiliensis* Pinto, 1918) in *Scolopendra* sp. (Chilopodasida) intestine.
2. *C. dujardini* (Schneider, 1875) Levine, 1979 (syns., *Actinocephalus dujardini* Schneider, 1875; *Acanthocephalus dujardini* [Schneider, 1875] Balbiani, 1884) in *Lithobius forficatus* (Chilopodasida) intestine.
3. *C. stella* (Léger, 1899) Levine, 1979 (syns., *Rhopalonia stella* Léger, 1899; *Actinocephalus stella* [Léger, 1899] Ormières, 1967) in *Geophilus* (syn., *Himantarium gabrielis* and *G. longicornis* (Chilopodasida) digestive tract.
4. *C. striata* (Léger and Duboscq, 1903) Levine, 1979 (syn., *Actinocephalus striatus* Léger and Duboscq, 1903) (TYPE SPECIES) in *Scolopendra cingulata* (Chilopodasida) intestine.

Genus *Crucocephalus* Sarkar, 1984. Young gamonts ovoid, becoming cylindrical with globular epimerite having six to eight broad vertical ridges set upon a short, thick-walled neck with a dilated base; mature gamonts fusiform to cylindroconical, solitary; gametocysts spherical, dehiscing by simple rupture; sporocysts biconical, with sharply pointed ends. TYPE SPECIES *C. dufouri* Sarkar, 1984.

1. *C. dufouri* Sarkar, 1984 (TYPE SPECIES) in *Dermestes* sp. larva (Coleopterorida) midgut.

Genus *Harendraia* Sarkar, 1984. Young gamonts ovoid or almost fusiform, with highly complex, bowl-like epimerite with a bulb-like round base, narrow elongated neck, and truncated apex set with four short, slender, symmetrical filaments or spines; mature gamonts solitary, cylindrical. Gametocysts spherical, dehiscing by simple rupture. Sporocysts ellipsoidal, released in lateral chain. TYPE SPECIES *H. intricata* Sarkar, 1984.

1.   *H. intricata* Sarkar, 1984 (TYPE SPECIES) in *Ptinus* sp. (Coleopterorida) midgut.

Genus *Levinea* Kori, 1985. Early gamonts ("sporadins") solitary; epimerite cup-like at the apex of a short neck, with numerous peripheral digitiform processes; dehiscence of gametocysts by simple rupture; oocysts cylindrobiconical; in odonate insects. TYPE SPECIES *L. agriocnemidis* Kori, 1985.

1.   *L. agriocnemidis* Kori, 1985 (TYPE SPECIES) in *Agriocnemis pygmaea* (Odonatorida) midgut.

Subfamily ACANTHOSPORINAE *Léger, 1892*

Oocysts with spines or thickenings at their poles, sometimes at the equator and also along their edges (this is the only difference between this subfamily and the Actinocephalinae); parasites of carnivorous insects (especially aquatic ones), of opilionids, and of chilopods.

Genus *Acanthospora* Léger, 1892. Mucron (epimerite?) a conical papilla with an obtuse point; or simple, knob-shaped, papilla-like; or crateriform, with partial septa around it. Oocysts biconical or ellipsoidal, with equatorial spines and terminal tufts; in insects. TYPE SPECIES *A. pileata* Léger, 1892.

1.   *A. bengalensis* Sarkar and Haldar, 1981 in *Ceriagrion cerinorubellum* (Odonatorida) intestine.
2.   *A. crypturgi* Geus, 1969 in *Crypturgus pusillus* (Coleopterorida) midgut.
3.   *A. elongata* Geus, 1969 in *Cerylon ferrugineum* (Coleopterorida) intestine.
4.   *A. pileata* Léger, 1892 (TYPE SPECIES) in *Cistelides* sp. and *Omoplus* sp. (Coleopterorida) intestine.
5.   *A. polymorpha* Léger, 1896 in *Hydrochara* (syns., *Hydrophilus, Hydrous) caraboides* and *Copelatus ruficollis* (Coleopterorida) intestine.
6.   *A. psychodae* Geus, 1969 in *Psychoda severini* (Dipterorida) intestine.

Genus *Grenoblia* Hasselmann, 1927. Mucron (epimerite?) small, simple; protomerite conical; gametocysts and oocysts unknown. TYPE SPECIES *G. legeri* Hasselmann, 1927.

1.   *G. legeri* Hasselmann, 1927 (TYPE SPECIES) in *Hydrophilus* sp. (Coleopterorida) intestine.

Genus *Corycella* Léger, 1892. Mucron (epimerite?) with a central button bearing a crown of strong hooks; oocysts biconical, with a tuft of fine hairs at the ends; in insects. TYPE SPECIES *C. armata* Léger, 1892.

1.   *C. armata* Léger, 1892 (TYPE SPECIES) in *Gyrinus natator* (Coleopterorida) alimentary canal.
2.   *C. orthomorpha* Hasselmann, 1918 in *Aeshna* sp. (Odonatorida) hindgut.

Genus *Ancyrophora* Léger, 1892. Mucron (epimerite?) composed of a head furnished with flexible or rigid appendages forming hooks which may curve backward; oocysts biconical, with polar tuft and six equatorial bristles; in intestine of insects. TYPE SPECIES *A. gracilis* Léger, 1892.

1.   *A. balazyi* Lipa, 1967 in *Carabus coriaceus* and Carabidae gen. sp. (Coleopterorida) intestine.
2.   *A. codreanui* Balcescu-Codreanu, 1973 in *Dictyogenus* sp. and *Isoperla* sp. (Plecopterorida) intestine.
3.   *A. cornuta* Baudoin and Mouthon, 1976 in *Dinocras megalocephala* (Plecopterorida) intestine.
4.   *A. gigantea* H. Hoshide, 1953 in *Calopteryx atrata* (Odonatorida) intestine.
5.   *A. gracilis* Léger, 1892 (syns., *Gregarina acus* von Stein, 1848 [?]; *Actinocephalus acus* von Stein, 1848) (TYPE SPECIES) in *Carabus* sp., *C. auratus*, *C. violaceus*, *C. arcensis*, *C. auronitens*, *C. cancellatus*, *C. glabratus*, *C. hortensis*, *C. insulicola*, *C. intricatus*, *C. nemoralis*, *C. nitens*, *C. ullrichi*, *Oeceoptoma* (syn., *Silpha*) *thoracica*, *Pristonychus* (= *Laemostenos*) *algerinus*, *Pterostichus niger*, *P. vulgaris*, *Hadrocarabus problematicus*, *Abax* sp., and Silphidae gen. sp. (Coleopterorida) digestive tract.
6.   *A. hispani* Massot and Ormières, 1979 in *Chrysocarabus (Chrysotribax) hispanus* (Coleopterorida) intestine.
7.   *A. ischnurae* Sarkar and Haldar, 1981 in *Ischnura senegalensis* (Odonatorida) midgut.
8.   *A. mutabilis* K. Hoshide, 1975 in *Copera annulata* (Odonatorida) intestine.
9.   *A. noteri* Rauchalles in Geus, 1969 in *Noterus clavicornis* and *N. crassicornis* (Coleopterorida) presumably intestine.
10. *A. obtusa* Balcescu-Codreanu, 1973 in *Dictyogenus* sp. and *Isoperla* sp. (Plecopterorida) intestine.
11. *A. octacantha* (von Frantzius, 1848) Baudoin, 1967 (syns., *Actinocephalus octacanthus* von Frantzius, 1848; *Gregarina heeri* von Frantzius, 1848) in *Phryganea striata* (Trichopterorida) intestine.
12. *A. ovoides* Sarkar and Haldar, 1981 in *Ischnura delicata* (Odonatorida) midgut.
13. *A. penetreti* Massot and Ormières, 1979 in *Penetretus rufipennis* (Coleopterorida) intestine.
14. *A. philonthi* Lipa, 1967 in *Philonthus laevicollis* (Coleopterorida) intestine.
15. *A. puytoraci* Baudoin, 1967 (syn., *Ramicephalus puytoraci* [Baudoin, 1967] Tuzet, Ormières, and Théodoridès, 1968) in *Trichostegia minor* (Trichopterorida) intestine.
16. *A. rostrata* Baudoin, 1971 in *Pterostichus cristatus femoratus* (Coleopterorida) intestine.
17. *A. similis* (Foerster, 1938) Geus, 1969 (syn., *Gregarina similis* Foerster, 1938) in *Soronia punctatissima*, *S. grisea*, *Epurea terminalis*, *Rhizophagus bipustulatus* (Coleopterorida), and *Amphigerontia bifasciata* (Corrodentiorida) intestine.
18. *A. stelliformis* (Schneider, 1875) Wellmer, 1911 (syn., *Actinocephalus stelliformis* Schneider, 1875) in *Carabus auratus*, *C. violaceus*, *Compsolacon crenicollis*, *Goerius similis*, *Ontholestes tesselatus*, *Philonthus carbonarius*, *Philonthus* sp., *Pterostichus vulgaris*, *Rhizotrogus* sp., *Staphylinus caesareus*, *S. erythropterus*, *S. olens*, Staphylinidae gen. sp., and *Silpha* sp. (Coleopterorida) intestine.
19. *A. tuzetae* Théodoridès and Jolivet, 1959 in *Systolocranius* sp. and *Chlaenites aruwinius* (Coleopterorida) intestine.
20. *A. uleiotae* Foerster, 1938 in *Uleiota planata* and *Rhizophagus nitidulus* (Coleopterorida) intestine.
21. *A. umbelliformis* Baudoin and Maillard, 1972 in *Hydrochara caraboides* (Coleopterorida) intestine.
22. *A. uncinata* Léger, 1892 in *Colymbetes* sp., *Dytiscus* sp., *D. marginalis*, *Noterus*

clavicornis, Rhantus exoletus, and *Hadrocarabus problematicus* (Coleopterorida), and also *Phryganea grandis, P. rhombica, Phryganea* sp., *Mystacides longicornis, Plectrocnemia conspersa,* and *Sericostoma* sp. (Trichopterorida) intestine.

Genus *Rhizionella* Baudoin, 1971. Mucron (epimerite?) a button bearing ascending filamentous appendages; oocysts biconical, with polar knobs and equatorial spines. TYPE SPECIES *R. tenuis* Baudoin, 1971.

1. *R. tenuis* Baudoin, 1971 (TYPE SPECIES) in *Procrustes purpurescens* and *Silpha atrata* (Coleopterorida) intestine.

Genus *Cometoides* Labbé, 1899. Mucron (epimerite?) composed of a spherical or flattened papilla surrounded by broad, thin, flexible filaments; gametocysts dehisce by simple rupture; oocysts cylindroconical, with polar spines and two rows of equatorial spines; in intestine of insects. TYPE SPECIES *C. crinitus* (Léger, 1892) Labbé, 1899.

1. *C. capitatus* (Léger, 1892) Labbé, 1899 (syn., *Pogonites capitatus* Léger, 1892) in *Hydrous* sp. and *Hydrochara caraboides* (Coleopterorida) intestine.
2. *C. crinitus* (Léger, 1892) Labbé, 1899 (syn., *Pogonites crinitus* Léger, 1892) (TYPE SPECIES) in *hydrobius* sp. Coleopterorida) intestine.
3. *C. pechumani* Anderson and Magnarelli, 1978 in *Chrysops fuliginosus* and *C. atlanticus* (Dipterorida) intestine.
4. *C. pileatus* Baudoin and Maillard, 1972 in *Hydrophilus piceus* (Coleopterorida) intestine.

Genus *Prismatospora* Ellis, 1914. Mucron (epimerite?) subglobular, with laterally recurved hooks; oocysts hexagonal, with one row of spines at each end; in dragonfly naiads. TYPE SPECIES *P. evansi* Ellis, 1914.

1. *P. evansi* Ellis, 1914 (TYPE SPECIES) in *Tramea lacerata* and *Sympetrum rubicundulum* (Odonatorida) intestine.

Genus *Tetraedrospora* Tschudovskaya, 1928. Mucron (epimerite?) a flattened disk, bordered by 14 to 16 hooks; gametocysts develop and oocysts emerge in host gut; oocysts elongate tetrahedral, with sides bearing a row of spines; in fungus gnats. TYPE SPECIES *T. sciarae* Tschudovskaya, 1928.

1. *T. sciarae* Tschudovskaya, 1928 (TYPE SPECIES) in *Sciara* sp., *S. militaris,* and *S. thomae* (Dipterorida) intestine.

Genus *Ramicephalus* Obata, 1955. Mucron (epimerite?) dish-like, with many upward-projecting processes (thicker than in *Cometoides*) at its periphery; gamonts solitary; gametocysts dehisce by simple rupture; oocysts biconical, with one row of polar spines and one row of six equatorial spines; in intestine of insects. TYPE SPECIES *R. ozakii* Obata, 1955.

1. *R. albertianus* (Théodoridès, Ormières, and Jolivet, 1958) Théodoridès, Desportes, and Mateu, 1976 (syn., *Actinocephalus albertianus* Théodoridès, Ormières, and Jolivet, 1958) in *Abisares viridipennis, Acrida* sp., *Atractomorpha aberrans, A. gerstaeckeri, Atractomorpha* sp., *Catantops quadratus, Chirista* sp., *Chondracris* sp., *Cyphocerastis* sp., *Duronia* sp., *Epistaurus succineus, Eyprepocnemis ibadana, Maura* sp., *Odontomelus* sp., *Orbillus coeruleus, Orthacris* sp., *Oxycatantops congoensis, Paracinema lucutenta, Paracoptacra cauta, Parapropacris rhodoptera, Parasphena*

ruandensis, Taphronota calliparea, Thisoicetrus guineensis, Trilophidia conturbata, and *Zonocerus variegatus* (Orthopterorida) intestine.

2. *R. amphoriformis* (Bush, 1928) Théodoridès, Desportes, and Jolivet, 1972 (syn., *Actinocephalus amphoriformis* Bush, 1928) in *Rudonia chloronata, Pseudochirista temporalis, Cataloipus cognatus, Acrotylus furcifer, Catantops humeralis, Cyrtacanthacris ruficornis lineata, Patanga succinata,* and *Trilophidia annulata* (Orthopterorida) intestine.

2a. *R. amphoriformis* var. *madecassa* (Théodoridès, 1962) Théodoridès, Desportes, and Jolivet, 1972 (syn., *Actinocephalus amphoriformis* var. *madecassa* Théodoridès, 1962) in *Paracinema* tricolor var. *madecassa* (Orthopterorida) intestine.

2b. *R. amphoriformis* var. *novaeguineae* Théodoridès, Desportes, and Jolivet, 1972 in *Eoscyllina inexpectata, Patanga succinata, Valanga papuasica, V. g. gohieri, Stenocatantops angustifrons, S. vitripennis, Heteropterus obscurella,* and *Acrida cheesmanae* (Orthopterorida) intestine.

3. *R. bodenheimeri* (Théodoridès and Jolivet, 1969) Théodoridès, Desportes, and Jolivet, 1972 (syn., *Actinocephalus bodenheimeri* Théodoridès and Jolivet, 1969) in *Arantia* sp., *Conocephalus maculatus, C. iris, Harposcepa* sp., *Homerocoryphus nitidulus, Phaneroptera nana sparsa, Poecilogramma* sp., *Pseudopyrhizia* sp., *Schulthessinia* sp., *Tylopsis* sp., and *Zeuneria* sp. (Orthopterorida) intestine.

4. *R. cecheni* Massot and Ormières, 1979 in *Cechenus (Iniopachys) pyrenaeus* (Coleopterorida) intestine.

5. *R. cervicornis* (Théodoridès, 1955) Tuzet, Ormières, and Théodoridès, 1968 (syn., *Ancyrophora cervicornis* Théodoridès, 1955) in *Silpha carinata* and *Hadrocarabus problematicus* (Coleopterorida) intestine.

6. *R. licini* (Tuzet and Théodoridès, 1951) Tuzet, Ormières, and Théodoridès, 1968 (syn., *Cometoides licini* Tuzet and Théodoridès, 1951) in *Licinus punctatosulcatus* (Coleopterorida) intestine.

7. *R. nebriae* Massot and Ormières, 1979 in *Nebria lafresnayei* (Coleopterorida) intestine.

8. *R. olivacus* Sarkar and Haldar, 1981 in *Ceriagrion olivacus* (Odonatorida) midgut.

9. *R. ozakii* Obata, 1955 (TYPE SPECIES) in *Chlaeniellus* (syn., *Chlaenius) inops* (Coleopterorida) intestine.

10. *R. rostratus* Massot and Ormières, 1979 in *Nebria lafresnayei* (Coleopterorida) intestine.

11. *R. tuzetae* (Théodoridès, and Jolivet, 1959) Tuzet, Ormières, and Théodoridès, 1968 (syn., *Ancyrophora tuzetae* Théodoridès, and Jolivet, 1959) in *Systolocranius* sp. and *Chlaenites aruwinius* (Coleopterorida) intestine.

12. *R. wellmeri* (Théodoridès, 1955) Tuzet, Ormières, and Théodoridès, 1968 (syn., *Cometoides wellmeri* Thédoridès, 1955) in *Carabus* sp., *Dolichus hatensis,* and *Pterostichus metallicus* (Coleopterorida) intestine.

13. *R. zophophili* Théodoridès, Desportes, and Jolivet, 1972 in *Zophophilus curticornis* (Coleopterorida) intestine.

Genus *Coronoepimeritus* H. Hoshide, 1959. Gamonts solitary; mucron (epimerite?) a crownlike, globular structure, situated on a short neck and furnished with many small, digitiform processes which may or may not be branched and which cover the surface of the crown; gametocysts dehisce by simple rupture; oocysts ellipsoidal or ovoid, with long, filamentlike polar spines; in insects. TYPE SPECIES *C. japonicus* H. Hoshide, 1959.

1. *C. japonicus* H. Hoshide, 1959 (TYPE SPECIES) in *Locusta migratoria danica, Oedaleus infernalis, Atractomorpha bedeli, Oxa japonica, O. velox,* and *Acrida lata* (Orthopterorida) intestine.

2. *C. monospinus,* H. Hoshide, 1959 in *Euconocephalus thunbergi* (Orthopterorida) intestine.

Genus *Dinematospora* Tuzet and Ormières, 1954. Development extracellular; mucron (epimerite?) hemispherical, flattened, then button-shaped, attached to the protomerite at the level of a chromophilic ring that persists until the time of syzygy; deutomerite with thick membrane; longitudinal myonemes well developed; moderate-sized amylopectin granules and black granules present; nucleus spherical or slightly ovoid, with nucleolus; young gamonts solitary; syzygy head to tail; gametocysts dehisce by simple rupture; oocysts with two long polar filaments; in insects. TYPE SPECIES *D. grassei* Tuzet and Ormières, 1954.

1. *D. grassei* Tuzet and Ormières, 1954 (TYPE SPECIES) in *Machilis tenuis* (Thysanurorida) intestine.

Genus *Doliospora* Ormières and Baudoin, 1969. Mucron (epimerite?) without ornamentation; oocysts asymmetrical, without terminal tufts, with two equatorial thickenings on the longitudinal cordons; in opilionids. TYPE SPECIES *D. repelini* (Léger, 1897) Ormières and Baudoin, 1969.

1. *D. repelini* (Léger, 1897) Ormières and Baudoin, 1969 (syn., *Acanthospora repelini* Léger, 1897) (TYPE SPECIES) in *Phalangium opilio, Opilio parietinus, O. grossipes, Oligolophus tridens, Platybunus bucephalus,* and *Liobunum rotundum* (Opilionasida) intestine.
2. *D. troguli* (Geus, 1969) Levine, 1980 (syn., *Acanthospora troguli* Geus, 1969) in *Trogulus tricarinatus* (Opilionasida) intestine.

Genus *Acanthosporidium* Georgévitch, 1951. Middle part of anterior end of mucron (epimerite?) in form of a somewhat tapered snout; a short neck separates the mucron (epimerite?) from the protomerite; gametocysts and oocysts unknown. TYPE SPECIES *A. gammari* Georgévitch, 1951.

1. *A. gammari* Georgévitch, 1951 (TYPE SPECIES) in unnamed host.
1a. *A. gammari* var. *phryganeus* Georgévitch, 1951 in Trichopterorida gen. sp. (Trichopterorida) intestine.
1b. *A. gammari* var. *ochridensis* Georgévitch, 1951 in Trichopterorida gen. spp. (Trichopterorida) intestine.
2. *A. ochridensis* Georgévitch, 1951 in Trichopterorida gen. spp. (Trichopterorida) intestine.

Genus *Quadruspinospora* Sarkar and Chakravarty, 1969. Mucron (epimerite?) spherical or subspherical, with variable number of stumpy, digitiform processes directed laterally; gametocysts thick-walled, spherical, dehiscing by simple rupture; oocysts ovoid, with two very long spines at each pole; gamonts solitary; in insects. TYPE SPECIES *Q. aiolopii* Sarkar and Chakravarty, 1969.

1. *Q. acridae* Haldar and Chakraborty, 1979 emend. (syn., *Q. acridii* Haldar and Chakraborty, 1979) in *Acrida exultata* (Orthopterorida) hepatic ceca and midgut.
2. *Q. aiolopii* Sarkar and Chakravarty, 1969 emend. Kundu and Haldar, 1983 (syn., *Q. aeolopii* Sarkar and Chakravarty, 1969) (TYPE SPECIES) in *Aiolopus* sp. (Orthopterorida) midgut.
3. *Q. atractomorphae* (Haldar and Chakraborty, 1978) Levine, 1985 (syn., *Q. atractomorphii* Haldar and Chakraborty, 1978) in *Atractomorpha crenulata* (Orthopterorida) midgut and hepatic ceca.
4. *Q. chakravartyi* Haldar and Chakrabory, 1975 emend. (syn., *Q. chakravartyei* Haldar and Chakraborty, 1975) in *Spathosternum* sp. (Orthopterorida) intestine.

5.   *Q. dichotoma* Kundu and Haldar, 1983 in *Spathosternum p. prasmiferum* (Orthopterorida) midgut and hepatic ceca.
6.   *Q. indoaiolopi* Haldar and Chakraborty, 1979 emend. (syn. *Q. indoaiolopii* Haldar and Chakrabory, 1979) in *Aiolopus* sp. and *A. thalassimus temulus* (Orthopterorida) hepatic ceca.
7.   *Q. megaspinosa* Haldar and Chakraborty, 1979 in *Trilophidia annulata* (Orthopterorida) hepatic ceca and gut.

Genus *Contospora* Devdhar and Amoji, 1978. Mucron (epimerite?) a conical knob, dentate at the base, with about 20 vertical lamellae; in arthropods. TYPE SPECIES *C. opalniae* Devdhar and Amoji, 1978.

1.   *C. opalniae* Devdhar and Amoji, 1978 (TYPE SPECIES) in *Opalnia* sp. (Opilionorida) midgut and ceca.

Genus *Tetractinospora* Sarkar and Haldar, 1981. With globular mucron (epimerite?) with about 16 laminate, vertical plates and a short neck; gamonts solitary; gametocysts dehisce by simple rupture; oocysts biconical, bent in the middle, with 2 sharp, stout spines at each end. TYPE SPECIES *T. victoris* Sarkar and Haldar, 1981.

1.   *T. victoris* Sarkar and Haldar, 1981 (TYPE SPECIES) in *Ceriagrion coromandelianum* (Odonatorida) midgut.

Genus *Echinoocysta* Levine, 1984. Epimerite simple, a globular or spherical knob; protomerite dome-shaped or hemispherical, with striated rim around its base, sitting on a short, cylindrical collar; oocysts biconical, with a row of eight to ten slender spines at each end; oocysts released from gametocyst in chains of two or three by simple rupture. TYPE SPECIES *E. phalangii* (Amoji and Devdhar, 1979) Levine, 1984.

1.   *E. phalangii* (Amoji and Devdhar, 1979) Levine, 1984 (syn., *Echinospora phalangii* Amoji and Devdhar, 1979) (TYPE SPECIES) in *Opalnia* sp. (Opilionorida) intestine.

Genus *Mukundaella* Sarkar, 1981. Mucron (epimerite?) cup-shaped, with numerous striations; oocysts diamond-shaped, with polar and meridional spines; in insects. TYPE SPECIES *M. undulatus* Sarkar, 1981.

1.   *M. gulbargaensis* Kori and Amoji, 1984 in *Copera* sp. (Odonatorida).
2.   *M. undulatus* Sarkar, 1981 (TYPE SPECIES) in *Enallagma* sp. (Odonatorida).

Genus *Tetrameridionospinispora* Kori and Amoji, 1985. Oocysts with rows of meridional spines. TYPE SPECIES *T. karnataki* Kori and Amoji, 1985.

1.   *T. ceriagrioni* Nazeer Ahmed and Narasimhamurti, 1979) Kori and Amoji, 1985 in *Ceriagrion coromandelianum* (Odonatorida) midgut.
2.   *T. karnataki* Kori and Amoji, 1985 (TYPE SPECIES) in *Agriocnemis* sp. (Odonatorida).

Subfamily MENOSPORINAE *Léger, 1892*
    Mucron (epimerite?) a large cup bordered with hooks, with a long neck; gamonts solitary; gametocysts without sporoducts; oocysts crescentic, smooth; in arthropods.

Genus *Menospora* Léger, 1892. Mucron (epimerite?) very persistent, with a long neck terminated by a cupule bordered by hooks; oocysts bent, banana-shaped; in insects. TYPE SPECIES *M. polyacantha* Léger, 1892.

1. *M. nonacantha* Devdhar and Deshpande, 1976 (syn., *M. nonacontha* Devdhar and Deshpande, 1976 *lapsus calami*) in *Agriocnemis* sp. (Odonatorida) intestine.
2. *M. polyacantha* Léger, 1892 (TYPE SPECIES) in *Agrion puella, A. pulchellum, Agrion* sp., *Coenagrion hastulatum, Enallagma cyathigerum, Erythroma najas, Lestes sponsa, L. viridis, Lestes* sp., *Platycnemis pennipes, Pyrrhosoma nymphila, Sympecma fusca,* and rarely *Ischnura elegans* (Odonatorida) digestive tract.

Genus *Hoplorhynchus* Carus, 1863. Mucron (epimerite?) long, terminating in a flattened bulb bordered by digitations anteriorly; oocysts ellipsoidal, a little curved. TYPE SPECIES *H. oligacanthus* (von Siebold, 1839) Carus, 1863.

1. *H. bourviensis* Hukui, 1952 in *Otocryptops rubiginosus* (Chilopodasida) intestine.
2. *H. gracilis* H. Hoshide, 1954 in *Aciagrion hisopa* (Odonatorida) intestine.
3. *H. hexacanthus* Obata, 1955 in *Coeagrion quadrigerum* (Odonatorida) intestine.
4. *H. magnus* H. Hoshide, 1959 in *Crocothemis servilia* (Odonatorida) intestine.
5. *H. miyanensis* K. Hoshide, 1977 in *Ceriagrion melanorum* (Odonatorida) intestine.
6. *H. oligacanthus* (von Siebold, 1839) Carus, 1863 (syns., *Gregarina oligacantha* von Siebold, 1839; *G. sieboldi* von Kölliker, 1848; *Stylorhynchus oligacanthus* [von Siebold, 1839] von Stein, 1848; *Acanthocephalus sieboldi* von Frantzius, 1848; *Actinocephalus sieboldi* [von Kölliker, 1848] von Frantzius, 1848; *A. oligacantha* [von Siebold, 1839] Bütschli, 1882) (TYPE SPECIES) in *Boyenia irene, Agrion* sp., *Calopteryx virgo* (syn., *Agrion forcipula*), *C. splendens, Calopteryx* sp., *Coenagrion hastulatum, Enallagma cyatherigerum, Lestes sponsa, Onychogamphus unchtus, Pyrrhosoma nymphula, Sympetrum danae, S. pedemontanum,* and *S. vulgatum* (Odonatorida) intestine.
7. *H. orthetri* H. Hoshide, 1953 in *Orthetrum albistylum speciosum* (Odonatorida) intestine.
8. *H. ozakii coreica* Théodoridès, Desportes, and Jolivet, 1979 in *Otocryptops rubiginosus* (Chilopodasida) intestine.
9. *H. polyhamatus* K. Hoshide, 1977 in *Munais strigata* (Odonatorida) intestine.
10. *H. ramidigitus* Sarkar and Haldar, 1980 in *Agriocnemis pygmaea* (Odonatorida) midgut.

Genus *Odonaticola* Sarkar and Haldar, 1981. Mucron (epimerite?) hat-shaped, with petaloid spines on the margin, with long neck; gamonts solitary; gametocysts dehisce by simple rupture; development extracellular; oocysts smooth, boat-shaped. TYPE SPECIES *O. hexacantha* Sarkar and Haldar, 1981.

1. *O. crocothemis* Kori and Amoji, 1983 in *Crocothemis servilia* (Odonatorida) midgut.
2. *O. haldari* Kori and Amoji, 1984 in *Trithemis aurora* (Odonatorida) fore- and midgut.
3. *O. hexacantha* Sarkar and Haldar, 1981 (TYPE SPECIES) in *Brachythemis contaminata* (Odonatorida) midgut.
4. *O. longicollare* Sarkar and Haldar, 1981 in *Diplacodes trivalis* (Odonatorida) midgut.
5. *O. orthetri* Sarkar and Haldar, 1981 in *Orthetrum sabina* (Odonatorida) midgut.
6. *O. rodgii* Sarkar and Haldar, 1981 in *Neurothemis t. tullia* (Odonatorida) midgut.

Family BRUSTIOSPORIDAE *Kundu and Haldar, 1981*
   Gamonts solitary; mucron (epimerite?) small, variously formed, with fine bristles, with short but distinct neck; gametocysts dehisce by simple rupture; oocysts spherical, set with brush borders and chained with fine, filamentous processes.

Genus *Brustiospora* Kundu and Haldar, 1981. With the characters of the family. TYPE SPECIES *B. indicola* Kundu and Haldar, 1981.

1. *B. indicola* Kundu and Haldar, 1981 (TYPE SPECIES) in *Stethorus* sp. (Coleopterorida) intestine.

Family ACUTIDAE *Stejskal, 1965*
   Gamonts solitary; mucron (epimerite?) simple, changing shape during development; gametocysts without sporoducts or residuum, dehiscing by opening in wall; oocysts ellipsoidal; in insects.

Genus *Acuta* Stejskal, 1965. Mucron (epimerite?) simple, growing continually, at first spherical, then cylindrical and drop-shaped and breaking off, leaving a scar; oocysts dehisce through a simple, irregular place in the gametocyst wall; in intestine of bees. TYPE SPECIES *A. rousseaui* Stejskal, 1965.

1. *A. rousseaui* Stejskal, 1965 (TYPE SPECIES) in *Apis mellifera* (Hymenopterorida) intestine.

Genus *Apigregarina* Stejskal, 1965. Mucron (epimerite?) simple, at first large, becoming ovoid, then spherical, and finally conical in the course of development; oocysts dehisce through a round opening in the gametocyst wall; in intestine of bees. TYPE SPECIES *A. stammeri* Stejskal, 1965.

1. *A. stammeri* Stejskal, 1965 (TYPE SPECIES) in *Apis mellifera* (Hymenopterorida) intestine.

Family MONOICIDAE *Geus, 1969*
   Autogamy present.

Genus *Monoica* Stejskal, 1964. Mucron (epimerite?) simple; young gamonts solitary; each gamont forms gametes of both sexes (i.e., autogamy present); gametes anisogamous; gametocysts open by simple rupture; gametocyst residuum present; oocysts ellipsoidal; in bees. TYPE SPECIES *M. apis* Stejskal, 1964.

1. *M. apis* Stejskal, 1964 (TYPE SPECIES) in *Apis mellifera* (Hymenopterorida) intestine.

Superfamily FUSIONICAE *Stejskal, 1965*
   Homoxenous; upon syzygy the nucleus and entocyte of the satellite go into the primite, where they fuse; anisogamous; gametocysts and oocysts unknown.

Family FUSIONIDAE *Stejskal, 1965*
   With the characters of the superfamily.

Genus *Fusiona* Stejskal, 1965. With the characters of the family; in insects. TYPE SPECIES *F. geusi* Stejskal, 1965.

1. *F. geusi* Stejskal, 1965 (TYPE SPECIES) in *Pycnoscelus surinamensis* (Orthopterorida) intestine.

Chapter 7

THE GREGARINES: NEOGREGARINES

Order **NEOGREGARINORIDA** Grassé, 1953
Merogony present, presumably acquired secondarily; gamonts septate; all in insects, in malpighian tubules, intestine, hemocoele, or fat tissues.

Family GIGADUCTIDAE *Filipponi, 1948*
Mucron (epimerite?) absent; gametocysts with single sporoduct; development intracellular.

Genus *Gigaductus* Crawley, 1903. Oocysts cylindrical, large. TYPE SPECIES *G. parvus* Crawley, 1903.

1.   *G. africanus* Théodoridès, Desportes, and Mateu, 1976 in *Plangiopsis* sp. (Orthopterorida) intestine.
2.   *G. agoni* Dederichs and Scholtyseck, 1977 in *Agonus sexpunctatum* (Coleopterorida) intestine.
3.   *G. americanus* Corbel, 1968 in *Gryllus capitatus* (Orthopterorida) intestine.
4.   *G. anchi* Tuzet and Ormières, 1966 in *Anchus ruficornis* (Coleopterorida) ceca and occasionally malpighian tubules.
5.   *G. brachyni* Dederichs and Scholtyseck, 1977 in *Brachynus crepitans* (Coleopterorida) intestine.
6.   *G. elongata* (Moriggi, 1943) Filipponi, 1948 (syn., *Endocryptella elongata* Moriggi, 1943) in *Calathus fuscipes* var. *latus* (Coleopterorida) intestine.
7.   *G. exiguus* Wellmer, 1911 (syns., *Endocryptella ghidini* Moriggi, 1943; *Gregarina exigua* [Wellmer, 1911] Watson, 1922; *G. [Gigaductus] exiguus* [Wellmer, 1911] Kamm, 1922) in *Pterostichus niger, P. melas, P. vulgaris,* and *Amara similita* (Coleopterorida) intestine.
8.   *G. macrospora* Filipponi, 1948 in *Pristonychus* (syn., *Laemostinus) algerinus* (Coleopterorida) intestine.
9.   *G. parvus* Crawley, 1903 (syn., *Gregarina parva* [Crawley, 1903] M. Watson, 1916) (TYPE SPECIES) in *Harpalus caliginosus* and *H. pennsylvanicus* (Coleopterorida) intestine.
10.   *G. podurae* (Léger, 1892) Crusz, 1960 (syns., *Clepsidrina podurae* Léger, 1892; *Gregarina podurae* [Léger, 1892] Labbé, 1899) in *Orchesella* (syn., *Podura) villosa* and *Orchesella* sp. (Collembolorida) intestine.
11.   *G. steropi* Massot and Ormières, 1979 in *Steropus madidus* (Coleopterorida) intestine.

Family OPHRYOCYSTIDAE *Léger and Duboscq, 1908*
"Monocystid"; development extracellular; meronts conical, attached to malpighian tubules by numerous rootlets or pseudopods; young gamonts vermicular; gamonts monogametic, isogamic, forming a single oocyst per gametocyst; oocysts with eight sporozoites; all in malpighian tubules.

Genus *Ophryocystis* Schneider, 1883. Multiplication by binary or multiple fission. TYPE SPECIES *O. buetschlii* Schneider, 1883.

1.   *O. buetschlii* Schneider, 1883 (TYPE SPECIES) in *Blaps* sp. (Coleopterorida) malpighian tubules.

2.   *O. caulleryi* Léger, 1900 in *Scaurus tristis* (Coleopterorida) malpighian tubules.
3.   *O. dendroctoni* Weiser, 1970 in *Dendroctonus pseudotsugae* (Coleopterorida) malpighian tubules.
4.   *O. duboscqi* Léger, 1907 in *Otiorhynchus meridionalis, O. ligustica,* and *O. fuscipes* (Coleopterorida) malpighian tubules.
5.   *O. elekroscirrha* McLaughlin and Myers, 1970 in *Danaus plexippus* and *D. gilippus berenice* (Lepidopterorida) hypodermal tissue.
6.   *O. francisci* Schneider, 1885 in *Akis algeriana* and *A. acuminata* (Coleopterorida) malpighian tubules.
7.   *O. hagenmuelleri* Léger, 1900 in *Olocrates gibbus* (Coleopterorida) malpighian tubules.
8.   *O. hessei* Léger, 1907 in *Omophlus brevicollis* (Coleopterorida) malpighian tubules.
9.   *O. hylesini* Purrini and Ormières, 1981 in *Hylesinus fraxini* (Coleopterorida) malpighian tubules.
10.   *O. hylobii* Purrini and Ormières, 1982 in *Hylobius abietis* (Coleopterorida) malpighian tubules.
11.   *O. mesnili* Léger, 1900 in *Tenebrio molitor* and *Trogoderma simplex* (Coleopterorida) malpighian tubules.
12.   *O. perezi* Léger, 1907 in *Dendarus tristis* (Coleopterorida) malpighian tubules.
13.   *O. schneideri* Léger and Hagenmüller, 1900 in *Blaps magica* (Coleopterorida) malpighian tubules.

Family SCHIZOCYSTIDAE *Léger and Duboscq, 1908*
   Development extracellular; mucron (epimerite?) made up first of small pseudopods and then consisting of a sort of sucker; nuclear multiplication proceeds in pairs with growth, merozoites (schizozoites) being arranged in clusters and becoming freed by dropping off; young gamonts vermiform; gamonts, gametocysts, and oocysts of the actinocephalid type.

Genus *Schizocystis* Léger, 1900. Mature trophozoites multinucleate, ovoid or cylindrical with differentiated anterior end; merogony by multiple fission; young gamonts become associated in syzygy; gametocyst with up to 32 oocysts; oocysts with 8 sporozoites; in Dipterorida. TYPE SPECIES *S. gregarinoides* Léger, 1900.

1.   *S. gregarinoides* Léger, 1900 (TYPE SPECIES) in *Ceratopogon solstialis* (Dipterorida) intestine.
2.   *S. legeri* Keilin, 1923 in *Systenus* sp. (Dipterorida) intestine.

Genus *Machadoella* Reichenow, 1935. Young gamonts look like those of *Monocystis*, with a thick cuticle creased by deep, longitudinal striations; body flexible but not contractile; nuclear multiplication begins in nematoid individuals; merozoites individualize when the meront is shortened and has taken an irregular form; division occurs without definite order, first isolating cytoplasmic islets without nuclei and finally producing ovoid merozoites with a single nucleus; gamonts appear like large nematoid gregarines; syzygy head to head; gametes appear similar — spheres with a short, conical mucron; gametocysts with three to six oocysts; oocysts navicular, with polar thickenings, with eight sporozoites; in malpighian tubules of insects. TYPE SPECIES *M. triatomae* Reichenow, 1935.

1.   *M. spinigeri* (Machado, 1913) Reichenow, 1935 (syn., *Schizocystis spinigeri* Machado, 1913) in *Spiniger* sp. (Homopterorida) intestine.
2.   *M. triatomae* Reichenow, 1935 (TYPE SPECIES) in *Triatoma dimidiata* and *Eutriatoma maculata* (Homopterorida) malpighian tubules.

Genus *Lymphotropha* Ashford, 1965. Young gamonts ovoid, with terminal projection and with faint longitudinal striae; gamonts mononucleate on association; oocysts with eight sporozoites; in insects. TYPE SPECIES *L. tribolii* Ashford, 1965.

1. *L. tribolii* Ashford, 1965 (TYPE SPECIES) in *Tribolium castaneum* (Coleopterorida) hemocoele.

Family CAULLERYELLIDAE *Keilin, 1914*

Development and merogony extracellular; meront globular, with nuclei localized in a single zone and merozoites grouped in a bouquet; young gamont more or less flask-shaped, with the neck an organ of fixation (pseudomerite); isogamy probable; gametocysts produce 4 to 11 oocysts; in digestive tract of dipteran larvae.

Genus *Caulleryella* Keilin, 1914. Gametocyst with eight oocysts; oocysts with eight typical sporozoites with very thin membrane which is difficult to see. TYPE SPECIES *C. aphiochaetae* Keilin, 1914.

1. *C. annulatae* Bresslau and Buschkiel, 1919 in *Culiseta annulata* and *Aedes rusticus* (Dipterorida) intestine.
2. *C. anophelis* Hesse, 1918 in *Anopheles claviger* (syn., *A. bifurcatus*) and *A. maculipennis* (Dipterorida) intestine.
3. *C. aphiochaetae* Keilin, 1914 (TYPE SPECIES) in *Aphiochaeta rufipes* (Dipterorida) intestine.
4. *C. maligna* Godoy and Pinto, 1922 in *Anopheles albitarsus* (syn., *Cellia allopha*) (Dipterorida) intestine.
5. *C. pipientis* Buschkiel, 1919 in *Culex pipiens* (Dipterorida) intestine.

Genus *Tipulocystis* Kramář, 1950. Sporozoites become mycetoid meronts attached to intestinal epithelial cells by a pseudomerite; these produce merozoites which become mononucleate gregarinoid meronts which are also attached to intestinal epithelial cells by a pseudomerite; they produce gamonts which fuse, and the resultant gametocyst forms a single oocyst containing eight sporozoites. TYPE SPECIES *T. maximae* Kramář, 1950.

1. *T. maximae* Kramář, 1950 (TYPE SPECIES) in *Tipula maxima* (Dipterorida) intestine.

Family SYNCYSTIDAE *Schneider, 1886*

Trophozoite aseptate, piriform or ovoid; merogony simple, ending in formation of 150 small merozoites; isogamy probable; gametocyst contains about 150 navicular oocysts with spines at the end; in hemocoele of insects.

Genus *Syncystis* Schneider, 1886. Merogony and sporogony extra- or intracellular; young gamonts elongate, amoeboid; mature meronts more or less spherical; gametocysts spherical. TYPE SPECIES *S. mirabilis* Schneider, 1886.

1. *S. aeshnae* Tuzet and Manier, 1953 in *Aeshna* sp. (Odonatorida) hemocoele.
2. *S. mirabilis* Schneider, 1886 (TYPE SPECIES) in *Nepa cinerea* (Hemipterorida) hemocoele.

Family LIPOTROPHIDAE *Grassé, 1953*

Development intracellular; merogony more or less regular; gametes isogamous; oocysts more or less numerous, navicular, with heavily staining polar thickenings; in insects.

Genus *Lipotropha* Keilin, 1923. Nuclear multiplication begins in the meronts while they are still small and continues as they grow; merozoites formed by a sort of beading; all merozoites disposed radially around a central cytoplasmic residuum; some merozoites grow without karyokinesis, become spherical, and turn into gamonts which associate in pairs, contracting and forming tiny (10-μm) gametocysts surrounded by a thin pellicle; isogamy appears complete, there being 32 gametes; each gametocyst forms 16 navicular oocysts, each with 8 sporozoites; in fat body of arthropods. TYPE SPECIES *L. macrospora* Keilin, 1923.

1.   *L. calliphorae* Chatton, 1937 in *Calliphora erythrocephala* (Dipterorida) fat body.
2.   *L. dorci* Ormières et al., 1969 in *Dorcus parallelopipedus* (Coleopterorida) fat tissues.
3.   *L. macrospora* Keilin, 1923 (TYPE SPECIES) in *Systenus adpropinquans* (Dipterorida) fat body.
4.   *L. microspora* Keilin, 1923 in *Systenus adpropinquans* and probably also *Dasyhelea obscura* and *Rhipidia ctenophora* but not developing completely in them (Dipterorida) fat body.
5.   *L. milloti* Tuzet and Manier, 1958 in *Spirostreptus madagascariensis* (Diplopodasida) fat body (?).

Genus *Menzbieria* Bogoyavlenskaya, 1922. Merogony either in body cavity or intestinal mucosa; first-generation merozoites elongate, second-generation merozoites ameboid; gamogony resembles that of *Lipotropha*; oocysts apparently of actinocephalid type; in intestine of arthropods. TYPE SPECIES *M. hydrachnae* Bogoyavlenskaya, 1922.

1.   *M. chalcographi* Weiser, 1955 in *Pityogenes chalcographus* (syn., *Ips typographus*) (Coleopterorida).
2.   *M. hydrachnae* Bogoyavlenskaya, 1922 (TYPE SPECIES) in *Hydrachna* sp. (Acarinasida) intestine.

Genus *Mattesia* Naville, 1930. Development intracellular, producing merozoites of 2 types, 1 with small nuclei and the other with large ones; merogony to the small-nucleate form appears to be the first, and merogony to the large-nucleate form appears to be the second; meronts about 40 μm, forming 15 to 60 elongate merozoites; gamonts, formed by the last agamic generation, generally fall into the hemocoele, couple, and continue to grow; gametocysts about 12 μm; first progamic division ends with formation of 2 unequal nuclei which divide asynchronously in their turn; the 2 small daughter nuclei are somatic, the 2 large ones sexual; each gamont forms 2 gametes; all gametes similar; gametocysts with 1 to 2 oocysts; oocysts with 8 sporozoites; haploid number of chromosomes 2; in insects. TYPE SPECIES *M. dispora* Naville, 1930.

1.   *M. bombi* Liu, Macfarlane, and Pengelly, 1974 in *Bombus affinis, B. bimaculatus, B. fervidus, B. griseocollis, B. impatiens, B. perplexus, B. terricola,* and *B. vagans* (Hymenopterorida) fat body.
2.   *M. dispora* Naville, 1930 (syn., *Coelogregarina ephestiae* Ghélélovitch, 1948) (TYPE SPECIES) in *Anagasta* (syn., *Ephestia*) *kuehniella, Ephestia cautella, Ephestia* sp., *Cryptolestes pusillus*, and *Plodia interpunctella* (Lepidopterorida) fat body, and perhaps *Laemophloeus ferrugineus* and *L. minutus* (Coleopterorida).
3.   *M. geminata* Jouvenaz and Anthony, 1979 in *Solenopsis geminata* (Hymenopterorida) hypodermis.
4.   *M. grandis* McLaughlin, 1965 in *Anthonomus grandis* (Coleopterorida) fat tissues.
5.   *M. orchopiae* Das Gupta, 1958 in *Orchopeas wickhami* (Siphonapterorida) fat body and malpighian tubules.

6.   *M. oryzaephili* Ormières, Louis, and Kuhl, 1971 in *Oryzaephilus surinamensis* (Coleopterorida) fat tissues.
7.   *M. povolnyi* Weiser, 1952 in *Homeosoma nebulellum* (Lepidopterorida) malpighian tubules and intestine.
8.   *M. schwenkei* Purrini, 1977 (syn., *M. schwenki* Purrini, 1977 of *Zool. Rec.* [1977] *lapsus calami*) in beetle *Dryocoetes autographus* (Coleopterorida) intestine.
9.   *M. trogodermae* Canning, 1964 in *Trogoderma granarium, T. simplex, T. sternale, T. variabile, T. inclusum, T. glabrum, T. grassmani,* and *T. ornatum* (Coleopterorida) fat body.

Genus *Lipocystis* Grell, 1938. Meronts give rise to numerous fusiform merozoites; this micronucleate merogony is followed, as in *Mattesia*, by a macronucleate merogony in which the merozoites seem to produce new meronts and finally a last generation, the gamonts; the young gamonts apparently continue to grow, a pair measuring a maximum of 55 μm; they are never surrounded by a membrane; the gamonts become multinucleate and divide into cytomeres at the edges of which the gametes develop; a large residuum containing somatic nuclei remains; gametes apparently isogamous; 100 to 300 oocysts per gametocyst; oocysts navicular with thickened ends, with 8 sporozoites; in fat body of insects. TYPE SPECIES *L. polyspora* Grell, 1938.

1.   *L. polyspora* Grell, 1938 (TYPE SPECIES) in *Panorpa communis* (Mecopterorida) fat body.

Genus *Farinocystis* Weiser, 1953. Development intracellular; first merogony produces 2 to 60 merozoites with small nuclei; these grow into new meronts with larger nuclei which develop into gamonts which produce gametes; gametes develop 8 nuclei before copulation; in gametic copulation one gamete surrounds the other; schizogony of zygote produces a 30-nucleate stage, and a spherical sporoblast forms around each nucleus; sporoblasts leave gametocyst, scatter through body of host, and each develops into a boat-shaped oocyst with a hyaline, button-like plug at each end; oocysts with 8 sporozoites each; in insects. TYPE SPECIES *F. tribolii* Weiser, 1953.

1.   *F. tenebrioides* Purrini and Ormières, 1979 in *Tenebrioides mauretanicus* (Coleopterorida) fat body.
2.   *F. tribolii* Weiser, 1953 (syn., *Triboliocystis garnhami* Dissanaike, 1955) (TYPE SPECIES) in *Tribolium castaneum, T. confusum, Tenebrioides mauretanicus,* and *Palembus acularis* (Coleopterorida) fat body.

Incertae Sedis
Genus *Sawayella* Marcus, 1939. Gametocyst apparently swallowed whole by host; 800 to 8000 sporozoites released which enter intestinal epithelial cells and turn into spherical meronts which grow and fall into intestinal lumen, in which mitosis (?) of an unusual type occurs; the meronts, 15 μm in diameter, undergo nuclear fragmentation, the nuclear membrane disappears, and the cytoplasm also fragments; gamonts couple either in a cell or in intestinal lumen; they form gametes (?) and then navicular and unusually small oocysts (4 to 7 × 1 to 2 μm) containing 8 spherical sporozoites; in midgut of marine Bryozoa. TYPE SPECIES *S. polyzoorum* Marcus, 1939.

1.   *S. polyzoorum* Marcus, 1939 (TYPE SPECIES) in *Thalamoporella* sp. and *Watersipora cucullata* (Bryozoa) intestine.

Chapter 8

THE COCCIDIA: AGAMOCOCCIDIORIDA, PROTOCOCCIDIORIDA, AND IXORHEORIDA

Subclass COCCIDIASINA Leuckart, 1879

Gamonts ordinarily present; mature gamonts small, typically intracellular; conoid not modified into mucron or epimerite; syzygy generally absent, if present involves gametes; anisogamy marked; life cycle characteristically consists of merogony, gamogony, and sporogony; most species in vertebrates.

Order AGAMOCOCCIDIORIDA Levine, 1979

Merogony and gamogony both absent; sporogony present.

Family RHYTIDOCYSTIDAE *Levine, 1979*

With characters of the order; in marine annelids.

Genus *Rhytidocystis* Henneguy, 1907. With characters of the family; oocyst with *n* sporocysts, each with two sporozoites. TYPE SPECIES *R. opheliae* Henneguy, 1907.

1. *R. henneguyi* de Beauchamp, 1912 in *Ophelia neglecta* (Polychaetasida) coelom, intestine, connective tissues, etc.
2. *R. opheliae* Henneguy, 1907 (TYPE SPECIES) in *Ophelia bicornis* (Polychaetasida) coelom.
3. *R. sthenelais* (Porchet-Henneré, 1972) Levine, 1979 (syn., *Dehornia sthenelais* Porchet-Henneré, 1972) in *Sthenelais boa* (Polychaetasida) coelom.

Order IXORHEORIDA Levine, 1984

Gamogony absent; merogony and sporogony present.

Family IXORHEIDAE *Levine, 1984*

With characters of the order, in marine holothurians.

Genus *Ixorheis* Massin, Jangoux, and Sibuet, 1978. With characters of the family; oocyst with one sporocyst; sporocyst with two sporozoites. TYPE SPECIES *I. psychropotae* Massin, Jangoux, and Sibuet, 1978.

1. *I. psychropotae* Massin, Jangoux, and Sibuet, 1978 (TYPE SPECIES) in *Psychropotes longicauda* (Holothurasida) intestinal blood vessels, under intestinal serosa and even coelom.

Order PROTOCOCCIDIORIDA Kheisin, 1956

Without merogony; most often extracellular, the intracellular phase being reduced to the penetration of and a period of latency in the host; trophozoite with permanent three-membraned wall; with sporocysts in oocyst; sporozoites with several dozen subpellicular microtubules; in annelids.

Family GRELLIIDAE *Levine, 1973*

Development extra- or intracellular; oocysts extraintestinal, usually in coelom; gamonts symmetrical; in annelids.

Genus *Grellia* Levine, 1973. Extracellular; gamogony produces 12 to 50 concave-disk-shaped or hemispherical microgametes; biflagellate oocysts with variable number of sporocysts, the number depending on the size of the macrogamont; sporocysts with 5 to 14 sporozoites each; in coelom of archiannelids and polychetes. TYPE SPECIES *G. dinophili* (Grell, 1953) Levine, 1973.

1. *G. dinophili* (Grell, 1953) Levine, 1973 (syn., *Eucoccidium dinophili* Grell, 1953) (TYPE SPECIES) in *Dinophilus gyrociliatus* and *D. notoglandulata* (Archiannelidasida) coelom.
2. *G. ophryotrochae* (Grell, 1960) Levine, 1973 (syn., *Eucoccidium ophryotrochae* Grell, 1960) in *Ophryotrocha puerilis*, *O. labronica*, *O. macrovifera*, *O. notoglandulata*, *O. robusta*, *O. hartmanni*, *O. adherens*, *O. diadema* (Polychaetasida), and *Dinophilus gyrociliatus* (Archiannelidasida) coelom.

Genus *Coelotropha* Henneré, 1963. Extracellular; microgamonts produce several hundred flagellated microgametes; oocysts with 6 to 60 sporocysts, each with about 20 to 50 or more sporozoites, sporocyst wall composed of two valves; young gamont with or without retractile rostrum; in coelom of polychetes. TYPE SPECIES *C. vivieri* Henneré, 1963.

1. *C. durchoni* (Vivier, 1963) Vivier and Henneré, 1964 (syns., *Eucoccidium durchoni* Vivier, 1963; *Caelotropha durchoni* Vivier of Vivier [1963] *lapsus calami*) in *Nereis diversicolor* (Polychaetasida) coelom.
2. *C. vivieri* Henneré, 1963 (TYPE SPECIES) in *Notomastus latericeus* (Polychaetasida) coelom.

Family MYRIOSPORIDAE *Grassé, 1953*
Male gamonts subdivided into several microgametoblasts which form the microgametes.

Genus *Myriospora* Lermantoff, 1913. Oocysts with 8 to several hundred sporocysts each with 24 or 32 sporozoites; male gamont helicoid or ellipsoidal; perhaps heteroxenous but probably homoxenous; known stages in polychetes. TYPE SPECIES *M. trophoniae* Lermantoff, 1913.

1. *M. gopalai* Ganapati, 1941 in *Cirratulus filiformis* (Polychaetasida) subepithelial tissues of intestine.
2. *M. petaloprocti* Ormières, 1975 in *Petaloproctus terricola* (Polychaetasida) coelom.
3. *M. polydorae* Ganapati, 1952 in *Polydora ciliata* (Polychaetasida) coelom.
4. *M. trophoniae* Lermantoff, 1913 (TYPE SPECIES) in *Trophonia plumosa* (Polychaetasida) heart.

Genus *Myriosporides* Henneré, 1966. Early development in intestinal wall; gamonts in coelom; gamont forms a number of microgametoblasts within it, and each microgametoblast forms quite a few biflagellate microgametes; ripe gamonts of both sexes shed their outer wall before uniting; oocysts form about 20 sporocysts, each with several tens of sporozoites; in polychetes. TYPE SPECIES *M. amphiglenae* Henneré, 1966.

1. *M. amphiglenae* Henneré, 1966 (TYPE SPECIES) in *Amphiglena mediterranea* (Polychaetasida) coelom and intestinal wall.

Genus *Mackinnonia* Janiszewska, 1963. Oocysts contain many sporocysts; sporocysts each contain two sporozoites; in oligochetes. TYPE SPECIES *M. tubificis* Janiszewska, 1963.

1.   *M. lumbricilli* Siau and Ormières, 1970 in *Lumbricillus lineatus* (Oligochaetasida) coelom.
2.   *M. tubificis* Janiszewska, 1963 (TYPE SPECIES) in *Tubifex tubifex* (Oligochaetasida) coelom.

Family ANGEIOCYSTIDAE *Léger, 1911*
Microgamonts crescent-shaped, undivided; macrogamonts spherical; oocysts form four ovoid sporoblasts, each of which forms the sporozoites.

Genus *Angeiocystis* Brasil, 1904. Oocysts with 4 sporocysts, each with about 16 sporozoites; gamonts at first sausage-shaped; possibly heteroxenous; known stages in heart of polychete. TYPE SPECIES *A. audouiniae* Brasil, 1904.

1.   *A. audouiniae* Brasil, 1904 (TYPE SPECIES) in *Cirriformia* (syn., *Audouinia*) *tentaculata* (Polychaetasida) heart.

Family ELEUTHEROSCHIZONIDAE *Chatton and Villeneuve, 1936*
Development extracellular; attached to epithelium in intestinal lumen of marine polychetes; gamonts helmet-shaped.

Genus *Eleutheroschizon* Brasil, 1906. With the characters of the family. TYPE SPECIES *E. duboscqi* Brasil, 1906.

1.   *E. duboscqui* Brasil, 1906 (TYPE SPECIES) in *Scoloplos armiger* (Polychaetasida) intestine.
2.   *E. murmanicum* Averinzev, 1908 in *Ophelia limacina* (Polychaetasida) intestine.

Chapter 9

THE COCCIDIA: ADELEINORINA

Order EUCOCCIDIORIDA Léger and Duboscq, 1910
Merogony, gamogony, and sporogony present; in vertebrates or invertebrates.

Suborder ADELEORINA Léger, 1911
Macrogamete and microgamont usually associated in syzygy during development; microgamont produces one to four microgametes; sporozoites enclosed in envelope; endodyogeny absent; homoxenous or heteroxenous.

Family ADELEIDAE *Mesnil, 1903*
Zygote inactive; sporocysts formed in oocyst; chiefly in invertebrates.

Genus *Adelea* Schneider, 1875. Oocysts ellipsoidal or ovoid, with thin wall; oocysts with 6 to 48 flattened sporocysts, each with 2 sporozoites; in chilopods and mollusks. TYPE SPECIES *A. ovata* Schneider, 1875.

1. *A. hyalospora* Narasimhamurti, 1960 in *Rhysida longipes* (Chilopodasida) intestine.
2. *A. ovata* Schneider, 1875 (TYPE SPECIES) in *Lithobius forficatus* (Chilopodasida) intestine.
3. *A. pachelebrae* de Mello, 1921 in *Pachelebra moesta* (Molluska) digestive gland and intestine.

Genus *Adelina* Hesse, 1911. Oocysts spherical or subspherical, with thick wall; oocysts with 3 to about 20 spherical or ellipsoidal sporocysts, each with 2 sporozoites; in annelids, chilopods, and insects. TYPE SPECIES *A. octospora* Hesse, 1911.

1. *A. acarinae* Purrini, 1984 in *Nothrus silvestris* (Acarina, Oribatei) fat body.
2. *A. akidium* (Léger, 1898) Hesse, 1911 (syn., *Adelea akidium* Léger, 1898) in *Akis algerina, Phylan* sp., *Asida* sp., and *Morica hybrida* (Coleopterorida) intestine.
3. *A. collembolae*, Purrini, 1984 in *Neanura muscorum* (Collembolorida) fat body.
4. *A. cryptocerci* Yarwood, 1937 in *Cryptocercus punctulatus* (Orthopterorida) intestine.
5. *A. deronis* Hauschka and Pennypacker, 1942 in *Dero limosa* (Oligochaetasida) mesodermal peritoneal cells.
6. *A. dimidiata* (Schneider, 1885) Hesse, 1911 (syns., *Klossia dimidiata* Schneider, 1885; *Adelea dimidiata* [Schneider, 1885] Labbé, 1896) in *Rhysida longipes, Scolopendra morsitans, S. cingulata, S. subspinipes,* and *S. camidens* (Chilopodasida) intestine.
6a. *A. dimidiata* var. *coccidioides* Léger and Duboscq, 1903 in *Scolopendra* spp. (Chilopodasida) intestine.
7. *A. enchytraei* Siau and Ormières, 1970 in *Enchytraeus albidus* (Oligochaetasida) periintestinal cells, coelomocytes, and coelom.
8. *A. hyalospora* Narasimhamurti, 1960 in *Rhysida longipes* (Chilopodasida) intestine.
9. *A. legeri* Levine, 1977 (syn., *Adelea* sp. Léger, 1900) in *Olocrates abbreviatus* (Coleopterorida) coelom.
10. *A. melolonthae* Tuzet et al., 1966 in *Melolontha melolontha* (Coleopterorida) coelom — fat tissue.
11. *A. mesnili* (Perez, 1899) Hesse, 1911 (syn., *Adelea mesnili* Perez, 1899) in *Tineola bisilliella, Anagasta* (syn., *Ephestia) kuehniella,* and *Plodia interpunctella* (Lepidopterorida) coelom.

12. *A. octospora* Hesse, 1911 (TYPE SPECIES) in *Slavina appendiculata* (Oligochaetasida) coelom.
13. *A. rayi* Narasimhamurti, 1977 in *Rhysida longipes* (Chilopodasida) midgut.
14. *A. riouxi* Levine, 1977 (syn., *Adelina* sp. Rioux et al., 1972) in *Sergentomyia m. minuta*, *Phlebotomus perniciosus*, and *P. ariasi* (Dipterorida) fat body (?).
15. *A. schellacki* Ray and Das Gupta, 1940 in *Cormocephalus dentipes* (Chilopodasida) intestine.
16. *A. sericesthis* Weiser and Beard, 1959 in *Sericesthis pruinosa* (Coleopterorida) fat body.
17. *A. simplex* (Schneider, 1885) Hesse, 1911 (syns., *Klossia simplex* Schneider, 1885; *Adelea simplex* [Schneider, 1885] Labbé, 1896) in *Gyrinus* sp. (Coleopterorida) intestine.
18. *A. tenebrionis* Sautet, 1930 in *Tenebrio molitor* and probably also *Cylindronotus laevioctostriatus* (Coleopterorida) fat body and sometimes epidermis.
19. *A. tipulae* (Léger, 1898) Hesse, 1911 (syn., *Adelea tipulae* Léger, 1898) in *Tipula* sp. (Dipterorida) intestine.
20. *A. transita* (Léger, 1904) Hesse, 1911 (syn., *Adelea transita* Léger, 1904) in *Embia solieri* (Embiopterorida) coelom.
21. *A. tribolii* Bhatia, 1937 in *Tribolium ferrugineum*, *T. confusum*, *T. castaneum*, *T. destructor*, *Ptinum pusillus*, *P. clavipes*, and *P. calyipes* (Coleopterorida) fat body and other organs.
22. *A. zonula* (Moroff, 1906) Hesse, 1911 (syn., *Adelea zonula* Moroff, 1906) in *Blaps mortisaga* and *Blaps* sp. (Coleopterorida) fat body, intestine.

Genus *Klossia* Schneider, 1875. Oocysts with quite numerous spherical sporocysts, each with four sporozoites; in invertebrates and perhaps vertebrates. TYPE SPECIES *K. helicina* Schneider, 1875.

1. *K. aphodii* Larsson, 1985 in *Aphodius fimetarius* and *A. contaminatus* (Coleopterorida) fat tissues.
2. *K. (?) bigemina* (Labbé, 1896) Labbé, 1899 (syn., *Eimeria bigemina* Labbé, 1896) in *Cryptops punctatus* (Chilopodasida) intestine.
3. *K. helicina* Schneider, 1875 (TYPE SPECIES) in *Helix hortensis*, *H. hispida*, *H. arbustorum*, *H. fruticum*, *H. umbrosa*, *Cepaea* (syn., *Helix*) *nemoralis*, *Succinea putris*, *S. pfeifferi*, and *S. gigantea* (Molluska) kidney.
4. *K. loossi* Nabih, 1938 in *Arion* spp. and *Limax* spp. (Molluska) kidney.
5. *K. musabaevae* Dzerzhinskii, 1982 in *Dryomys nitedula* (Mammalia) feces.
6. *K. pachyleparon* Colley and Else, 1975 in *Varanus nebulosus* (Reptilia) feces.
7. *K. perplexens* Levine, Ivens, and Kruidenier, 1955 in *Peromyscus maniculatus* (Mammalia) feces.
8. *K. soror* Schneider, 1881 in *Theodoxus* (syn., *Neritina*) *fluviatilis* (Molluska) kidney.
9. *K. tellinae* Buchanan, 1979 in *Tellina tenuis* (Molluska) renal organ.
10. *K. variabilis* Levine, Ivens, and Kruidenier, 1955 in *Corynorhinus rafinesqii* (Mammalia) feces.
11. *K. vitrina* Moroff, 1911 in *Vitrina eliptica* (?) (Molluska) kidney.

Genus *Orcheobius* Schuberg and Kunze, 1906. Oocysts with 25 to *n* sporocysts, each with 4 sporozoites; in annelids. TYPE SPECIES *O. herpobdellae* Schuberg and Kunze, 1906.

1. *O. carinii* (Pinto, 1926) Levine, 1981 (syn., *Cariniella carinii* Pinto, 1926) in *Leptodactylus ocellatus* (Amphibia) feces.
2. *O. cruzi* Carini and Pinto, 1930 in an unidentified oligochete (Oligochaetasida) genital region.

3. *O. herpobdellae* Schuberg and Kunze, 1906 (TYPE SPECIES) in *Herpobdella ato-maria* (syn., *Nephelis vulgaris*) (Hirudinasida) testis.

Genus *Chagasella* Machado, 1911. Oocysts with three sporocysts; in intestine of insects. TYPE SPECIES *C. hartmanni* (Chagas, 1910) Machado, 1911.

1. *C. alydi* Machado, 1913 in *Alydus* sp. (Hemipterorida) intestine.
2. *C. ganapatii* Narasimhamurti and Kalavati, 1968 in *Odontotermes obesus* (Isopterorida) intestine.
3. *C. gibbsi* Levine, 1977 (syn., *Chagasella* sp. Gibbs, 1944) in *Cenaeus carnifex* (Hemipterorida) salivary glands.
4. *C. hartmanni* (Chagas, 1910) Machado, 1911 (syns., *Adelea hartmanni* Chagas, 1910; *Chagasia hartmanni* [Chagas, 1910] Léger, 1911) (TYPE SPECIES) in *Dysdercus ruficollis* (Hemipterorida) intestine.

Genus *Ithania* Ludwig, 1947. Oocysts with 1 to 4 sporocysts, each with 9 to 33 sporozoites; in insects. TYPE SPECIES *I. wenrichi* Ludwig, 1947.

1. *I. wenrichi* Ludwig, 1947 (TYPE SPECIES) in *Tipula abdominalis* (Dipterorida) intestine.

Genus *Rasajeyna* Beesley, 1977. Oocysts with up to 18 sporocysts, each with 1 sporocyte; in insects. TYPE SPECIES *R. nannyla* Beesley, 1977.

1. *R. nannyla* Beesley, 1977 (TYPE SPECIES) in *Tipula paludosa* and *T. vittata* (Dipterorida) midgut cells.

Genus *Ganapatiella* Kalavati, 1977. Gamonts associated from an early stage, developing close association; microgamonts produce 8 comma-shaped microgametes; oocysts ovoid, with 1 side flattened; they form 15 to 20 sporoblasts, each of which becomes a spherical sporocyst containing 2 elongate sporozoites. TYPE SPECIES *G. odontotermitis* Kalavati, 1977 emend.

1. *G. odontotermitis* Kalavati, 1977 emend. (syn., *G. odontotermi* Kalavati, 1977) (TYPE SPECIES) in *Odontotermes obesus* (Isopterorida) fat tissue.

Genus *Gibbsia* Levine, 1986. Oocysts spherical, with four elongate sporocysts; sporocysts with one sporozoite each; merogony, gamogony, and sporogony occur in blood cells; microgametes not flagellated; in diplopods. TYPE SPECIES *G. archiuli* Levine, 1986.

1. *G. archiuli* Levine, 1986 (TYPE SPECIES) in *Archiulus moreleti* (Diplopodasida) blood cells.

Family LEGERELLIDAE *Minchin, 1903*
    Zygote inactive; no sporocysts formed in oocysts; in diplopods, nematodes, or insects.

Genus *Legerella* Mesnil, 1900. With the characters of the family. TYPE SPECIES *L. nova* (Schneider, 1881) Mesnil, 1900.

1. *L. grassii* Splendore, 1920 in *Nosopsyllus* (syn., *Ceratophyllus*) *fasciatus* (Siphonapterorida) Malpighian tubules.

2.   *L. helminthorum* Canning, 1962 in *Mononchus composticola* (Nematoda) intestinal cells.
3.   *L. hydropori* Vincent, 1927 in *Hydroporus palustris* (Coleopterorida) Malpighian tubules.
4.   *L. nova* (Schneider, 1881) Mesnil, 1900 (syns., *Eimeria nova* Schneider, 1881; *Eimeriella nova* [Schneider, 1881] Stiles, 1902) (TYPE SPECIES) in *Glomeris ornata, G. guttata*, and *G. marginata* (Diplopodasida) Malpighian tubules.
5.   *L. parva* Nöller, 1914 in *Echidnophaga* (syn., *Ceratophyllus) gallinae* and *E.* (syn., *C.) columbae* (Siphonapterorida) Malpighian tubules.
6.   *L. testiculi* Cuénot, 1902 in *Glomeris marginata* (Diplopodasida) testes.

Family HAEMOGREGARINIDAE *Léger, 1911*

Zygote active (ookinete), secreting a flexible membrane which is stretched during development; heteroxenous, life cycle involving 2 hosts, 1 vertebrate and the other invertebrate; merogony in various cells of vertebrates; gamonts in vertebrate blood cells, with about 70 to 80 subpellicular microtubules; sporogony in invertebrates.

Genus *Haemogregarina* Danilewsky, 1885. Vertebrate hosts reptiles, amphibia, and fish; invertebrate hosts leeches, isopods, and arthropods; gamonts primarily in erythrocytes; merogony ordinarily in vertebrate internal organs; sporogony in invertebrates; no sporokinetes; oocysts with eight or more naked sporozoites; infection of vertebrate host by bite of invertebrate. TYPE SPECIES *H. stepanowi* Danilewsky, 1885.

1.   *H. (Hepatozoon?) acanthoclini* Laird, 1953 in *Acanthoclinus quadridactylus* (Osteichthys) erythrocytes.
2.   *H. acipenseri* Nawrotzky, 1914 emend. Levine, 1985 (syns., *H. ascipenseri* Nawrotzky, 1914 *lapsus calami; Leucocytogregarina* sp. Perekropov, 1928) in *Acipenser ruthenus* (Osteichthys) erythrocytes.
3.   *H. aeglefini* H. Henry, 1913 (syn., *H. urophysis* Fantham, Porter, and Richardson, 1942) in *Melanogrammus aeglefinus* (syn., *Gadus aeglefinus*), *Pollachius virens, Gadus morhua* (syn., *G. callarias*), *Urophycis tenuis*, and *U. chuss* (Osteichthys) erythrocytes.
4.   *H. aegyptia* Mohammed and Mansour, 1963 in *Bufo regularis* (Amphibia) erythrocytes.
5.   *H. agamae* Laveran and Pettit, 1909 in *Agama agama* (syn., *A. colonorum*) and *A. ruppelli* (Reptilia) erythrocytes and also leukocytes. (This is probably *Schellackia*.)
6.   *H. aguai* Phisalix, 1930 (syns., *Karyolysus aquai [sic]* [Phisalix, 1950] Scorza, Dagert, and Iturriza Arocha, 1956; *H. minima* Chaussat of Léger [1918]) in *Bufo marinus* (Amphibia) erythrocytes.
7.   *H. algiri* Manceaux, 1908 in *Zamenis hippocrepis* and *Z. algirus* (Reptilia) erythrocytes.
8.   *H. ameivae* Carini and Rudolph, 1912 in *Ameiva surinamensis* (Reptilia) erythrocytes.
9.   *H. amethystina* Johnston, 1909 in *Liasis* (syn., *Python) amethystinus kinghorni* and *Morelia spilotes variegata* (Reptilia) erythrocytes.
10.   *H. amphisbaena* Pessoa, 1968 in *Amphisbaena alba* (Reptilia) blood.
11.   *H. anarhichadis* H. Henry, 1912 emend. Fantham, Porter, and Richardson, 1942 (syn., *H. anarrhichabis* H. Henry, 1912) in *Anarhicas lupus* (Osteichthys) erythrocytes.
12.   *H. ancistrodoni* Zmeev, 1939 in *Ancistrodon halys* (Reptilia) erythrocytes.
13.   *H. andamanensis* R. Ray, 1980 in *Rana cyanophlyctis* (Amphibia) erythrocytes.
14.   *H. aragaoi* di Primio, 1925 in *Paroaria capitata* (Aves).
15.   *H. arantesi* Pessoa, 1967 (syns., *H. butantanensis* Arantes, 1932; *Hepatozoon arantesi* [Pessoa, 1967] Pessoa, de Biasi, and Puorto, 1974) in *Xenodon* (syn., *Ophis) merremii* (Reptilia) erythrocytes.
16.   *H. aspidomorphi* Mackerras, 1961 in *Aspidomorphus harriettae* and *Notechis scutatus* (Reptilia) and perhaps *Stegonotus pulumbeus* (Osteichthys) erythrocytes.

17. *H. assiuticus* Abdel-Rahman et al., 1978 in *Bufo regularis* (Amphibia) erythrocytes.
18. *H. aurorae* Lehmann, 1960 in *Rana a. aurora* (Amphibia) erythrocytes.
19. *H. australis* Mackerras, 1961 in *Pseudechis australis* (Reptilia) erythrocytes.
20. *H. babudierii* Ricci, 1954 in *Lacerta sicula patrizii, L. muralis,* and *L. viridis* (Reptilia) erythrocytes.
21. *H. bagensis* Ducloux, 1904 in *Emys leprosa* (Reptilia) erythrocytes.
22. *H. balli* Paterson and Desser, 1976 in *Chelydra s. serpentina* (Reptilia) erythrocytes.
23. *H. bancrofti* Johnston and Cleland, 1912 emend. Mackerras, 1961 (syn., *H. [Karyolysus] bancrofti* Johnston and Cleland, 1912) in *Pseudechis guttatus* (syn., *P. mortonensis*), *P. australis,* and *Notechis scutatus* (Reptilia) erythrocytes.
24. *H. baueri* C. D. Becker, 1968 (syn., *H. cotti* Bauer, 1948) in *Cottus sibiricus* (Osteichthys) erythrocytes.
25. *H. berestneffi* Castellani and Willey, 1905 in *Rana tigrina, R. limnocharis, R. hexadactyla,* and *R. cyanophlyctis* (Amphibia) erythrocytes.
26. *H. berestnewi* (Finkelstein, 1908) Andrushko and Markov, 1955 (syn., *Karyolysus berestnewi* Finkelstein, 1908) in *Lacerta viridis* (syn., *L. muralis*) (Reptilia) erythrocytes.
27. *H. bertonii* Schouten, 1941 in *Lepidosiren paradoxa* (Osteichthys) erythrocytes.
28. *H. bettencourti* França, 1908 in *Anguilla* sp. (Osteichthys) erythrocytes.
29. *H. bicapsulata* França, 1912 in *Lacerta muralis* (Reptilia) erythrocytes.
30. *H. bicarinati* Pessoa and Cavalheiro, 1969 emend. (syns., *H. bicarinata* Pessoa and Cavalheiro, 1969; *Hepatozoon bicarinatus* [Pessoa and Cavalheiro, 1969] Pessoa, de Biasi, and Puorto, 1974) in *Chironius bicarinatus* and *C. laevicolis* (Reptilia) erythrocytes.
31. *H. bigemina* Laveran and Mesnil, 1901 in *Blennius pholis, B. gattorugine, Abudufduf saxatilis, Acanthopargus bifasciatus, Amblygobius albumaculatus, Argyrops spinifer, Artedius fenestralis, Auxis thazard, Bairdella chrysura, Balistes vetula, Bathygobius saporator, Calamus bajanado, Caranx bartholomaei, C. crysops, C. hippos, C. ruber, Centropristis striatus, Cephalopholis hemistictus, C. miniatus, Chelinius trilobatus, Chrysophrys haffara, Clinus perspicillatus, Coryphaena hippurus, Ctenochaetus strigosus, Diplectrum formosum, Epinephelus adscensionis, E. fasciatus, E. fuscoguttatus, E. guttatus, E. morio, E. striatus, E. summana, E. tauvina, Ericentrus rubrus, Eucinostomus gula, Gerres cinereus, Gymnothorax funebris, Haemulon album, H. aurolineatum, H. flavolineatum, H. plumieri, H. sciurus, Halichoeres bivittata, Hemiramphus brasiliensis, Hyporhamphus unifasciatus, Istiophorus americanus, Kyphosus bigibbus, Lagodon rhomboides, Lethrinus machsena, L. nebulosus, L. variegatus, Lutianus griseus, Lutjanus apodus, L. bohar, L. synagris, Malacanthus plumeri, Menticirrhus littoralis, Mugil trichodon, Mulloidichthys auriflamma, Mycteroperca bonaci, M. microlepis, Notoelinus fenestratus, Ocyurus chrysurus, Oliverichtus melobesia, Perapercis hexophthalma, Plectopomus maculatus, Pomacanthus maculosus, Pteragogus opercularis, Sarcus croicensis, S. sordidus, Scomberomorus regalis, S. cavalia, Seriola dumerili, Sparisoma aurofrenatum, Sphraena barracuda, Strongylura notata, Synodus japonicus, Thalassoma bifasciatum, T. purpureum, Tripterygion medium, T. rufopileum, T. varium, Upeneus tragula, Variola louti,* and *Zonichthys falcatus* (Osteichthys) erythrocytes and leukocytes.
32. *H. billeti* Simond, 1901 in *Trionyx stellatus* (syn., *Chitra indica*) (Reptilia) erythrocytes.
33. *H. bipileata* Svahn, 1955 in *Naja tripudians* (Reptilia) erythrocytes.
34. *H. bitis* Fantham, 1925 (syn., *H. arietans* Fantham, 1925 of Hoare [1932]) in *Bitis arietans* (Reptilia) erythrocytes.
35. *H. blanchardi* Brumpt and Lebailly, 1904 in *Gobius niger* (Osteichthys) erythrocytes.
36. *H. boigae* Mackerras, 1961 in *Boiga fusca* and *Dendrophis punctulatus* (Reptilia) erythrocytes.

37. *H. boodoni* Phisalix, 1914 in *Boaedon* (syn., *Boodon) fuliginosus* (Reptilia) erythrocytes.
38. *H. bornandi* Galli-Valerio, 1916 in *Natrix* (syn., *Tropidonotus) natrix* (Reptilia) erythrocytes.
39. *H. borreli* Nicolle and Comte, 1906 in *Varanus griseus* (Reptilia) erythrocytes.
40. *H. boskiani* Catouillard, 1909 in *Acanthodactylus boskianus* (Reptilia) erythrocytes.
41. *H. bothi* Lebailly, 1905 in *Bothus rhombus* (Osteichthys) erythrocytes.
42. *H. boueti* França, 1911 (syn., *H. boneti* França, 1925 of Tuzet and Grjebine [1957] *lapsus calami*) in *Bufo regularis* and *B. melanosticus* (?) (Amphibia) erythrocytes.
43. *H. boyli* Lehmann, 1959 in *Rana b. boyli* (Amphibia) erythrocytes and erythroblasts.
44. *H. bradfordi* Sambon, 1909 in *Tropidonotus fasciatus* (Reptilia) erythrocytes.
45. *H. brendae* Sambon and Seligmann, 1907 in *Psammophis sibilans* (Reptilia) erythrocytes.
46. *H. brevoortiae* Saunders, 1964 in *Brevoortia tyrannus* (Osteichthys) erythrocytes and erythroblasts.
47. *H. brumpti* Sambon, 1907 in *Coluber melanoleucus* (Reptilia) erythrocytes.
48. *H. bungari* (Billet, 1895) Labbé, 1899 (syn., *Laverania bungari* Billet, 1895) in *Bungarus fasciatus* (Reptilia) erythrocytes.
49. *H. butantanensis* Pessoa, 1928 (syn., *Hepatozoon butantanensis* [Pessoa, 1928] Pessoa, de Biasi, and Puorto, 1974) in *Philodryas aestivus* (Reptilia) erythrocytes.
50. *H. calligaster* Lewis, 1913 emend. Mackerras, 1961 (syn., *H. calligasger* Lewis, 1913) in *Dendrophis calligaster* (Reptilia) erythrocytes.
51. *H. callionymi* Brumpt and Lebailly, 1904 in *Callionymus dracunculus* (Osteichthys) erythrocytes.
52. *H. camarai* Santos Dias, 1954 in *Varanus albigularis* (Reptilia) erythrocytes.
53. *H. cantliei* Sambon, 1907 in *Eryx conicus, E. miliaris,* and *E. tataricus* (Reptilia) erythrocytes.
54. *H. capsensis* Conor, 1909 in *Acanthodactylus pardalis* and *A. scutellatus* (Reptilia) erythrocytes.
55. *H. capsulata* Phisalix, 1931 in *Crotalus terrificus* and *C. durissus terrificus* (Reptilia) erythrocytes.
56. *H. carchariasi* Laveran, 1908 in *Carcharias* sp. (Elasmobranchii) erythrocytes.
57. *H. carlosi* Levine, 1982 (syns., *H. minuta* França, 1909; *Schellackia minuta* [França, 1909] Reichenow, 1953) in *Lacerta ocellata* (Reptilia) erythrocytes.
58. *H. carpionis* Franchini and Saini, 1923 in *Cyprinus carpio* (Osteichthys) digestive tract.
59. *H. catesbianae* Stebbins, 1904 in *Rana catesbiana* (Amphibia) erythrocytes.
60. *H. catostomi* C. D. Becker, 1962 in *Catostomus macrocheilus* and *C. columbianus* (Osteichthys) erythrocytes.
61. *H. cayennensis* Leger, 1918 in *Bufo marinus* (Amphibia) erythrocytes.
62. *H. cenchridis* Phisalix, 1931 (syn., *H. cenchris* Phisalix of Pessoa and Cavalheiro [1969]) in *Epicrates cenchris* and *E. cerebris* (Reptilia) erythrocytes.
63. *H. chamaeleonis* Franchini, 1933 (syn., *H. chamaeleonis* Rousselot, 1953) in *Chamaeleo vulgaris* and *C. basilicus* (Reptilia) erythrocytes.
64. *H. chartusica* Zmeev, 1936 in *Coluber carelini* (Reptilia) erythrocytes.
65. *H. cheissini* Ovezmukhammedov, 1969 in *Agama caucasica* (Reptilia) erythrocytes.
66. *H. chelodinae* Laveran and Pettit, 1909 (syns., *H. clelandi* Johnston, 1909; *H. dentata* Lewis, 1913) in *Chelodina longicollis, C. oblonga, C. expansa, Emydura krefftii, E. macquarii, E. latisternum,* and *Elseya dentata* (Reptilia) erythrocytes.
67. *H. chironii* Pessoa, 1967 emend. (syns., *H. chironiusi* Pessoa, 1967; *Hepatozoon chironiusi* [Pessoa, 1967] Pessoa, de Biasi, and Puorto, 1974) in *Chironius flavolineatus* (Reptilia) erythrocytes.
68. *H. chodukini* Markov and Bogdanov, 1961 in unnamed snake (Reptilia) erythrocytes.

69. *H. clamatae* (Stebbins, 1905) Lehmann, 1960 emend. Levine, 1985 (syns., *Karyolysus clamatae* Stebbins, 1905; *H. [K.] clamatae* [Stebbins, 1905] Lehmann, 1960) in *Rana clamitans* (syn., *R. clamatae*), *R. pipiens*, and *R. catesbiana* (Amphibia) erythrocytes and also leukocytes.
70. *H. clavata* Neumann, 1909 in *Solea lutea* (Osteichthys) erythrocytes.
71. *H. clemmydis* von Prowazek, 1910 in *Clemmys japonica* (Reptilia) erythrocytes.
72. *H. cnemidophori* Carini, 1941 in *Cnemidophorus l. lemniscatus* (Reptilia) erythrocytes.
73. *H. coelorhynchi* Laird, 1952 in *Coelorhynchus australis* and *Physiculus bachus* (Osteichthys) erythrocytes.
74. *H. colisa* Mandal et al., 1983 in *Colisa fasciatus* (Osteichthys) erythrocytes.
75. *H. colubri* Börner, 1901 (syn., *Hepatozoon colubri* [Börner, 1901] Pessoa, de Biasi, and Puorto, 1974) in *Erythrolampus aesculapii, Coluber carelini, C. tyria*, and *Elaphe longissimus* (Reptilia) erythrocytes.
76. *H. columbae* Franchini, 1927 in *Columba livia* (Aves) erythrocytes.
77. *H. corallus* Pessoa and Cavalheiro, 1970 in *Corallus hortulanus* and *C. enydris* (Reptilia) erythrocytes.
78. *H. coronellae* França, 1912 in *Coronella girundica* (Reptilia) erythrocytes.
79. *H. cotti* Brumpt and Lebailly, 1904 (syn., *H. cotti* Bauer, 1948) in *Cottus bubalis, C. gobio, and C. scorpius* (Osteichthys) erythrocytes.
80. *H. crocodilinorum* Börner, 1901 in *Crocodilus frontatus* and *Alligator mississippiensis* (Reptilia) erythrocytes, spleen, and bone marrow.
81. *H. crotali* Laveran, 1902 (syn., *Karyolysus crotali* [Laveran, 1902] Lühe of Pessoa, de Biasi, and Puorto [1974]) in *Crotalus a. atrox* and *C. durissus cascavella* (Reptilia) erythrocytes.
82. *H. crotaphopeltis* Hoare, 1932 in *Crotaphopeltis hotamboeia* (Reptilia) erythrocytes.
83. *H. ctenosaurae* Sokoloff and Mooser, 1943 in *Ctenosaura pectinata* (Reptilia) erythrocytes.
84. *H. cunninghami* Mackerras, 1961 in *Egernia cunninghami* and *E. striolata* (Reptilia) erythrocytes.
85. *H. curvirostris* Billet, 1904 in *Lacerta ocellata* (Reptilia) erythrocytes.
86. *H. cyprini* Smirnova, 1971 in *Cyprinus carpio* (Osteichthys) erythrocytes.
87. *H. dakarensis* A. Léger and M. Leger, 1920 in *Diagramma mediterraneum* (Osteichthys) erythrocytes.
88. *H. darlingi* Leger, 1918 in *Bufo marinus* (Amphibia) erythrocytes.
89. *H. darwiniensis* Lewis, 1913 in *Pseudechis australis* (Reptilia) erythrocytes.
90. *H. dasyatis* Saunders, 1958 in *Dasyatis americana* (Elasmobranchii) erythroblasts and erythrocytes.
91. *H. delagei* Laveran and Mesnil, 1902 in *Raja punctata, R. mosaica, R. erinacea, R. radiata, R. senta*, and *Squalus acanthias* (Elasmobranchii) erythrocytes and erythroblasts.
92. *H. dendrophidis* Johnston and Cleland, 1910 emend. Mackerras, 1961 (syn., *H. [Karyolysus] dendrophidis* Johnston and Cleland, 1910) in *Dendrophis punctulatus* (Reptilia) erythrocytes.
93. *H. denisoniae* Mackerras, 1961 in *Denisonia pallidiceps, D. signata, Pseudechis porphyriacus*, and *Notechis scutatus* (Reptilia) erythrocytes.
94. *H. digueti* Phisalix, 1914 in *Sistrurus catenatus* (Reptilia) erythrocytes.
95. *H. dimorphon* Brimont, 1909 in *Testudo tabulata* (Reptilia) erythrocytes.
96. *H. dogieli* Hoare, 1920 in *Bitis gabonica* (Reptilia) erythrocytes.
97. *H. dolichopyrena* Zmeev, 1938 (syn., *H. dolychopyrena* Zmeev, 1939 of Ovezmukhammedov and Shammakov [1969]) in *Gymnodactylus caspius* (Reptilia) erythrocytes.
98. *H. drymobii* Marullaz, 1912 (syn., *Hepatozoon drymobii* [Marullaz, 1912] Pessoa, de Biasi, and Puorto, 1974) in *Mastigodryas bifossatus* (Reptilia) erythrocytes.

99. *H. echisi* Mohiuddin, Pal, and Warsi, 1967 in *Echis carinatus* (Reptilia) erythrocytes and lung capillaries.
100. *H. egerniae* Mackerras, 1961 in *Egernia cunninghami* and *E. striolata* (Reptilia) erythrocytes.
101. *H. eidsvoldensis* Mackerras, 1961 in *Pseudechis australis* (Reptilia) erythrocytes.
102. *H. emydae* von Prowazek, 1910 in *Emyda japonica* (Reptilia) erythrocytes.
103. *H. enswerae* Hoare, 1932 in *Naja melanoleuca* (Reptilia) erythrocytes.
104. *H. epuluensis* van den Berghe, 1942 in *Rana oxyrhynchus* (Amphibia) erythrocytes.
105. *H. eremiae* Zmeev, 1936 in *Eremias lineolata, E. grammica, E. regeli, E. scripta, E. intermedia, E. velox*, and *Eremias* spp. (Reptilia) erythrocytes.
106. *H. eristavi* Krasil'nikov, 1964 in *Vipera lebetina* (Reptilia) erythrocytes.
107. *H. esocis* Nawrotzky, 1914 in *Esox* sp. (Osteichthys) erythrocytes.
108. *H. eumecei* Khodukin and Sofiev, 1940 in *Eumeces schneideri* (Reptilia) erythrocytes.
109. *H. faiyumensis* Mansour and Mohammed, 1966 in *Bufo regularis* (Amphibia) erythrocytes.
110. *H. fitzsimonsi* Santos Dias, 1953 in *Kinixys belliana zuluensis* (Reptilia) erythrocytes.
111. *H. flesi* Lebailly, 1904 in *Flesus vulgaris* (Osteichthys) erythrocytes.
112. *H. fragilis* Fantham, 1930 in *Blennius cornutus* (Osteichthys) erythrocytes.
113. *H. francai* Abdel-Rahman et al., 1978 in *Bufo regularis* (Amphibia) erythrocytes.
114. *H. franchinii* Yakimov and Rastegaeva, 1930 in *Python molurus* (Reptilia) erythrocytes.
115. *H. froilanoi* França, 1925 in *Bufo regularis* (Amphibia) erythrocytes.
116. *H. fuscus* Lewis, 1913 in *Liasis fuscus* (Reptilia) erythrocytes.
117. *H. gangetica* Misra, 1976 (syn., *H. simondi* Misra et al., 1974) in *Trionyx gangeticus* (Reptilia) erythrocytes.
118. *H. gigas* Pessoa and Cavalheiro, 1970 in *Hydrodynastes gigas* (Reptilia) erythrocytes.
119. *H. gightiensis* Cuenod, 1909 in *Lithorhynchus diadema* (Reptilia) erythrocytes.
120. *H. gilbertiae* Mackerras and Mackerras, 1925 in *Ellerkeldia* (syn., *Gilbertia*) *semicincta* and *E. annulata* (Osteichthys) erythrocytes.
121. *H. gilruthi* Mackerras and Mackerras, 1961 in *Varanus tristis orientalis* (Reptilia) erythrocytes.
122. *H. gobii* Brumpt and Lebailly, 1904 in *Gobius niger* (Osteichthys) erythrocytes.
123. *H. gobionis* Franchini and Saini, 1923 in *Gobio fluviatilis* (Osteichthys) erythrocytes.
124. *H. gouldii* Johnston and Cleland, 1912 emend. Mackerras, 1961 (syn., *H. [Karyolysus] gouldii* Johnston and Cleland, 1912) in *Varanus gouldii* (Reptilia) erythrocytes.
125. *H. gracilis* Wenyon, 1909 (syn., *H. [Karyolysus (?)] gracilis* Wenyon, 1909 of Wenyon [1926]) in *Mabuia quinquetaeniata* (Reptilia) erythrocytes.
126. *H. hamata* Garnham, 1950 in *Lacerta jacksoni* and possibly *Mabuya maculilabris* (Reptilia) erythrocytes.
127. *H. hankini* Simond, 1901 in *Gavialis gangeticus* (Reptilia) erythrocytes.
128. *H. hartochi* Kohl-Yakimoff and Yakimoff, 1915 in *Gobius aurantus* (Osteichthys) erythrocytes.
129. *H. hassleri* Schouten, 1938 in *Apostolepis ambinigra* (Reptilia) erythrocytes.
130. *H. hemiscyllii* Mackerras and Mackerras, 1961 in *Hemiscyllium ocellatum* (Elasmobranchii) erythrocytes.
131. *H. heterodonti* von Prowazek, 1910 in *Heterodontus japonicus* (Elasmobranchii) erythrocytes.
132. *H. heteronotae* Mackerras, 1961 in *Heteronota binoei* (Reptilia) erythrocytes.
133. *H. hinuliae* Johnston and Cleland, 1910 in *Sphenomorphus* (syn., *Hinulia*) *quoyii* (Reptilia) erythrocytes.
134. *H. hoplichthys* Laird, 1952 in *Hoplichthys haswelli* (Osteichthys) erythrocytes.
135. *H. hortai* Brumpt, 1928 in *Rana esculenta* (Amphibia) erythrocytes.

136. *H. hydromedusae* Carini, 1942 in *Hydromedusa tectifera* (Reptilia) erythrocytes.
137. *H. hyperolii* Hoare, 1932 in *Hyperolius* sp. (Amphibia) erythrocytes.
138. *H. ibera* Tartakovskii, 1913 in *Testudo graeca* (Reptilia) erythrocytes.
139. *H. iguanae* Laveran and Nattan-Larrier, 1912 (syn., *H. iguanae* Ducceschi, 1914) in *Iguana tuberculata, I. nudicollis,* and *Tupinambis teguixin* (Reptilia) erythrocytes.
140. *H. imperatoris* Seidelin, 1911 in *Boa imperator* (Reptilia) erythrocytes.
141. *H. irkalukpiki* Laird, 1961 in *Salvelinus alpinus, S. malma,* and perhaps *Caregonus clupeaformis* (Osteichthys) erythrocytes.
142. *H. jakimovi* Khodukin and Sofiev, 1940 (syn., *H. helioscope* Khodukin and Sofiev, 1940) in *Phrynocephalus helioscopus, P. reticulatus, P. interscapularis,* and *P. mystaceus* (Reptilia) erythrocytes.
143. *H. jararacussu* Pessoa, 1968 (syns., *Haemogregarina* sp. Migone, 1916; *Hepatozoon jararacussu* [Pessoa, 1968] Pessoa, de Biasi, and Puorto, 1974) in *Bothrops jararacussu* and *B. neuwiedi* (Reptilia) erythrocytes.
144. *H. joannoni* (Hagenmüller, 1898) Laveran, 1901 (syn., *Danilewskya joannoni* Hagenmüller, 1898) in *Macroprotodon cucullatus* (Reptilia) erythrocytes.
145. *H. johnstoni* Mackerras, 1961 in *Varanus v. varius* and *V. gouldii* (Reptilia) erythrocytes.
146. *H. juxtanuclearis* Carini, 1947 (syn., *H. [Karyolysus (?)] juxtanuclearis* Carini, 1947) in *Boa constrictor* (Reptilia) erythrocytes.
147. *H. kaloulae* R. Ray, 1980 in *Kaloula pulchra taprobanica* (Amphibia) erythrocytes.
148. *H. labbei* Börner, 1901 in *Clemmys elegans* and *Platymys* sp. (Reptilia) erythrocytes.
149. *H.laevicolis* Pessoa, 1967 (syn., *Hepatozoon laevicollis* [Pessoa, 1967] Pessoa, de Biasi, and Puorto, 1974) in *Chironius laevicollis* and *C. foveatus* (Reptilia) erythrocytes.
150. *H. lahillei* Brumpt, 1928 in *Natrix* (syn., *Tropidonotus) natrix* and *N. taxipilota* (Reptilia) erythrocytes.
151. *H. laternae* Lebailly, 1904 in *Platophrys laternae* (Osteichthys) erythrocytes.
152. *H. laverani* Simond, 1901 in *Cryptopus* (syn., *Emyda) granosus* (Reptilia) erythrocytes.
153. *H. lavieri* Tuzet and Grjebine, 1957 in *Bufo regularis* (Amphibia) erythrocytes.
154. *H. lebetina* Sergent, 1918 in *Vipera lebetina* (Reptilia) erythrocytes.
155. *H. legeri* Scorza, Dagert, and Iturriza Arocha, 1956 (syn., *Haemogregarina* sp. Leger, 1918) in *Bufo marinus* (Amphibia) erythrocytes.
156. *H. lepidosirenis* Jepps, 1927 in *Lepidosiren paradoxa* (Osteichthys) erythrocytes.
157. *H. leptoscopi* Laird, 1952 in *Leptoscopus macropygus* (Osteichthys) erythrocytes.
158. *H. lermensis* Martinez, 1941 in *Thamnophis marostemma* (Reptilia) erythrocytes.
159. *H. lignieresi* Laveran, 1906 in *Anguilla* sp. (Osteichthys) erythrocytes.
160. *H. lobianci* Yakimov and Kohl-Yakimov, 1912 emend. Levine, 1985 (syn., *H. lo bianci* Yakimov and Kohl-Yakimov, 1912) in *Torpedo marmoratus* (Osteichthys) erythrocytes.
161. *H. londoni* Yakimov and Kohl-Yakimov, 1912 in *Blennius trigloides* (Osteichthys) erythrocytes.
162. *H. luisieri* França, 1910 in *Zamenis hippocrepis* (Reptilia) erythrocytes.
163. *H. lusitanica* França, 1912 in *Psammodromus algirus* (Reptilia) erythrocytes.
164. *H. lutzi* Hartmann and Chagas, 1910 (syn., *Hepatozoon lutzi* [Hartmann and Chagas, 1910] Pessoa, de Biasi, and Puorto, 1974) in several unidentified species of snakes (including *Eunectes murinus [?]*) (Reptilia) erythrocytes.
165. *H. mabuiae* Nicolle and Comte, 1906 in *Mabuia vittata* (Reptilia) erythrocytes.
166. *H. macroscinca* Laveran, 1907 in unnamed lizard (macroskink) (Reptilia) erythrocytes.
167. *H. maculatus* R. Ray, 1980 in *Rhacophorus maculatus* (Amphibia) erythrocytes.
168. *H. magna* (Grassi and Feletti, 1891) Labbé, 1899 (syns., *H. ranarum* Kruse, 1890 in part; *Drepanidium magnum* Grassi and Feletti, 1891; *D. krusei* Labbé, 1892; *Danilewskya krusei* [Labbé, 1892] Labbé, 1894) in *Rana esculenta, R. ridibunda, R. pipiens, R. limnocharis,* and *R. tigrina* (Amphibia) erythrocytes.

169. *H. malabarica* de Mello, 1932 in *Emyda* (syn., *Cryptopus) granosus vittata* (Reptilia) erythrocytes.
170. *H. malacitensis* Alvarez Calvo, 1975 in *Lacerta hispanica vaucheri* (Reptilia) erythrocytes.
171. *H. manceauxi* França, 1912 in *Zamensis hippocrepis* (Reptilia) erythrocytes.
172. *H. mansoni* Sambon and Seligmann, 1907 in *Zamenis flagelliformis* and *Masticophis f. flagellum* (Reptilia) erythrocytes.
173. *H. maputensis* Santos Dias and de Sousa, 1952 in *Pelusios sinuatus zulensis* (Reptilia) erythrocytes.
174. *H. marceaui* França, 1910 in *Lacerta muralis* (Reptilia) erythrocytes.
175. *H. marzinowskii* Yakimov and Kohl-Yakimov, 1912 in *Gobius jozo* (Osteichthys) erythrocytes.
176. *H. mavori* Laird and Bullock, 1969 in *Macrozoarces americanus* (Osteichthys) erythrocytes.
177. *H. megalocystis* Gilruth, Sweet, and Dodd, 1910 in *Morelia spilotes variegata* (Reptilia) erythrocytes.
178. *H. mellisselensis* Georgévitch, 1940 in *Lacerta serpa* var. *galvagnii* (Reptilia) erythrocytes.
179. *H. mesnili* Simond, 1901 in *Emys tectum* (Reptilia) erythrocytes.
180. *H. metachiri* Regendanz and Kikuth, 1928 in *Metachirus quica* (Mammalia) erythrocytes.
181. *H. minuta* Neumann, 1909 in *Gobius minutus* (Osteichthys) erythrocytes.
182. *H. mirabilis* Castellani and Willey, 1904 in *Natrix* (syn., *Tropidonotus) piscator, N. mairii,* and *Boiga fusca* (?) (Reptilia) erythrocytes.
183. *H. missoni* Carini, 1909 in *Tupinambus teguixin* (Reptilia) erythrocytes.
184. *H. modesta* Pessoa and Cavalheiro, 1969 (syn., *Hepatozoon modesta* [Pessoa and Cavalheiro, 1969] Pessoa, de Biasi, and Puorto, 1974) in *Helicops modesta* (Reptilia) erythrocytes.
185. *H. moloensis* Hoare, 1920 in *Bufo* sp. (Amphibia) erythrocytes.
186. *H. moreliae* Johnston, 1909 in *Morelia* (syn., *Python) spilotes variegata* (Reptilia) erythrocytes.
187. *H. moringa* Pessoa and de Biasi, 1975 in *Gymnothorax moringa* (Osteichthys) erythrocytes.
188. *H. mugili* Carini, 1932 in *Mugil brasiliensis, M. cephalus, Mugil* sp., *Awaous ocellaris,* and *Stenogobius genivittatus* (Osteichthys) erythrocytes.
189. *H. musotae* Hoare, 1932 in *Boaedon lineatus* and *B. geometricus* (Reptilia) erythrocytes.
190. *H. myoxocephali* Fantham, Porter, and Richardson, 1942 in *Myoxocephalus octodecemspinosus* and *M. scorpius* (Osteichthys) erythrocytes.
191. *H. najae* Laveran, 1902 in *Naja tripudians, N. haje, N. nigricollis, N. nivea, N. oxiana,* and *N. mosembica* (Reptilia) erythrocytes.
192. *H. nana* França, 1912 in *Lacerta muralis* and *L. sicula patrizii* (Reptilia) erythrocytes.
193. *H. nasuta* Eisen, 1895 in *Eclipidrilus frigidus* (Oligochaetasida) blood cells.
194. *H. neireti* Laveran, 1905 in *Rana* sp. (Amphibia) erythrocytes.
195. *H. nicollei* França, 1909 in *Lacerta ocellata* (Reptilia) erythrocytes.
196. *H. nicoriae* Castellani and Willey, 1904 in *Nicoria trijuga* (Reptilia) erythrocytes.
197. *H. nili* Wenyon, 1909 in *Ophriocephalus obscurus* (Osteichthys) erythrocytes.
198. *H. ninakohlyakimovae* (Yakimov, 1916) Wenyon, 1926 emend. Levine, 1985 (syns., *Leucocytogregarina ninae kohl-yakimovae* Yakimov, 1916; *L. ninae kohl-yakimovi* Yakimov, 1917; *Leucocytozoon ninae kohl-yakimovae* Yakimoff, 1917; *Hepatozoon ninae kohl-jakimoff* [Jakimoff, 1915] Bykhovskaya-Pavlovskaya et al., 1962) in *Barbus* sp. (Osteichthys) leukocytes.
199. *H. nobrei* França, 1912 in *Lacerta muralis* (Reptilia) erythrocytes.

200. *H. nucleobisecans* Shortt, 1917 in *Bufo melanostictus* and *B. andersoni* (Amphibia) erythrocytes.
201. *H. obscura* Mackerras, 1961 in *Egernia cunninghami* (Reptilia) erythrocytes.
202. *H. ophisauri* Tartakovskii, 1913 in *Ophisaurus apodus* (Reptilia) erythrocytes.
203. *H. pallida* França, 1908 in *Psammodromus algirus* (Reptilia) erythrocytes and mononuclear blood cells.
204. *H. palmeri* Mackerras, 1961 in *Physignathus lesueurii* (Reptilia) erythrocytes.
205. *H. paradoxa* Santos Dias, 1954 in *Varanus albigularis* (Reptilia) erythrocytes.
206. *H. parasiluri* Bykhovskaya-Pavlovskaya et al., 1962 in *Parasilurus asotus* (Osteichthys) erythrocytes.
207. *H. parmae* Mackerras and Mackerras, 1925 in *Parma microlepis* (Osteichthys) erythrocytes.
208. *H. parvula* Santos Dias, 1953 in *Kinixys belliana zuluensis* (Reptilia) erythrocytes.
209. *H. pattoni* Ray, 1960 in *Rana hexadactyla* (Amphibia) erythrocytes.
210. *H. pavlovskyi* Zmeev, 1935 in *Gymnodactylus fedtschenkoi, G. russowi,* and *Teratoscincus scincus* (Reptilia) erythrocytes.
211. *H. pellegrini* Laveran and Pettit, 1910 in *Damonia subtrijuga* (Reptilia) erythrocytes.
212. *H. percae* Franchini and Saini, 1923 in *Perca fluviatilis* (Osteichthys) scrapings from stomach and intestine.
213. *H. percomsi* Shortt, 1922 in *Agama nupta* var. *fusca* (Reptilia) erythrocytes.
214. *H. perfilievi* Zmeev, 1939 (syn., *H. repfilievi* Zmeev of Markov and Bogdanov [1961] *lapus calami*) in *Echis carinatus* (Reptilia) erythrocytes.
215. *H. perinucleophilum* R. Ray, 1980 in *Rana tigrina* (Amphibia) erythrocytes.
216. *H. perrieri* Phisalix, 1913 in *Lachesis* (syn., *Bothrops) neuwidii* (Reptilia) erythrocytes.
217. *H. pestanae* França, 1911 (syn., *H. pistanea* França, 1910 of Tuzet and Grjebine [1957] *lapsus calami*) in *Bufo regularis* (Amphibia) erythrocytes.
218. *H. petrishtchewae* Zmeev, 1937 in *Phrynocephalus interscapularis* (Reptilia) erythrocytes.
219. *H. phisalix* Yakimov and Rastegaeva, 1930 in *Python molurus* (Reptilia) erythrocytes.
220. *H. phyllodactyli* Laveran, 1917 in *Phyllodactylus gerrhopygus* (Reptilia) erythrocytes.
221. *H. phylodriasi* Carini, 1910 (syns., *H. phyllodryae* Gonzalez, 1926; *H. phylodryae* Gonzalez, 1926 of Pessoa, de Biasi, and Puerto [1974]; *H. butantanensis* Pessoa, 1928 in part; *Hepatozoon philodryasi* [sic] [Carini, 1910] Pessoa, de Biasi, and Puorto, 1974; *Haemogregarina, philodryasi* Carini, 1910 of Pessoa, de Biasi, and Puorto [1974] [*lapsus calami*]) in *Phylodrias schotti, P. olfersi, P. patagoniensis, P. aestivus, P. natteri, P. psammophideus,* and *P. baroni* (Reptilia) erythrocytes.
222. *H. pigmentata* Markov and Bogdanov, 1961 in *Tephrometopon lineolatum* (Reptilia) erythrocytes.
223. *H. pituophis* Laveran and Pettit, 1909 in *Pituophis melanoleucus* (syn., *Coluber melanoleucus*) and *P. s. sayi* (Reptilia) erythrocytes.
224. *H. platessae* Lebailly, 1904 (syn., *H. achiri* Saunders, 1955) in *Trinectes maculatus* (syn., *Achirus fasciatus), Glyptocephalus cynoglossus, Pleuronectes platessa, Paralichthys dentatus, Pseudopleuronectes americanus,* and *Scophthalmus aquosus* (Osteichthys) erythrocytes.
225. *H. platydactyli* Billet, 1900 in *Platydactylus mauritanicus* (Reptilia) erythrocytes.
226. *H. pococki* Sambon and Seligmann, 1907 in *Python molurus* and *Morelia spilotes variegata* (Reptilia) erythrocytes.
227. *H. poecilogyrus* Pessoa, 1967 in *Leimadophis poecilogyrus* (Reptilia) erythrocytes.
228. *H. polypartita* Neumann, 1909 in *Gobius paganellus* (Osteichthys) erythrocytes.
229. *H. procteri* Shortt, 1922 in *Phyllodactylus elisae* (Reptilia) erythrocytes.
230. *H. prolata* Tartakovskii, 1913 in *Coluber ravergieri* (Reptilia) erythrocytes.

231. *H. psammodromi* Soulié, 1904 in *Psammodromus algirus* (Reptilia) erythrocytes.
232. *H. pseudechis* Johnston, 1909 in *Pseudechis porphyriacus* and *P. australis* (Reptilia) erythrocytes.
233. *H. pseudemydis* Acholonu, 1974 in *Pseudemys scripta elegans, P. decorata, P. floridana hoyi, P. stejneri, Chelydra s. serpentina, Graptemys kohni, Kinosternon subrubrum hippocrepis, Terrepene c. carolina, T. c. triunguis,* and *Trionyx spinifer* (Reptilia) erythrocytes and also leukocytes.
234. *H. pseudoboae* Pessoa, 1967 in *Pseudoboa nigra* (Reptilia) erythrocytes.
235. *H. pullatus* Pessoa, 1968 (syns., *Haemogregarina* sp. Migone, 1916; *Hepatozoon pullatus* [Pessoa, 1968] Pessoa, de Biasi, and Puorto, 1974) in *Spilotes pullatus* (Reptilia) erythrocytes.
236. *H. pythonis* (Billet, 1895) Labbé, 1899 (syns., *Danilewskya pythonis* Billet, 1895; *H. colubri* Börner, 1901; *H. shattocki* Sambon and Seligmann, 1907 in part; *H. amethystina* Johnston, 1909 in part; *H. reichenowi* Yakimov and Rastegaeva, 1930) in *Python reticulatus, P. molurus, P. sebae, Morelia spilotes variegata, M. s. spilotes, Liasis amethystinus kinghorni,* and *Chondropython viridis* (Reptilia) erythrocytes.
237. *H. quadrigemina* Brumpt and Lebailly, 1904 in *Callionymus lyra* (syn., *C. dracunculus*) (Osteichthys) erythrocytes.
238. *H. rara* Laveran and Mesnil, 1902 in *Damonia reevesii* (Reptilia) erythrocytes.
239. *H. refringens* Sambon and Seligmann, 1907 in *Pseudaspis cana* (Reptilia) erythrocytes.
240. *H. reichenowi* Schubotz, 1913 in *Cycloderma aubryi* (Reptilia) erythrocytes.
241. *H. robertsonae* Sambon, 1909 (syn., *Trypanosoma pythonis* Robertson, 1906) in *Python regius* or *P. sebae* (Reptilia) erythrocytes.
242. *H. rojasi* Jörg, 1931 in *Lachesis alternatus* and *Xenodon merreni* (Reptilia) erythrocytes.
243. *H. romani* Phisalix, 1931 in *Crotalus terrificus* and *C. durissus terrificus* (Reptilia) erythrocytes.
244. *H. rovignensis* Minchin and Woodcock, 1910 in *Trigla lineata* (Osteichthys) erythrocytes.
245. *H. rubrimarensis* Saunders, 1960 in *Acanthurus sokal, A. nigricans, Scarus sordidus, S. ghobban, S. harid, S. guttatus,* and *Chlorurus* sp. (Osteichthys) erythrocytes.
246. *H. sachai* Kirmse, 1978 in *Scophthalmus maximus* (Osteichthys) white and red blood cells.
247. *H. salariasi* Laird, 1951 in *Salarias periophthalmus* (Osteichthys) erythrocytes.
248. *H. salimbeni* Leger, 1919 in *Tupinambis nigropunctatus* (Reptilia) erythrocytes.
249. *H. salvelini* (Fantham, Porter, and Richardson, 1942) Hsu, Campbell, and Levine, 1973 (syn., *Leucocytozoon salvelini* Fantham, Porter, and Richardson, 1942) in *Salvelinus fontinalis* (Osteichthys) leukocytes.
250. *H. samboni* Giordano, 1907 in *Vipera aspis* and *V. libertina* (Reptilia) erythrocytes.
251. *H. schaudinni* França, 1908 (syn., *H. schaudinnae* var. *africana* França, 1909) in *Lacerta ocellata* (Reptilia) erythrocytes.
252. *H. scheini* Mathis and Leger, 1911 in *Rana tigrina* (Amphibia) erythrocytes.
253. *H. scorpaenae* Neumann, 1909 in *Scorpaena ustulata* (Osteichthys) erythrocytes.
254. *H. sebai* Laveran and Pettit, 1909 in *Python seba* (Reptilia) erythrocytes.
255. *H. seligmanni* Sambon, 1907 in *Bothrops* (syn., *Lachesis*) *mutus* (Reptilia) erythrocytes.
256. *H. sergentium* Nicolle, 1904 in *Gongylus ocellata* (Reptilia) erythrocytes.
257. *H. serpentium* (Lutz, 1901) Sambon, 1907 (syns., *Drepanidium serpentium* Lutz, 1901; *Karyolysus serpentium* [Lutz, 1901] Lühe, 1906; *Hepatozoon serpentium* [Lutz, 1901] Pessoa, de Biasi, and Puorto, 1974) in *Eunectes murinus, Boa constrictor, Drymobius bifossatus (Coryphodon pantherinus), Coluber corais, Spilotes pullatus, Xenodon neuwiedii, Rhabdinaea merremii, Philodryas olfersii, Herpetodryas carinata, Curotalus* spp., *Bothrops* spp., and the following species reported without a parasite

species name by Patton (1908): *Bungarus coeruleus (candidus), Dryophis mycterizans, Dendrophys pictus, Eryx johnii, Gongylophis conicus, Naja tripudians, Python molurus, Natrix piscator, Tropidonotus stolatus, Vipera russellii,* and *Zamenis mucosus* (Reptilia) erythrocytes.

258. *H. serrei* Phisalix, 1914 in *Caiman trigonatus* (syns., *Jacaretinga trigonatus, Crocodilus trigonatus, C. palpebrosus, Alligator trigonatus*) (Reptilia) erythrocytes.

259. *H. seurati* Laveran and Pettit, 1911 in *Cerastes cornutus* (Reptilia) erythrocytes.

260. *H. shattocki* Sambon and Seligmann, 1907 (syn., *H. amethystina* Johnston, 1909 in part) in *Morelia (Python) spilotes, Liasis amethystina kinghorni, Notechis scutatus,* and *Pseudechis guttatus* (?) (Reptilia) erythrocytes.

261. *H. sheppardi* Santos Dias, 1952 in *Crocodilus niloticus* (Reptilia) erythrocytes.

262. *H. shirikenimori* Miyata, 1977 in *Triturus pyrrhogaster ensicauda* (Amphibia) erythrocytes.

263. *H. simondi* Laveran and Mesnil, 1901 in *Solea solea* (Osteichthys) erythrocytes.

264. *H. sinimbui* Carini, 1942 in *Iguana iguana* (Reptilia) erythrocytes.

265. *H. sonomae* (Lehmann, 1959) Levine, 1984 (syn., *Karyolysus sonomae* Lehmann, 1959) in *Rana b. boylii* (Amphibia) erythrocytes.

266. *H. stegonoti* Mackerras, 1961 in *Stegonotus plumbeus* (Reptilia) erythrocytes.

267. *H. stepanowi* Danilewsky, 1885 (syns., *Drepanidium stepanowi* [Danilewsky, 1885] Labbé, 1892; *Danilewskya stepanowi* [Danilewsky, 1885] Labbé, 1894; *H. bruneti* Commes, 1919; *H. tyromixus* Thiroux, 1911) (TYPE SPECIES) in *Emys orbicularis* (syns., *Emys lutaria, Cistudo europaea*), *Amyda triunguis, Chelydra serpentinaea, Chrysemys picta dorsalis, C. belli marginata, Emydoides* (syn., *Emys*) *blandingi, Kinixys homeara, Kinosternon subrubrum, K. cruentatum, Pseudemys concinna, P. floridanus, P. scripta, Sternotherus carinatus, S. minor, S. odoratus, Testudo esculenta, T. marginata* (syn., *T. campanulata*), *Trionyx ferox, T. muticus, T. spiniferus,* and *Trionyx* sp., (Reptilia) erythrocytes.

268. *H. stepanowiana* Laveran and Mesnil, 1902 in *Damonia reevesii* (Reptilia) erythrocytes.

269. *H. sternothoeri* França, 1912 in *Pelusios* (syn. *Stenothoerus*) *derbianus* and *S. sinuatus* (Reptilia) erythrocytes.

270. *H. striata* Schubotz, 1913 in *Cycloderma aubryi* (Reptilia) erythrocytes.

271. *H. taeniolati* Mackerras, 1961 in *Sphenomorphus taeniolatus* (Reptilia) erythrocytes.

272. *H. temporariae* (Nöller, 1920) Levine and Nye, 1977 (syn., *Nematopsis temporariae* Nöller, 1920) in *Rana temporaria* (Amphibia) erythrocytes.

273. *H. terzii* Sambon and Seligmann, 1907 in *Boa constrictor* (Reptilia) erythrocytes.

274. *H. testudinis* Laveran and Nattan-Larrier, 1912 in *Testudo emys* (Reptilia) erythrocytes.

275. *H. tetraodontis* Mackerras and Mackerras, 1961 in *Tetraodon hispidus* (Osteichthys) erythocytes.

276. *H. thamnophis* Roudabush and Coatney, 1937 (syn., *H. thamnophium* Sokoloff and Mooser, 1940) in *Thamnophis megalops, T. radix,* and *T. sirtalus* (Reptilia) erythrocytes.

277. *H. theileri* Laveran, 1905 (syn., *Pseudohaemogregarina ranae* Averinzew, 1941) in *Rana angolensis* (syn., *R. nuttii*) (Amphibia) erythrocytes.

278. *H. thomsoni* Minchin, 1908 in *Agama tuberculata* and *A. sanguinolenta* (Reptilia) erythrocytes.

279. *H. thyrosoideae* de Mello and Valles, 1936 in *Thyrosoidea macrurus* (Osteichthys) erythrocytes.

280. *H. tigrinae* Hoare, 1918 in *Rhabdophis t. tigrinus* (syn., *Natrix tigrina*) (Reptilia) and also possibly *Elaphe quadrivirgata* and *Agkistrodon halys* (Reptilia) erythrocytes.

281. *H. tilapiae* A. Léger and M. Leger, 1914 in *Tilapia lata* (Osteichthys) erythrocytes.

282. *H. tiliquae* Johnston and Cleland, 1912 in *Tiliqua scincoides* (Reptilia) erythrocytes.

283. *H. tincae* Levine, 1982 (syn., *H. laverani* Franchini and Saini, 1923) in *Tinca tinca* (Osteichthys) erythrocytes, stomach, and intestinal epithelium.

284. *H. toddi* Wolbach, 1914 in *Varanus niloticus* (Reptilia) erythrocytes and endothelial cells.
285. *H. tonkinensis* Mathis and Leger, 1911 of Leger (1918) in *Bufo melanostictus* (Amphibia) erythrocytes.
286. *H. torpedinis* Neumann, 1909 in *Torpedo ocellata* (Osteichthys) erythrocytes.
287. *H. travassosi* di Primio, 1925 in *Taraba m. major* (Aves) leukocytes.
288. *H. trionyxis* Thiroux, 1911 in *Trionyx triunguis* (Reptilia) erythrocytes.
289. *H. trumata* Tartakovskii, 1913 in *Coluber najadum* (Reptilia) erythrocytes.
290. *H. tuatarae* Laird, 1950 in *Sphenodon punctatus* (Reptilia) erythrocytes and leukocytes.
291. *H. tucumanensis* Senez, 1918 in *Lachesis alternatus* and *Phyllodryas baroni* (Reptilia) erythrocytes.
292. *H. tunisiensis* Nicolle, 1904 in *Bufo mauritanicus* and *B. regularis* (Amphibia) erythrocytes.
293. *H. turcomanica* Khodukin and Sofiev, 1940 (syn., *H. agamae* Khodukin and Sofiev, 1940) in *Agama caucasica, A. lehmanni, A. sanguinolenta,* and *Phrynocephalus reticulatus* (Reptilia) erythrocytes.
294. *H. turkestanica* Yakimov and Shokhor, 1917 in *Silurus glanis* and *Silurus* sp., (Osteichthys) erythrocytes.
295. *H. varani* Laveran, 1905 in *Varanus exanthematicus, V. griseus, V. niloticus,* and *V. komodoensis* (Reptilia) erythrocytes.
296. *H. varanicola* Johnston and Cleland, 1910 (syn., *H. [Karyolysus] varanicola* Johnston and Cleland, 1910) in *Varanus varius, V. tristis orientalis, V. gouldii,* and *V. salvator* (Reptilia) erythrocytes.
297. *H. veloxi* Zakharyan, 1972 in *Eremias velox* (Reptilia) erythrocytes.
298. *H. viperini* Billet, 1904 (syn., *Karyolysus viperini* [Billet, 1904] Lühe, 1906) in *Tropidonotus viperinus* (Reptilia) erythrocytes.
299. *H. vubirizi* Hoare, 1932 in *Simocephalus butleri* (Reptilia) erythrocytes.
300. *H. wardi* Sambon, 1907 in *Coronella getula* (Reptilia) erythrocytes.
301. *H. weissi* Conor, 1912 in *Naja haje* (Reptilia) erythrocytes.
302. *H. wladimirovi* Yakimov and Kohl-Yakimov, 1912 in *Gobius cruentatus* (Osteichthys) erythrocytes.
303. *H. xavieri* de Mello, 1932 in *Emyda granosa* (Reptilia) erythrocytes.
304. *H. yakimovikohli* Wladimiroff, 1910 emend. Levine, 1985 (syn., *H. yakimovi-kohli* Wladimiroff, 1910) in *Gobius capito* (Osteichthys) erythrocytes.
305. *H. zambiensis* Peirce, 1984 in *Dispholidus typus* (Reptilia) erythrocytes.
306. *H. zamenis* Laveran, 1902 in *Zamenis hippocrepis, Z. mucosus,* and *Chrysopelea ornata* (Reptilia) erythrocytes.
307. *H. zumpti* Santos Dias, 1952 in *Dendroaspis polyleps* (Reptilia) erythrocytes.

Genus *Karyolysus* Labbé, 1894. Vertebrate hosts reptiles; invertebrate hosts mites; gamonts primarily in erythrocytes; merogony in vertebrate internal organs; syngamy, meiosis, and sporogony in mite; oocysts form many motile sporoblasts ("sporokinetes") which enter new cells to form sporocysts which develop 20 to 30 sporozoites each; infection of vertebrate host by ingestion of mite. TYPE SPECIES *K. lacertae* (Danilewsky, 1886) Reichenow, 1913.

1. *K. berestnewi* Finkelstein, 1907 in *Lacerta muralis* (Reptilia) erythrocytes.
2. *K. bicapsulatus* (França, 1910) Reichenow, 1921 (syn., *Haemogregarina bicapsulata* França, 1910; *H. bicoprulata* França, 1912 *lapsus calami*) in *Lacerta muralis* (Reptilia) erythrocytes.
3. *K. biretortus* (Nicolle, 1904) Reichenow, 1921 (syn., *Haemogregarina biretorta* Nicolle, 1904) in *Lacerta viridis* (syn., *L. ocellata* var. *pater*).

4. *K. lacazei* (Labbé, 1894) Reichenow, 1921 (syns., *Drepanidium danilewskii* Labbé, 1892 (?); *Danilewskya lacazei* Labbé, 1894; *Haemogregarina lacazei* [Labbé, 1894] Labbé, 1899; *Haemocytozoon clavatum* [?] Danilewsky, 1886) in *Lacerta agilis* and *L. muralis* (Reptilia) erythrocytes.

5. *K. lacertae* (Danilewsky, 1886) Reichenow, 1913 (syns., *Haemogregarina lacertae* Danilewsky, 1886; *H. lacertarum* [Danilewsky, 1886] Labbé, 1894; *Drepanidium lacertarum* [Danilewsky, 1886] Labbé, 1894; *Caryolysus lacertarum* [Danilewsky, 1886] Labbé, 1899) (TYPE SPECIES) in *Lacerta agilis, L. muralis, L. viridis, L. vivipara,* and *L. ocellata* (Reptilia) erythrocytes.

6. *K. latus* Svahn, 1975 in *Lacerta agilis* and *L. vivipara* (Reptilia) erythrocytes and viscera.

7. *K. minor* Svahn, 1975 in *Lacerta agilis* (Reptilia) erythrocytes and viscera.

8. *K. octocromosomi* Alvarez Calvo, 1975 in *Lacerta hispanica* (Reptilia) erythrocytes.

9. *K. subtilis* Ricci, 1954 in *Lacerta sicula patrizii* (Reptilia) erythrocytes.

10. *K. zuluetai* Reichenow, 1920 in *Lacerta muralis* (Reptilia) erythrocytes.

Genus *Hepatozoon* Miller, 1908. Vertebrate hosts amphibia, reptiles, birds, and mammals; invertebrate hosts mites, ticks, insects, and leeches; gamonts in erythrocytes or leukocytes; merogony in vertebrate internal organs; sporogony in invertebrate host; no sporokinetes; oocysts enormous, with n sporocysts , each with 4 to 16 or more sporozoites; infection of vertebrate host by ingestion of invertebrate host. TYPE SPECIES *H. muris* (Balfour, 1906) Wenyon, 1926.

1. *H. acomys* Mohammed and Saoud, 1972 in *Acomys cahirinus* and *A. dimidiatus* (Mammalia) erythrocytes.

2. *H. adiei* Hoare, 1924 in Indian eagle (species undetermined) (Aves) leukocytes.

3. *H. aegypti* Bashtar, Boulos, and Mehlhorn, 1984 in *Spalerosophis diadema* (Reptilia) lung capillary endothelial cells and erythrocytes. A vector is mosquito *Culex pipiens molestus.*

4. *H. akodoni* (Carini and Maciel, 1915) Wenyon, 1926 (syn., *Haemogregarina akodoni* Carini and Maciel, 1915; *Leucocytozoon akodoni* [Carini and Maciel, 1915] Coatney, 1937) in *Akodon fuliginosus* (Mammalia) leukocytes.

5. *H. alactagae* Zasukhin, 1936 (syn., *H. alactaguli* Brumpt, 1946) in *Alactagulus pumilio* (syn., *A. acontion), Allactaga major, A. sibirica, Stylodipus telum,* and *Dipus sagitta* (Mammalia) erythrocytes.

6. *H. albatrossi* Peirce and Prince, 1980 in *Diomedea exulans* (TYPE HOST), *D. melanophris,* and *D. chrysostoma* (Aves) mononuclear leukocytes.

7. *H. argantis* Garnham, 1954 in *Argas brumpti* (Arthropoda) and probably an agamid lizard or *Tarentola* (Reptilia).

8. *H. arvalis* (Martoglio, 1913) Wenyon, 1926 (syns., *Leucocytogregarina arvalis* Martoglio, 1913; *H. orvalis* [Martoglio] Schwetz and Collart, 1920 *lapsus calami*) in unknown rodent (perhaps *Arvicanthis niloticus*) (Mammalia) leukocytes.

9. *H. arvicanthis* Schwetz and Collart, 1930 (syn., *H. arvicanthidis* (Schwetz and Collart, 1930) in *Arvicanthis niloticus* (syn., *A. abyssinicus*) (Mammalia) leukocytes.

10. *H. arvicolae* Brumpt, 1946 (syn., *Karyolysis [sic]* sp. Miyairi, 1934) in *Arvicola hatanezumi* (Mammalia) unspecified blood cells.

11. *H. atticorae* (Aragão, 1911) Hoare, 1924 (syn., *Haemogregarina atticorae* Aragão, 1911) in *Notochelidon gyanoleucus* (Aves) leukocytes.

12. *H. balfouri* (Laveran, 1905) Wenyon, 1926 (syns., *Haemogregarina balfouri* Laveran, 1905; *H. jaculi* Balfour, 1905) in *Jaculus gordoni, J. orientalis, J. johnstoni,* and *J. jaculus* (Mammalia) erythrocytes.

13. *H. brachyspizae* (Aragão, 1911) Hoare, 1924 (syn., *Haemogregarina brachyspizae* Aragão, 1911) in *Zonotrichia capensis* (Aves) leukocytes.

14. *H. breinli* (Mackerras, 1961) Mackerras, 1962 (syn., *Haemogregarina breinli* Mackerras, 1961) in *Varanus tristis orientalis, V. varius,* and *V. gouldii* (Reptilia) erythrocytes.

15. *H. burneti* Lavier and Callot, 1938 in *Tarentola mauritanica* (Reptilia) erythrocytes.

16. *H. caimani* (Carini, 1909) Pessoa, de Biasi, and de Souza, 1972 (syns., *Haemogregarina caimani* Carini, 1909; *H. brasiliensis* di Primio, 1925; *Haemogregarina* sp. Migone, 1916 in part) in *Caiman latirostris* and *C. crocodilus* (Reptilia) erythrocytes.

17. *H. canis* (James, 1905) Wenyon, 1926 (syns., *Leucocytozoon* sp. Dutton et al., 1907; *L. canis* James, 1905; *Haemogregarina canis* [James, 1905] Wenyon, 1906; *H. rotundata* Patton, 1910; *H. chattoni* Léger, 1912; *H. canis-adusti* Nuttall, 1910; *Hepatozoon canis-adusti* [Nuttall, 1910] Wenyon, 1926; *H. chattoni* [Léger, 1912] Wenyon, 1926; *H. pattoni* Léger, 1912 of Krampitz et al. [1968] *lapsus calami; Leucocytogregarina rotundata canis familiaris* Martoglio, 1913; *Leucocytogregarina* sp. Rodhain et al., 1913) in *Canis familiaris, C. aureus, C. adustus, Vulpes vulpes, Crocuta crocuta,* and *Paradoxurus hermaphroditus* (Mammalia) leukocytes.

18. *H. carinicauda* Pessoa and Cavalheiro, 1969 in *Helicops carinicaudus* (Reptilia) erythrocytes.

19. *H. chabaudi* Brygoo, 1963 in *Chamaeleo brevicornis, C. oustaleti,* and *C. pardalis* (Reptilia) erythrocytes.

20. *H. citellicolum* (Wellman and Wherry, 1910) Wenyon, 1926 (syn., *Leucocytozoon citellicola* Wellman and Wherry, 1910) in *Spermophilus beecheyi* (Mammalia) leukocytes.

21. *H. criceti* (Nöller, 1912) Nöller, 1920 (syn., *Leucocytogregarina criceti* Nöller, 1912) in *Cricetus frumentarius* (Mammalia) leukocytes.

22. *H. cricetomysi* Brumpt, 1936 (syn., *Hepatozoon* sp. Rodhain, 1915) in *Cricetomys gambianus* (Mammalia) erythrocytes.

23. *H. cuniculi* (Sangiorgi, 1914) Wenyon, 1926 (syn., *Leucocytogregarina cuniculi* Sangiorgi, 1914) in *Oryctolagus cuniculus* (Mammalia) leukocytes.

24. *H. cyclagrasi* (Arantes, 1934) Pessoa, Sacchetta, and Cavalheiro, 1970 (syn., *Haemogregarina cyclagrasi* Arantes, 1934) in *Hydrodynastes gigas* (Reptilia) erythrocytes.

25. *H. dasyuri* (Welsh, Dalyell, and Burfitt, 1910) Wenyon, 1926 (syn., *Haemogregarina dasyuri* Welsh, Dalyell, and Burfitt, 1910) in *Dasyurus viverrinus* (Mammalia) erythrocytes.

26. *H. dendromi* Brumpt, 1946 emend. (syns., *Hepatozoon* sp. Klein, 1910; *H. dendromysi* Brumpt, 1946; *H. dendromusi* Brumpt, 1946 emend. Killick-Kendrick, 1974) in *Dendromus insignis* (Mammalia) erythrocytes.

27. *H. didelphydis* (d'Utra e Silva and Arantes, 1916) Wenyon, 1926 (syn., *Haemogregarina didelphydis* d'Utra e Silva and Arantes, 1916) in *Didelphys didelphys aurita, Metadirus nudicaudatus,* and *Philander opossum* (Mammalia) erythrocytes.

28. *H. dolichomorphon* Killick-Kendrick, 1984 in *Idiurus macrotis* (Mammalia) monocytes.

29. *H. domerguei* Landau, Chabaud, Michel, and Brygoo, 1970 in *Madagascarophis colubrina,* transmissible to *Lioheterodon modestus, Python sebae, Oplurus sebae, Lacerta muralis, L. sicula,* and *L. vivipara* (Reptilia) erythrocytes.

30. *H. epsteini* Kakabadze and Zasukhin, 1969 in *Rattus norvegicus* (Mammalia) leukocytes.

31. *H. erhardovae* *Krampitz,* 1964 in *Clethrionomys glareolus* (Mammalia) leukocytes.

32. *H. esoci* (Shapoval, 1950) Bykhovskaya-Pavlovskaya et al., 1962 (syn., *Leucocytogregarina esoci* Shapoval, 1950) in *Esox* sp. (Osteichthys) leukocytes and plasma cells.

33. *H. felis* Patton, 1908 (syns., *Haemogregarina felis-domesticae* Patton, 1908; *Hepatozoon felisdomesticae* [Patton, 1908] Wenyon, 1926; *Leucocytozoon felis-domestici* [Patton, 1908] Patton, 1908) in *Felis catus* (Mammalia) leukocytes.

34. *H. funambuli* (Patton, 1906) Wenyon, 1926 (syn., *Leucocytozoon funambuli* Patton, 1906) in *Funambulus pennanti* (Mammalia) mononuclear leukocytes.
35. *H. fusifex* Ball, Chao, and Telford, 1969 (syn., *H. fusiflex* Pessoa, de Biasi, and Puorto, 1974 *lapsus calami*) in *Boa constrictor* (Reptilia) erythrocytes.
36. *H. gaetulum* Sergent, 1921 (syn., *H. getulum* Sergent, 1921 of Killick-Kendrick [1974]) in *Atlantoxerus gaetulus* (Mammalia) leukocytes.
37. *H. gerbilli* (Christophers, 1905) Wenyon, 1926 (syn., *Haemogregarina gerbilli* Christophers, 1905) in *Tatera* (syn., *Gerbillus) indicus, Rhombomys opimus,* and *Meriones erythrourus* (Mammalia) erythrocytes.
38. *H. graomysi* Brumpt, 1946 emend. Killick-Kendrick, 1974 (syns., *Hepatozoon* sp. Romana, 1945; *H. groamysi* Brumpt, 1946) in *Graomys medius* (Mammalia).
39. *H. griseisciuri* Clark, 1958 in *Sciurus carolinensis* (Mammalia) leukocytes (monocytes).
40. *H. guangdongense* Li, 1982 in *Ptyas korros* and experimentally *Natrix piscator, Elaphe radita,* and *Enhydris chinensis* (Reptilia).
41. *H. hoogstraali* Uilenberg, 1970 in *Hemicentetes semispinosus* (Mammalia) erythrocytes.
42. *H. insectivorae* Levine, 1982 in *Sorex araneus* and *Crocidura leucodon* (Mammalia) leukocytes.
43. *H. krampitzi* Levine, 1982 (syns., *H. microti* [Coles, 1914] of Erhardová [1955] in part; *Hepatozoon* sp. Krampitz, 1964 from *Microtus oeconomus; Hepatozoon* sp. Ohbayashi, 1971) in *Microtus oeconomus* (Mammalia) leukocytes.
44. *H. leimadophis* (Pessoa, 1967) Pessoa and de Biasi, 1974 emend. (syns., *Haemogregarina leimadophisi* Pessoa, 1967; *Hepatozoon leimadophisi* [Pessoa, 1967] Pessoa and de Biasi, 1974) in *Leimadophis poecilogyrus* (Reptilia) erythrocytes.
45. *H. leporis* (Patton, 1908) Wenyon, 1926 (syn., *Leucocytozoon leporis* Patton, 1908) in *Lepus nigricollis* (Mammalia) leukocytes.
46. *H. leptodactyli* (Lesage, 1908) Pessoa, 1970 (syns., *Leucocytozoon ranarum* Carini, 1907; *Haemogregarina leptodactyli* Lesage, 1908; *H. heteronucleata* Carini, 1909; *Haemogregarina* sp. Carini, 1911; *H. ranarum* [Carini, 1907] Carini, 1945) in *Leptodactylus ocellatus* and *L. pentadactylus* (Amphibia) erythrocytes.
47. *H. leptosoma* Wood, 1962 in *Peromyscus maniculatus* and *P. boylii* (Mammalia) leukocytes.
48. *H. luehi* (Sambon, 1909) Pessoa, Cavalheiro, and de Souza, 1970 emend. Levine, 1985 (syns., *Haemogregarina luehi* Sambon, 1909; *Hepatozoon luhei* [Sambon, 1909] Pessoa, Cavalheiro, and de Souza, 1970 *lapsus calami*) in *Corallus cooki, C. caninus, C. enydris,* and *C. hortulanus* (Reptilia) erythrocytes.
49. *H. lusitanicum* Najera, 1937 in *Eliomys lusitanicus* (Mammalia) leukocytes.
50. *H. lygosomarum* (Doré, 1919) Allison and Desser, 1982 (syn., *Haemogregarina lygosomarum* Doré, 1919) in *Leiolopisma nigriplantare* (syn., *L. moco*) (Reptilia) erythrocytes, liver, and spleen; a vector is mite *Ophionyssus scincorum.*
51. *H. mauritanicum* (Sergent and Sergent, 1904) Michel, 1973 (syns., *Haemogregarina mauritanica* Sergent and Sergent, 1904; *Coelomoplasma hyalomma* Brumpt, 1938) in *Testudo graeca* and *T. mauritanica* (Reptilia) erythrocytes.
52. *H. mereschkowskii* Tartakovskii, 1913 in *Spermophilus guttatus* and *S. musicus* (Mammalia) leukocytes.
53. *H. mesnili* Robin, 1936 in *Gecko verticillatus* (Reptilia) erythrocytes.
54. *H. microti* (Coles, 1914) Wenyon, 1926 (syns., *Haemogregarina microti* Coles, 1914; *Hepatozoon micratai* Coles, 1914 of Batabaeva [1979]; *H. lavieri* [Brumpt, 1946] of Frank [1978]; *H. sylvatici* [Coles, 1914] Wenyon, 1926 of Lavier [1921]) in *Microtus agrestis* and perhaps *Allactaga severtzovi* and *A. evertsi* (Mammalia) leukocytes.
55. *H. migonei* (Schouten, 1934) Pessoa, Sacchetta, and Cavalheiro, 1970 (syn., *Haemogregarina migonei* Schouten, 1934) in *Hydrodynastes* (syn., *Cyclagras) gigas* (Reptilia) erythrocytes.

56. *H. miliaris* (Pessoa, 1968) Pessoa and Cavalheiro, 1969 (syn., *Haemogregarina miliaris* Pessoa, 1968) in *Liophis miliaris* (Reptilia) erythrocytes.
57. *H. minchini* Garnham, 1950 in *Crotaphopeltis degeni* (Reptilia) erythrocytes.
58. *H. mocassini* (Laveran, 1902) Nadler and Miller, 1984 (syn., *Haemogregarina mocassini* Laveran, 1902) in *Agkistrodon piscivorus* (Reptilia) erythrocytes.
59. *H. muris* (Balfour, 1906) Wenyon, 1926 (syns., *Leucocytozoon muris* Balfour, 1906; *Haemogregarina muris* [Balfour, 1906] Laveran, 1905; *Hepatozoon perniciosus* Miller, 1908; *L. ratti* Adie, 1906; *Haemogregarina ratti* [Adie, 1906] *auctores; Hepatozoon ratti* [Adie, 1906] Wenyon, 1926; *Leucocytogregarina innoxia* Kusama, Kasai, and Kobayashi, 1919; *L. muris* [Balfour, 1905] França and Pinto, 1912; *H. innoxia* [Kusama, Kasai, and Kobayashi, 1919] Wenyon, 1926) (TYPE SPECIES) in *Rattus norvegicus, R. rattus, R. alexandrinus, R. assimilis, R. conatius, R. frugivorus*, and *Praomys tullbergi* (Mammalia) leukocytes.
60. *H. musculi* (Porter, 1908) Wenyon, 1926 (syns., *Leucocytozoon musculi* Porter, 1908; *Leucocytogregarina musculi* [Porter, 1908] Sangiorgi, 1912) in *Mus musculus* (Mammalia) leukocytes.
61. *H. neophrontis* (Todd and Wolbach, 1912) Bray, 1964 (syns., *Leucocytogregarina neophrontis* Todd and Wolbach, 1912; *Leucocytozoon neophrontis* [Todd and Wolbach, 1912]; *Toxoplasma neophrontis* [Todd and Wolbach, 1912] Wenyon, 1926; *H. monachus* [Todd and Wolbach, 1912] of Baker et al. [1972] *lapsus calami*) in *Necrosyrtes* (syn., *Neophron) monachus* (Aves) leukocytes (mononuclear cells and rarely eosinophils and neutrophils).
62. *H. normani* Killick-Kendrick, 1984 (syn., *Hepatozoon* sp. Killick-Kendrick and Bellier, 1971) in *Idiurus macrotis* (Mammalia) monocytes.
63. *H. pallida* (Pessoa, Sacchetta, and Cavalheiro, 1971) Levine, 1982 (syn., *Haemogregarina pallida* Pessoa, Sacchetta, and Cavalheiro, 1971) in *Thamnodynastes pallidus nattereri* (Reptilia) erythrocytes.
64. *H. peramelis* (Welsh and Dalyell, 1910) Wenyon, 1926 (syn., *Haemogregarina peramelis* Welsh and Dalyell, 1910) in *Perameles nasuta* (Mammalia) blood plasma.
65. *H. peromysci* Levine, 1982 (syn., *H. muris* of Wood [1962]) in *Peromyscus b. boylii* and *P. truei gilberti* (Mammalia) leukocytes.
66. *H. petauri* (Welsh and Barling, 1910) Wenyon, 1926 (syn., *Haemogregarina petauri* Welsh and Barling, 1910) in *Petaurus* sp. and probably *P. sciureus* (Mammalia) erythrocytes.
67. *H. pettiti* (Thiroux, 1910) Hoare, 1932 (syn., *Haemogregarina pettiti* Thiroux, 1910) in *Crocodilus niloticus* (Reptilia) erythrocytes.
68. *H. pintoi* (?) (di Primio, 1925) Baker et al., 1972 (syn., *Haemogregarina pintoi* di Primio, 1925) in *Cathartes aura ruficollis* (Aves) leukocytes.
69. *H. pitymysi* Splendore, 1920 in *Pitymys savii* (Mammalia) leukocytes.
70. *H. plicatum* (Martoglio, 1913) Brumpt, 1946 (syns., *Leucocytogregarina plicatum marmotae* Martoglio, 1913; *H. plicatum marmotae* [Martoglio, 1913] Brumpt, 1946) in *Hyrax* sp. (?) or *Pectinator speke* (?) (Mammalia) leukocytes.
71. *H. plimmeri* (Sambon, 1909) Pessoa et al., 1971 (syn., *Haemogregarina plimmeri* Sambon, 1909) in *Bothrops jararaca* (syn., *Lachesis lanceolatus*) and *B. moojeni* (Reptilia) erythrocytes.
72. *H. poroariae* (Aragão, 1911) Hoare, 1924 (syn., *Haemogregarina poroariae* Aragão, 1911) in *Paroaria* [*sic*] *dominica* [?] (syn., *Poroaria larvata*) (Aves) leukocytes.
73. *H. procyonis* Richards, 1961 in *Procyon lotor* and *P. cancrivorus panamensis* (Mammalia) leukocytes.
74. *H. prodhoni* Landau, 1973 in *Oplurus quadrimaculatus* (Reptilia) leukocytes.
75. *H. rarefaciens* (Sambon and Seligmann, 1907) Ball, Chao, and Telford, 1967 (syn.,

Haemogregarina rarefaciens Sambon and Seligmann, 1907) in *Drymarchon* (syn., *Coluber) corais* and experimentally *Constrictor constrictor*, and in *Pituophis catenifer* (Reptilia) erythrocytes and also leukocytes.

76. *H. rhamphocoeli* (Aragão, 1911) Hoare, 1924 (syns., *Haemogregarina rhamphocoeli* Aragão, 1911; *H. remphocoeli* Aragão, 1911 of de Mello [1915] *lapsus calami*) in *Rhamphocoelus brasilius* (Aves) leukocytes.

77. *H. rhipicephali* (Brumpt, 1938) Levine, 1984 (syn., *Coelomoplasma rhipicephali* Brumpt, 1938) in *Rhipicephalus bursa* (Arthropoda) coelom.

78. *H. roulei* (Phisalix and Laveran, 1913) Pessoa, de Biasi, and de Souza, 1972 (syns., *Haemogregarina roulei* Phisalix and Laveran, 1913; *H. raulei* Phisalix and Laveran, 1913 of Pessoa [1968] *lapsus calami*) in *Bothrops* (syn., *Lachesis) alternatus* (Reptilia) erythrocytes.

79. *H. sauromali* Lewis and Wagner, 1964 in *Sauromalus hispidus, S. australis, S. obesus,* and *S. varius* (Reptilia) erythrocytes and sometimes leukocytes.

80. *H. sciuri* (Coles, 1914) Brumpt, 1946 (syn., *Haemogregarina sciuri* Coles, 1914) in *Sciurus vulgaris* (Mammalia) leukocytes.

81. *H. strigatus* (Pessoa, 1967) Pessoa, Cavalheiro, and de Souza, 1970 (syn., *Haemogregarina strigatus* Pessoa, 1967) in *Thamnodynastes strigatus* (Reptilia) erythrocytes.

82. *H. sylvatici* (Coles, 1914) Wenyon, 1926 (syn., *Haemogregarina sylvatici* Coles, 1914) in *Apodemus sylvaticus, A. flavicollis, A. mystacinus,* and (experimentally) *Clethrionomys glareolus* (Mammalia) leukocytes.

83. *H. tanagrae* (Aragão, 1911) Hoare, 1924 (syn., *Haemogregarina tanagrae* Aragão, 1911) in *Thraupis* (syn., *Tanagra) palmarium* (Aves) leukocytes.

84. *H. triatomae* (Osimani, 1942) Reichenow, 1953 (syn., *Haemogregarina triatomae* Osimani, 1942) in *Tupinambis teguixin* and probably *Tropidurus torquatus* (Reptilia) erythrocytes.

85. *H. tupinambis* (Laveran and Salimbeni, 1909) Pessoa, de Biasi, and Sacchetta, 1974 (syns., *Haemogregarina tupinambis* Laveran and Salimbeni, 1909; *H. tupinambisi* Carini, 1909; *H. caranii* Laveran, 1909) in *Tupinambis teguixin* (Reptilia) erythrocytes.

Genus *Cyrilia* Lainson, 1981. Oocysts with 20 or more naked sporozoites. TYPE SPECIES *C. gomesi* (Neiva and Pinto, 1926) Lainson, 1981.

1. *C. gomesi* (Neiva and Pinto, 1926) Lainson, 1981 (syn., *Haemogregarina gomesi* Neiva and Pinto, 1926) (TYPE SPECIES) in *Symbranchius marmoratus* (Osteichthys) erythrocytes.

2. *C. uncinata* (Khan, 1978) Lainson, 1981 (syn., *Haemogregarina uncinata* Khan, 1978) in *Lycodes lavalaei* and *L. vahlii* (Osteichthys) erythrocytes.

Family KLOSSIELLIDAE *Smith and Johnson, 1902*
Zygote inactive; typical oocyst not formed; a number of sporocysts, each with many sporozoites, develop within a membrane which is perhaps laid down by the host cell; two to four nonflagellated microgametes formed by microgamonts; homoxenous, with sporogony and merogony in different locations of the same host; in kidney and other organs of host.

Genus *Klossiella* Smith and Johnson, 1902. With the characters of the family; in vertebrates. TYPE SPECIES *K. muris* Smith and Johnson, 1902.

1. *K. bettongiae* Barker, Munday, and Hartley, 1984 in *Bettongia gaimardi* (Mammalia) kidneys.

2. *K. beveridgei* Barker, Munday, and Hartley, 1984 in *Lagorchestes conspicillatus* (Mammalia) kidneys.

3. *K. boae* Zwart, 1964 in *Boa constrictor* (Reptilia) kidneys.
4. *K. callitris* Barker, Munday, and Harrigan, 1975 in *Macropus fuliginosus melanops* (Mammalia) kidneys.
5. *K. cobayae* Seidelin, 1914 in *Cavia porcellus* (Mammalia) kidneys.
6. *K. convolutor* Barker, Munday, and Harrigan, 1975 in *Pseudocheirus peregrinus* (Mammalia) kidneys.
7. *K. equi* Baumann, 1946 (syns., *K. equi* Seibold and Thorson, 1955; *Eimeria utinensis* Selan and Vittorio, 1924) in *Equus caballus, E. asinus,* and zebra (scientific name not given — presumably *E. burchelli*) (Mammalia) kidneys.
8. *K. hydromyos* Winter and Watt, 1971 in *Hydromys chrysogaster* (Mammalia) kidneys.
9. *K. killicki* Boulard, 1975 in *Hipposideros caffer guineensis, H. caffer,* and presumably *Rhinolophus* sp. (Mammalia) kidneys.
10. *K. mabokensis* Boulard and Landau, 1971 in *Rattus (syn., Praomys) jacksoni* and experimentally *Mus musculus* (Mammalia) kidneys.
11. *K. muris* Smith and Johnson, 1902 (TYPE SPECIES) in *Mus musculus* (Mammalia) kidneys.
12. *K. quimrensis* Barker, Munday, and Harrigan, 1975 (syn., *Klossiella* sp. Derrick and Smith in Mackerras, Mackerras, and Sanders, 1953) in *Isoodon obesulus* and *Perameles gunnii* (Mammalia) kidneys.
13. *K. rufi* Barker, Munday, and Hartley, 1984 in *Macropus rufus* (Mammalia) kidneys.
14. *K. rufogrisei* Barker, Munday, and Hartley, 1984 in *Macropus rufogriseus* (Mammalia) kidneys.
15. *K. schoinobatis* Barker, Munday, and Hartley, 1984 in *Schoinobates volans* (Mammalia) kidneys.
16. *K. serendipensis* Barker, Munday, and Harrigan, 1975 in *Wallabia bicolor* (Mammalia) kidneys.
17. *K. tejerai* Scorza, Torrealba, and Dagert, 1957 in *Didelphis marsupialis* and *Marmosa cinerea demararae* (Mammalia) kidneys.
18. *K. thylogale* Barker, Munday, and Hartley, 1984 in *Thylogale billardierii* (Mammalia) kidneys.

Chapter 10

THE COCCIDIA: EIMERIORINA

Suborder EIMERIORINA Léger, 1911

Macrogamete and microgamont develop independently; syzygy absent (except perhaps in Dobellidae); microgamont typically produces many microgametes; zygote not motile; sporozoites typically enclosed in a sporocyst; with conoid so far as known; endodyogeny present or absent; homoxenous or heteroxenous.

Family SPIROCYSTIDAE *Léger and Duboscq, 1915*

Merogony present, within host intestinal cells; meronts vermicular, curved, with one end markedly narrowed; mature meronts coiled like a snail shell, with numerous nuclei; syzygy apparently absent; gametes dissimilar, nonflagellate, in chlorogogen cells, somatic and visceral peritoneum; one oocyst per gametocyst; gametocysts and oocysts only in chlorogogen cells; oocysts very thick-walled, ovoid or piriform, with micropyle; each oocyst contains one coiled, vermicular, naked sporozoite.

Genus *Spirocystis* Léger and Duboscq, 1911. With the characters of the family; in oligochetes. TYPE SPECIES *S. nidula* Léger and Duboscq, 1911.

1.   *S. nidula* Léger and Duboscq, 1911 (TYPE SPECIES) in *Lumbriculus variegatus* (Oligochaetasida) coelom, etc.

Family SELENOCOCCIDIIDAE *Poche, 1913*

Meronts develop as vermicules in host intestinal lumen; meronts with myonemes and a row of nuclei.

Genus *Selenococcidium* Léger and Duboscq, 1910. With the characters of the family; in lobster. TYPE SPECIES *S. intermedium* Léger and Duboscq, 1910.

1.   *S. intermedium* Léger and Duboscq, 1910 (TYPE SPECIES) in *Homarus gammarus* (Decapodasida) intestine.

Family DOBELLIDAE *Ikeda, 1914*

Male and female gamonts produced by micro- and macroschizogony, respectively; syzygy present.

Genus *Dobellia* Ikeda, 1914. With the characters of the family; in intestinal epithelium of sipunculids. TYPE SPECIES *D. binucleata* Ikeda, 1914.

1.   *D. binucleata* Ikeda, 1914 (TYPE SPECIES) (syn., *D. dimorphonucleata* [?] Ikeda, 1913) in *Petalostoma minutum* (Sipunculida) intestinal epithelium.

Family AGGREGATIDAE *Labbé, 1899*

Development in host cell proper; oocysts typically with many sporocysts; most genera heteroxenous, with merogony in one host and gamogony in another; syzygy absent.

Genus *Aggregata* Frenzel, 1885. Oocysts large, with many sporocysts; sporocysts with 3 to 28 sporozoites; heteroxenous, with merogony in a decapod crustacean and gamogony in a cephalopod mollusk; (the synonymy of the species in this genus is complicated and I make no pretense that I am positive what the early authors actually saw). TYPE SPECIES *A. octopiana* (Schneider, 1875) Frenzel, 1885.

1. *A. coelomica* Léger, 1901 (syn., *A. duboscqui* Moroff, 1908) in *Pinnotheres pisum* and *P. pinnotheres* (Decapodasida) gut wall and *Octopus* sp. (Molluska) spiral gut.
2. *A. eberthi* (Labbé, 1895) Léger and Duboscq, 1906 (syns., *Benedenia eberthi* Labbé, 1895; *Klossia eberthi* Labbé, 1896; *Legerina eberthi* [Labbé] Jacquement, 1903; *Eucoccidium eberthi* [Labbé] Lühe, 1903; *K. sepiana* Labbé, 1896; *A. portunidarum* Frenzel, 1885 in part; *A. arcuata* Moroff, 1908; *A. mingazzinii* Moroff, 1908; *A. minima* Moroff, 1908; *A. frenzeli* Moroff, 1908; *A. mamillana* Moroff, 1908) in *Macropipus* (syn., *Portunus*) *armatus, M. arcuatus, M. corrugatus, M. depurator, M. bolivari, M. holsatus, M. puber, M. tuberculatus,* and *M. vernalis* (Decapodasida) gut wall and coelom, and *Sepia officinalis* (Molluska) gut wall.
3. *A. inachi* G. W. Smith, 1905 emend. Léger and Duboscq, 1906 in *Inachus dorsettensis, I. communissimus,* and *I. scorpio* (Decapodasida) gut wall; molluskan host unknown.
4. *A. jacquemeti* (Moroff, 1906) (syn., *Eucoccidium jacquemeti* Moroff, 1906) in *Octopus* sp. (Molluska) location not stated; arthropod host unknown.
5. *A. kudoi* Narasimhamurti, 1979 in *Sepia elliptica* (Molluska) proximal gut epithelium, decapod host unknown.
6. *A. labbei* Moroff, 1908 in *Octopus* sp. (Molluska) spiral intestine; arthropod host unknown.
7. *A. leandri* Goodrich, 1950 in *Leander squilla, Solenocera membranacea, Acanthephyra eximia,* and *Gennadas elegans* (Decapodasida) gut; molluskan host unknown.
8. *A. legeri* Moroff, 1908 in *Octopus* sp. (Molluska) spiral intestine; arthropod host unknown.
9. *A. maxima* Théodoridès and Desportes, 1975 in *Sergestes robustus* (Decapodasida) outer wall of digestive tract; molluskan host unknown.
10. *A. octopiana* (Schneider, 1875) Frenzel, 1885 (syns., *Benedenia octopiana* Schneider, 1875; *Klossia octopiana* [Schneider, 1875] Labbé, 1896; *Legeria octopiana* [Schneider, 1875] Blanchard, 1900; *Eucoccidium octopiana* [Schneider, 1875] Lühe, 1902; *Legerina octopiana* [Schneider, 1875] Jacquement, 1903) (TYPE SPECIES) in *Octopus vulgaris* (Molluska) intestine; arthropod host unknown.
11. *A. ovata* Moroff, 1908 in *Octopus* sp. (Molluska) spiral intestine; arthropod host unknown.
12. *A. portunidarum* Frenzel, 1885 in *Portunus arcuatus* and *Carcinus maenas* (Decapodasida) intestine; molluskan host unknown.
13. *A. reticulosa* Moroff, 1908 in *Octopus* sp. (Molluska) spiral intestine; arthropod host unknown.
14. *A. schneideri* Moroff, 1908 in *Octopus* sp. (Molluska) spiral intestine; arthropod host unknown.
15. *A. siedleckii* Moroff, 1908 in *Octopus* sp. (Molluska) spiral intestine; arthropod host unknown.
16. *A. spinosa* Moroff, 1906 in *Octopus* sp. (Molluska) spiral intestine; arthropod host unknown.
17. *A. stellata* Moroff, 1908 in *Octopus* sp. (Molluska) spiral intestine; arthropod host unknown.
18. *A. vagans* Léger and Duboscq, 1903 in *Eupagurus prideauxi* (Decapodasida) intestine and coelom; molluskan host unknown.

Genus *Merocystis* Dakin, 1911. Oocyst with many sporocysts, each with one sporozoite; merogony unknown; presumably heteroxenous. TYPE SPECIES *M. kathae* Dakin, 1911.

1. *M. kathae* Dakin, 1911 (TYPE SPECIES) in *Buccinum undatum* (Molluska) kidney.

Genus *Pseudoklossia* Léger and Duboscq, 1915. Oocyst with zero or many sporocysts, each with two sporozoites (if sporocysts occur); merogony unknown; presumably heteroxenous; known stages in mollusks. TYPE SPECIES *P. glomerata* Léger and Duboscq, 1915.

1. *P. chitonis* Debaisieux, 1919 in *Acanthochites fascicularis* (Molluska) liver.
2. *P. glomerata* Léger and Duboscq, 1915 (TYPE SPECIES) in *Tapes floridus* and *T. virgineus* (Molluska) kidneys.
3. *P. patella* Debaisieux, 1919 in *Patella vulgaris* (Molluska) location in host not given.
4. *P. pectinis* Léger and Duboscq, 1917 in *Pecten maximus* (Molluska) kidney.
5. *P. pelseneeri* (Léger, 1897) Léger and Duboscq, 1915 (syn., *Hyaloklossia pelseneeri* Léger, 1897) in *Donax* sp. and *Tellina* sp. (Molluska) kidneys.
6. *P. tellinovum* (Buchanan, 1979) nov. comb. (syn., *Merocystis tellinovum* Buchanan, 1979) in *Tellina tenuis* (Molluska) primary germ cells and oogonia. (*Remarks. Pseudoklossia* species contain two sporozoites per sporocyst, while *Merocystis* species contain one, so this species must be shifted from *Merocystis* to *Pseudoklossia*.)

Genus *Grasseella* Tuzet and Ormières, 1960. Oocyst with many sporocysts, each with two sporozoites; in ascidians. TYPE SPECIES *G. microcosmi* Tuzet and Ormières, 1960.

1. *G. microcosmi* Tuzet and Ormières, 1960 (TYPE SPECIES) in *Microcosmus sulcatus* and *Pyura microcosmus* (Urochordata) epithelial cells of liver and more rarely cells of digestive tract.

Genus *Ovivora* Mackinnon and Ray, 1937. Oocysts with many sporocysts, each with up to 12 (?) sporozoites; homoxenous; in eggs of echiuroids. TYPE SPECIES *O. thalassemae* (Lankester, 1885) Mackinnon and Ray, 1937.

1. *O. thalassemae* (Lankester, 1885) Mackinnon and Ray, 1937 (TYPE SPECIES) (syn., *Monocystis thalassemae* Lankester, 1885) in *Thalassema neptuni* (Echiurida) eggs.

Genus *Selysina* Duboscq, 1917. Oocysts with no sporocysts but with a variable number of "heliospores" consisting of many sporozoites arranged in a circle around a residuum like the petals of a flower; in ascidians. TYPE SPECIES *S. perforans* Duboscq, 1917.

1. *S. duboscqui* Harant, 1931 in *Styela partita* and *Polycarpa pomaria* (Urochordata) mesenchyme of intestine and mantle epithelium.
2. *S. incerta* Duboscq and Harant, 1923 in *Amaroucium aerolatum* (Urochordata) connective tissue, especially near the stomach and genital organs.
3. *S. perforans* Duboscq, 1917 (TYPE SPECIES) in *Stolonica socialis* (Urochordata) statoblasts, epidermis, and peribranchial epithelium.

Family CARYOTROPHIDAE *Lühe, 1916*
Oocysts without definite wall, with about 20 sporocysts, each with 8 or 12 sporozoites; homoxenous; in annelids.

Genus *Caryotropha* Siedlecki, 1902. Sporocysts with 12 sporozoites each. TYPE SPECIES *C. mesnili* Siedlecki, 1902.

1. *C. mesnili* Siedlecki, 1902 (TYPE SPECIES) in *Polymnia nebulosa* (Polychaetasida) coelom, testes.

Genus *Dorisiella* Ray, 1930. Sporocysts each with eight sporozoites. TYPE SPECIES *D. scolelepidis* Ray, 1930.

1.   *D. scolelepidis* Ray, 1930 (TYPE SPECIES) in *Scolelepis fuliginosa* (Polychaetasida) intestine.

Family CRYPTOSPORIDIIDAE *Léger, 1911*
    Development just under surface membrane of host cell or within its brush border and not in cell proper; meronts with a knob-like attachment organelle at some point on their surface; oocysts with or without sporocysts, homoxenous; microgametes without flagella.

Genus *Cryptosporidium* Tyzzer, 1907. Oocysts either without sporocysts or with a single one, with sporozoites; in GI and/or respiratory tract of vertebrates. TYPE SPECIES *C. muris* Tyzzer, 1907.

1.   *C. crotali* Triffitt, 1925 (syn., *C. serpentis* Levine, 1981) in *Crotalus confluentis, Elaphe guttata, E. subocularis, Crotalus horridus, Sansinia madagascarensis,* and *Pseudechis porphyriacus* (Reptilia) intestine.
2.   *C. meleagridis* Slavin, 1955 (syns., *C. anserinum* Proctor and Kemp, 1974; *C. tyzzeri* Levine, 1961) in *Meleagris gallopavo, Gallus gallus, Pavo cristatus, Coturnix coturnix, Anser anser, Poephila cincta,* and *Amazona autumnalis* (Aves) GI and/or respiratory tracts.
3.   *C. muris* Tyzzer, 1907 (TYPE SPECIES) in *Mus musculus, Bos taurus,* and probably various other mammals (Mammalia) GI tract.
4.   *C. nasorum* Hoover et al., 1981 in *Naso lituratus* (Osteichthys) intestine.
5.   *C. parvum* Tyzzer, 1912 (syns. *C. agni* Barker and Carbonell, 1974; *C. bovis* Barker and Carbonell, 1974; *C. cuniculus* Inman and Takeuchi, 1979; *C. felis* Iseki, 1979; *C. garnhami* Bird, 1981; *C. rhesi* Levine, 1981; *C. wrairi* Vetterling et al., 1971) in *Mus musculus, Bos taurus,* presumably *Homo sapiens, Macaca mulatta, Ovis aries, Capra hircus, Cervus elaphus, Odocoileus hemionus, Gazella subgutturosa, Sus scrofa, Equus caballus, Canis familiaris, Felis catus, Procyon lotor, Oryctolagus cuniculus, Sylvilagus floridanus, Apodemus sylvaticus, Clethrionomys glareolus, Cavia porcellus, Sciurus carolinensis,* and experimentally *Rattus norvegicus, Mesocricetus auratus* (all Mammalia), and probably *Gallus gallus* (Aves) GI tract. (See Upton and Current, 1980, 1985.)

Family PFEIFFERINELLIDAE *Grassé, 1953*
Oocysts without sporocysts, with eight naked sporozoites; fertilization of macrogamete ordinarily through a "vaginal" tube; homoxenous.

Genus *Pfeifferinella* von Wasielewski, 1904. With the characters of the family; in Molluska. TYPE SPECIES *P. ellipsoides* von Wasielewski, 1904.

1.   *P. ellipsoides* von Wasielewski, 1904 (TYPE SPECIES) in *Planorbarius corneus* (Molluska) liver.
2.   *P. impudica* Léger and Hollande, 1912 in *Lehmannia* (syn., *Limax*) *marginata* (Molluska) liver.

Family EIMERIIDAE *Minchin, 1903*
    Development in host cell proper; without attachment organelle or "vaginal" tube; oocysts with zero, one, two, four, or more sporocysts, each with one or more sporozoites; sporocysts

univalved, without dehiscence line; homoxenous; merogony and gamogony within host, sporogony typically outside; microgametes with two or three flagella; without metrocytes; in vertebrates or invertebrates.

Genus *Tyzzeria* Allen, 1936. Oocysts without sporocysts, with eight naked sporozoites. TYPE SPECIES *T. perniciosa* Allen, 1936.

1. *T. allenae* Chakravarty and Basu, 1946 emend. Pellérdy, 1974 (syn., *T. alleni* Chakravarty and Basu, 1947) in *Nettapus coromandelianus* (Aves) rectum.
2. *T. chenicusae* Ray and Sarkar, 1967 (syn., *Eimeria chenicusae* Ray and Sarkar, 1967 of Pellérdy [1974] *lapsus calami*) in *Nettapus coromandelianus* (Aves) feces.
3. *T. galli* Fernando and Remmler, 1973 in *Gallus lafayettei* (Aves) feces.
4. *T. natrix* (Matubayasi, 1936) Matubayasi, 1937 (syns., *Koidzumiella natrix* Matubayasi, 1936; *T. [K.] natrix* [Matubayasi, 1936] Matubayasi, 1937) in *Rhabdophis t. tigrinus* (Reptilia) intestine.
5. *T. parvula* (Kotlán, 1933) Klimeš, 1963 (syns., *Eimeria anseris* Kotlán, 1932 in part; *E. parvula* Kotlán, 1933; *T. anseris* Nieschulz, 1947) in *Anser anser, A. albifrons frontalis, A. caerulescens, A. rossi, Branta bernicla hrota, B. canadensis,* and *Olor columbianus* (Aves) small intestine.
6. *T. pellerdyi* Bhatia and Pande, 1966 in *Anas strepera, A. acuta, A. americana, A. p. platyrhynchos,* and *Aythya nyroca* (Aves) feces.
7. *T. perniciosa* Allen, 1936 (TYPE SPECIES) in *Anas platyrhynchos domestica, A. acuta, Aythya affinis,* and *A. erythropus* (Aves) small intestine.
8. *T. peromysci* Levine and Ivens, 1960 in *Peromyscus maniculatus* and *P. leucopus* (Mammalia) intestinal contents.
9. *T. typhlopis* Ovezmukhamedov, 1968 emend. (syn., *T. typhlopisi* Ovezmukhamedov, 1968) in *Typhlops vermicularis* (Reptilia) feces.

Genus *Alveocystis* Bel'tenev, 1980. Oocysts with eight naked sporozoites; homoxenous; in invertebrates. TYPE SPECIES *A. macrocoronata* (Lüling, 1942) Levine, 1985.

1. *A. macrocoronata* (Lüling, 1942) Levine, 1985 (syns., *A. intestinalis* Bel'tenev, 1980; *Klossia* [?] *macrocoronata* [Lüling, 1942] Levine, 1977; *? macrocoronata* Lüling, 1942) (TYPE SPECIES) in *Priapulus caudatus* (TYPE HOST) and *Halicryptus spinulosus*) (Priapuloidea) intestinal epithelial cells.
2. *A. gugleri* (Wacha, 1981) Levine, 1985 (syn., *Pfeifferinella gugleri* Wacha, 1981) in *Triodopsis albolabris* (Gastropodasida) digestive gland.

Genus *Eimeria* Schneider, 1875. Oocysts with four sporocysts, each with two sporozoites; merogony intracellular; sporogony extracellular; in vertebrates and a few invertebrates. TYPE SPECIES *E. falciformis* (Eimer, 1870) Schneider, 1875.

1. *E. abdildaevi* Dzerzhinskii, 1982 in *Dryomys nitedula* (Mammalia) feces.
2. *E. abenovi* Svanbaev, 1979 (syn., *E. faurei* [Moussu and Marotel, 1902] Martin, 1909 of Svanbaev [1979] in part) in *Gazella subgutturosa* (Mammalia) feces.
3. *E. abidzhanovi* Davronov, 1973 in *Rhombomys opimus* (Mammalia) feces.
4. *E. ablephari* Cannon, 1967 in *Ablepharus boutonii* (Reptilia) midgut.
5. *E. abramovi* Svanbaev and Rakhmatullina, 1967 in *Anas p. platyrhynchos* (Aves) feces.
6. *E. absheronae* Musaev, 1970 emend. Musaev and Mamedova, 1981 (syns., *E. apsheronica* Musaev, 1970); *E. aemula* Yakimoff, 1931 in the goat; *E. faurei* [Moussu

and Marotel, 1902] Martin, 1909 in the goat) in *Capra hircus, C. ibex*, and probably *C. sibirica* (Mammalia) intestine.

7.   *E. abusalimovi* Musaev and Veisov, 1965 in *Dryomys nitedula* (Mammalia) feces.

8.   *E. abuschevi* Veisov, 1962 in *Microtus majori* (Mammalia) feces.

9.   *E. acanthodactyli* (Phisalix, 1930) Levine and Becker, 1933 (syn., *Coccidium acanthodactyli* Phisalix, 1933) in *Acanthodactylus scutellatus* (Reptilia) bile ducts and liver.

10.   *E. accipitris* Schwalbach, 1959 in *Accipiter nisus, Falco naumanni,* and *Hieraaetus pennatus* (Aves) feces.

11.   *E. acerinae* Pellérdy and Molnár, 1971 in *Acerina cernua* (Osteichthys) intestine.

12.   *E. acervulina* Tyzzer, 1929 (syns., *E. diminuta* Fernando and Remmler, 1973; *E. acervulina* var. *diminuta* [Fernando and Remmler, 1973] Long, 1974) in *Gallus gallus* and *G. lafayettei* (Aves) anterior small intestine.

13.   *E. achburunica* Musaev and Alieva, 1961 in *Meriones libycus* (syn., *M. erythrourus*) (Mammalia) large intestine.

14.   *E. adenoeides* Moore and Brown, 1951 in *Meleagris gallopavo* (Aves) posterior ileum, ceca, and rectum.

15.   *E. adiyamanensis* Sayin, 1981 in *Spalax ehrenbergi* (Mammalia) intestinal contents.

16.   *E. adleri* Yakimoff and Gousseff, 1936 in *Vulpes vulpes* (Mammalia) feces.

17.   *E. aeromysis* Colley and Mullin, 1971 in *Aeromys tephromelas* (Mammalia) intestine.

18.   *E. aesculapi* Carini, 1933 in *Erythrolamprus aesculapi* (Reptilia) intestine.

19.   *E. africana* Levine et al., 1959 in *Lophuromys s. sikapusi* (Mammalia) intestine.

20.   *E. africiensis* Musaev and Mamedova, 1981 in *Capra hircus* (Mammalia) feces.

21.   *E. agamae* (Laveran and Pettit, 1910) Reichenow, 1921 (syn., *Coccidium agamae* Laveran and Pettit, 1910) in *Agama agama colonorum* (Reptilia) liver (epithelial cells of bile ducts).

22.   *E. agrarii* Musaev and Veisov, 1965 in *Apodemus agrarius* (Mammalia) feces.

23.   *E. aguti* Carini, 1935 in *Dasyprocta aguti* (Mammalia) feces.

24.   *E. ahsata* Honess, 1942 in *Ovis c. canadensis, O. aries,* and *O. musimon* (Mammalia) small intestine.

25.   *E. ahtanumensis* Clark, 1970 in *Sceloporus occidentalis* (Reptilia) bile duct and gall bladder epithelium.

26.   *E. akeriana* Ismailov and Gaibova, 1983 in gerbil *Meriones blackleri* (Mammalia) small intestine.

27.   *E. alabamensis* Christensen, 1941 in *Bos taurus, B. indicus,* and *Bubalus bubalis* (Mammalia) posterior ileum.

28.   *E. alakuli* Rakhmatullina-Batyrshina and Svanbaev, 1972 in *Fulica atra* (Aves) feces.

29.   *E. albertensis* Hair and Mahrt, 1970 in *Ursus americanus* (Mammalia) feces.

30.   *E. albigulae* Levine, Ivens, and Kruidenier, 1957 in *Neotoma albigula* (Mammalia) intestinal contents.

31.   *E. alces* Arnastauskene, 1974 in *Alces alces* (Mammalia) feces.

32.   *E. alectoreae* Ray and Hiregoudar, 1959 (syns., *E. lyruri* Galli-Valerio, 1927 in *Alectoris graeca,* not in *Lyrurus tetrix; E. legionensis* Cordero del Campillo and Pla Hernandez, 1966) in *A. graeca* (Aves) feces.

33.   *E. alijevi* Musaev, 1970 (syns., *E. galouzoi* Yakimoff and Rastegaieff, 1930 in part; *E. kandilovi* Musaev, 1970; *E. parva* Kotlán, Mócsy, and Vajda, 1929 in the goat) in *Capra hircus, C. ibex,* and *C. sibirica* (Mammalia) small intestine.

34.   *E. alischerica* Musaev and Veisov, 1965 in *Rattus norvegicus* (Mammalia) feces.

35.   *E. allactagae* Iwanoff-Gobzem, 1934 in *Allactaga* major, *A. elater,* and *A. jaculus* (Mammalia) feces.

36.   *E. almataensis* Musaev, 1970 (syn., *E. debliecki* Douwes, 1921 of Svanbaev [1958] in *Sus scrofa* (Mammalia) feces.

37. *E. alpacae* Guerrero, 1967 in *Lama pacos* (Mammalia) feces.
38. *E. alpina* Supperer and Kutzer, 1961 in *Rupicapra rupicapra* (Mammalia) feces.
39. *E. amarali* Pinto, 1928 in *Bothrops neuwiedii* (Reptilia) feces.
40. *E. ambassi* Patnaik and Acharya, 1972 in *Barbus ambassis* (Osteichthys) abscess in shoulder.
41. *E. amburdariana* Musaev and Veisov, 1962 in *Mesocricetus auratus* (Mammalia) feces.
42. *E. ambystomae* Saxe, 1955 in *Ambystoma tigrinum, Desmognatha monticola,* and *D. quadramaculata* (Amphibia) intestine.
43. *E. ameivae* Lainson, 1968 in *Ameiva undulata* (Reptilia) feces.
44. *E. americana* Carvalho, 1943 in *Lepus townsendii campanius* (Mammalia) intestine.
45. *E. ammonis* Musaev, 1920 (syn., *E. faurei* [Moussu and Marotel, 1902] Martin, 1909 of Svanbaev [1958] in *Ovis ammon*) in *Ovis ammon* (Mammalia) feces.
46. *E. amreini* Pellérdy, 1974 (syns., *E. cystisfelleae* var. *americana* Amrein, 1952 in part; *E. noctisauris* Bovee and Telford, 1965 in part) in *Uta stansburiana* (Reptilia) gall bladder.
47. *E. amurensis* Akhmerov, 1959 in Dogel' and Akhmerov, 1959 in *Pseudorasbora parva* and *Sarcochilichtys sinensis* (Osteichthys) liver, kidney, and other organs.
48. *E. amydae* Roudabush, 1937 in *Amyda spinifera* (Reptilia) small and large intestine.
49. *E. anatis* Scholtyseck, 1955 in *Anas platyrhynchos* (Aves) small intestine.
50. *E. andamanensis* Mandal and Nair, 1973 in *Taphozous melanopogon* (Mammalia) small intestine.
51. *E. andrewsi* Yakimoff and Gousseff, 1935 in *Sciurus vulgaris* (Mammalia) small intestine.
52. *E. anekalensis* Rajasekariah et al., 1971 in *Felis pardus* (Mammalia) feces.
53. *E. angusta* Allen, 1934 in *Tetrastes bonasia, Canachites canadensis, Bonasa umbellus,* and also perhaps *Centrocercus urophasianus* and *Pediocetes phasianellus campestris* (Aves) intestine.
54. *E. anili* Haldar, Ray, and Mandal, 1982 in *Sturnus c. contra* (Aves) feces.
55. *E. ankarensis* Sayin, 1969 in *Bubalus bubalis* (Mammalia) feces.
56. *E. anseris* Kotlán, 1932 emend. Kotlán, 1933 in *Anser anser domesticus* and also *A. c. caerulescens* and *Branta canadensis hutchinsi* (Aves) mostly middle and posterior small intestine.
57. *E. antilocaprae* Huizinga, 1942 emend. Levine and Ivens, 1970 (syn., *E. antelocaprae* Huizinga, 1942) in *Antilocapra americana* (Mammalia) feces.
58. *E. antilocervi* Ray and Mandal, 1960 emend. Levine and Ivens, 1970 (syn., *E. antelocervi* Ray and Mandal, 1960) in *Antilope cervicapra* (Mammalia) feces.
59. *E. antonellii* Straneva and Gallati, 1980 in *Neotoma floridana* (Mammalia) feces.
60. *E. apionodes* Pellérdy, 1954 in *Apodemus flavicollis* (Mammalia) small intestine.
61. *E. apodemi* Pellérdy, 1954 in *Apodemus flavicollis, A. sylvaticus,* and *A. agrarius* (Mammalia) small intestine.
62. *E. arabiana* Veisov, 1961 in *Meriones vinogradovi* (Mammalia) small intestine.
63. *E. arasinaensis* Musaev and Veisov, 1965 in *Mus musculus* (Mammalia) feces.
64. *E. araxena* Musaev and Veisov, 1960 in *Meriones shawi* (syn., *M. tristrami*) (Mammalia) feces.
65. *E. arctica* Yakimoff, Matschoulsky, and Spartansky, 1939 in *Rangifer tarandus* (Mammalia) feces.
66. *E. arctomysi* Galli-Valerio, 1931 in *Marmota marmota* (Mammalia) feces.
67. *E. aristichthysi* Lee and Chen, 1964 in *Aristichthys nobilis* and *Hypophthalmichthys molotrix* (Osteichthys) intestine. (See Dyková and Lom, 1983.)
68. *E. arizonensis* Levine, Ivens, and Kruidenier, 1957 in *Peromyscus truei, P. maniculatus, P. eremicus,* and *P. leucopus* (Mammalia) feces.

69. *E. arkhari* Yakimov and Machulskii, 1937 in *Ovis ammon polii*, *O. polii sewerzowi*, and *O. vignei* (Mammalia) feces.
70. *E. arkutinae* Golemansky, 1978 (syns., *E. keilini* Yakimoff and Gousseff of Ryšavý [1954]; *Eimeria* sp. Golemansky and Yankova, 1973) in *Apodemus sylvaticus*, *A. flavicollis*, and *A. agrarius* (Mammalia) intestinal contents.
71. *E. arloingi* (Marotel, 1905) Martin, 1909 (syns., *Coccidium arloingi* Marotel, 1905; *C. caprae* Jaeger, 1921; *E. ahsata* Honess, 1942 of Chevalier [1942] from the goat; *E. crandallis* Honess, 1942 of Chevalier [1966] from the goat; *E. faurei* [Moussu and Marotel, 1902] of Tsygankov, Paichuk, and Balbaeva [1963] and of some other Russian authors from the goat; *E. hawkinsi* Ray, 1952 in part) in *Capra hircus*, *C. aegagrus*, *C. falconeri*, *C. ibex*, and *C. sibirica* (Mammalia) small intestine.
72. *E. arnaldoi* Pinto and Maciel, 1929 in *Thamnodynastes nattereri* (Reptilia) gall bladder.
73. *E. arundeli* Barker, Munday, and Presidente, 1979 in *Vombatus ursinus* (Mammalia) intestine.
74. *E. arusica* Musaev and Veisov, 1961 in *Cricetulus migratorius* (Mammalia) feces.
75. *E. arvicanthis* van den Berghe and Chardome, 1956 in *Arvicanthis abyssinicus rubescens* (Mammalia) probably intestine.
76. *E. arvicolae* (Galli-Valerio, 1905) Reichenow, 1921 (syns., *Coccidium arvicolae* Galli-Valerio, 1905; *E. arvalis* Iwanoff-Gobzem, 1934; *E. musculi* Yakimoff and Gousseff of Svanbaev [1956]) in *Microtus nivalis* and also *M. arvalis* (Mammalia) feces.
77. *E. asadovi* Musaev and Veisov, 1965 in *Dryomys nitedula* (Mammalia) feces.
78. *E. ascotensis* Levine and Ivens, 1965 (syn., *E. neosciuri* Prasad, 1960 of Webster [1960]) in *Sciurus carolinensis* and *S. niger rufiventer* (Mammalia) small intestine.
79. *E. asiatici* Levine and Ivens, 1965 (syn., *E. beecheyi* Henry of Tanabe and Okinami [1940]) in *Eutamias asiaticus* (Mammalia) cecal contents.
80. *E. assaensis* Levine and Ivens, 1965 (syn., *E. callosphermophili* [*sic*] Henry of Svanbaev [1962] in *Meriones tamariscinus*) in *Meriones tamariscinus* (Mammalia) feces.
81. *E. astrachanbazarica* Musaev and Veisov, 1960 in *Meriones shawi* (syn., *M. tristrami*) (Mammalia) large intestine contents.
82. *E. asturi* Galli-Valerio, 1935 in *Accipiter gentilis* (Aves) intestinal contents.
83. *E. asymmetrica* Supperer and Kutzer, 1961 in *Cervus elaphus* (Mammalia) feces.
84. *E. athabascensis* Samoil and Samuel, 1977 in *Lepus americanus* (Mammalia) feces.
85. *E. atheni* Chauhan and Jain, 1979 in *Athene brama* (Aves) feces.
86. *E. attenuata* Wacha and Christiansen, 1974 in *Thamnophis sirtalis parietalis* and also *Natrix s. sipedon* (Reptilia) feces.
87. *E. auburnensis* Christensen and Porter, 1939 (syns., *E. ildefonsoi* Torres and Ramos, 1939; *E. khurodensis* Rao and Hiregaudar, 1954) in *Bos taurus*, *B. indicus*, and *Bubalus bubalis* (Mammalia) middle and lower thirds of small intestine.
88. *E. audubonii* Duszynski and Marquardt, 1969 in *Sylvilagus audubonii* and *S. floridanus* (Mammalia) feces.
89. *E. aurata* Musaev and Veisov, 1962 in *Mesocricetus auratus* (Mammalia) feces.
90. *E. aurati* Hoffman, 1965 in *Carassius auratus* (Osteichthys) intestine.
91. *E. aurei* Bhatia et al., 1979 in *Canis aureus naria* (Mammalia) feces.
92. *E. auriti* Mirza, 1975 in *Hemiechinus auritus* (Mammalia) feces.
93. *E. austriaca* Supperer and Kutzer, 1961 in *Cervus elaphus* (Mammalia) feces.
94. *E. aythyae* Farr, 1965 in *Aythya affinis* (Aves) small intestine.
95. *E. azul* Wiggins and Rothenbacher, 1979 in *Sylvilagus floridanus* (Mammalia) feces.
96. *E. babaevi* Svanbaev, 1979 (syn., *E. ninakohlyakimovae* Yakimoff and Rastegaieff, 1930 of Svanbaev [1958] in *Capra sibirica*) in *Capra sibirica* (Mammalia) feces.
97. *E. babatica* Sugar, 1978 in *Lepus europaeus* (Mammalia) large intestine.
98. *E. bactriani* Levine and Ivens, 1970 (syns., *E. cameli* Nöller, 1933; *E. cameli* Iwanoff-

Gobzem, 1934; *E. nolleri* Reichenow, 1953 of Abdussalam and Rauf [1958]) in *Camelus bactrianus* (Mammalia) small intestine.

99. *E. badamlinica* Musaev and Veisov, 1963 in *Apodemus sylvaticus* (Mammalia) feces.

100. *E. badchisica* Glebezdin, 1969 in *Rhombomys opimus* (Mammalia) feces.

101. *E. baghdadensis* Mirza, 1975 in *Mus musculus* (Mammalia) feces.

102. *E. bailwardi* Glebezdin, 1971 in *Calomyscus bailwardi* (Mammalia) feces.

103. *E. baiomysis* Levine, Ivens, and Kruidenier, 1958 in *Baiomys taylori* and *B. musculus* (Mammalia) intestinal contents.

104. *E. bakanensis* Svanbaev and Rakhmatullina, 1971 in *Vulpes vulpes* (Mammalia) feces.

105. *E. balchanica* Glebezdin, 1978 in *Ochotona rufescens* (Mammalia) feces.

106. *E. balozeti* Yakimoff and Gousseff, 1938 in *Sturnus vulgaris* (Aves) feces.

107. *E. balphae* Ernst, Chobotar, and Anderson, 1967 in *Dipodomys ordii, D. agilis, D. merriami,* and *D. spectabilis* (Mammalia) feces.

108. *E. bandicota* Bandyopadhyay and Dasgupta, 1982 in *Bandicota bengalensis* (Mammalia) intestine.

109. *E. bandipurensis* Ray, Banik, and Mukherjea, 1965 in *Funambulus palmarum* and *F. tristriatus* (Mammalia) feces.

110. *E. banffensis* Lepp, Todd, and Samuel, 1973 in *Ochotona princeps, O. collaris,* and *O. hyperborea yesoensis* (Mammalia) colon contents.

111. *E. banyulensis* Lom and Dyková, 1982 in *Crenilabrus mediterraneus* (Osteichthys) middle part of intestine.

112. *E. barbeta* Kar, 1944 in *Cyanops asiatica* (Aves) intestinal contents.

113. *E. bareillyi* Gill, Chhabra, and Lal, 1963 (syn., *E. bubalis* Abdussalam and Rauf, 1956 of Yasin and Abdussalam [1958]) in *Bubalus bubalis* (Mammalia) small intestine.

114. *E. barleyi* Straneva and Gallati, 1980 in *Neotoma floridana* (Mammalia) feces.

115. *E. barretti* Lepp, Todd, and Samuel, 1972 in *Ochotona princeps* (Mammalia) feces.

116. *E. basilisci* Duszynski, 1969 in *Basiliscus basiliscus* (Reptilia) feces.

117. *E. baskanica* Nukerbaeva and Svanbaev, 1977 in *Mustela erminea* (Mammalia) feces.

118. *E. bassagensis* Svanbaev, 1979 (syn., *E. arvicolae* Galli-Valerio, 1905 of Svanbaev [1958] in *Alticola strelzovi*) in *Alticola strelzovi* (Mammalia) feces.

119. *E. batabatensis* Levine and Ivens, 1965 (syn., *E. arvicolae* Galli-Valerio of Musaev and Veisov [1960]) in *Arvicola terrestris* (Mammalia) large intestine contents.

120. *E. bateri* Bhatia, Pandey, and Pande, 1965 in *Coturnix c. coturnix* and *C. c. japonica* (Aves) small intestine.

121. *E. battakhi* Dubey and Pande, 1963 in *Anas platyrhynchos* (Aves) feces.

122. *E. bazi* Chauhan and Bhatia, 1970 in *Pseudibis papillosa* (Aves) feces.

123. *E. (?) beauchampi* Léger and Duboscq, 1917 in *Glossobalanus minutus* (Hemichordata) hepatic ceca.

124. *E. beckeri* Yakimoff and Sokoloff, 1935 (syn., *E. ussuriensis* Yakimoff and Sprinholtz-Schmidt, 1939) in *Spermophilus pygmaeus, S. eversmanni, S. maximus,* and *S. fulvus* (Mammalia) feces.

125. *E. beecheyi* Henry, 1932 in *Spermophilus beecheyi* (Mammalia) large intestine contents.

126. *E. belawini* Yakimoff, 1930 in *Hyla arborea* (Amphibia) feces.

127. *E. beldingii* Veluvolu, 1984 in Veluvolu and Levine, 1984 in *Spermophilus beldingi* (Mammalia) feces.

128. *E. bentongi* Colley and Mullin, 1971 in *Hylomys suillus* (Mammalia) intestinal contents.

129. *E. berkinbaevi* Abenov and Svanbaev, 1979 in *Spermophilus fulvus* (Mammalia) feces.

130. *E. betica* Martinez and Hernandez, 1973 in *Sus scrofa* (Mammalia) feces (?).

131. *E. beyerae* (Ovezmukhammedov, 1977) emend. Levine, 1985 (syn., *E. beyeri* Ovezmukhammedov, 1977) in *Ablepharus deserti* (Reptilia) feces.

132. *E. bhutanensis* Ray and Hiregoudar, 1959 in *Polyplectron bicoloratus bakeri* (Aves) feces.

133. *E. bicrustae* Veisov, 1962 in *Microtus majori* (Mammalia) feces.
134. *E. bijlikuli* Svanbaev, 1962 in *Hemiechinus auritus* (Mammalia) feces.
135. *E. bilamellata* Henry, 1932 (syn., *E. eubeckeri* Hall and Knipling, 1935) in *Spermophilus lateralis chrysodeirus, S. armatus, S. beecheyi, S. citellus, S. franklinii, S. tridecemlineatus, S. variegatus, S. columbianus,* and *S. richardsoni* (Mammalia) intestine.
136. *E. bistratum* Veisov, 1961 in *Meriones vinogradovi* (Mammalia) cecum and colon.
137. *E. bitis* Fantham, 1932 (syn., *E. annea* Fantham and Porter, 1954) in *Bitis lachesis* (Reptilia) gall bladder and bile ducts.
138. *E. blarinae* Todd, French, and Levine, 1986 in *Blarina brevicauda* (Mammalia) intestinal contents.
139. *E. boae* Prasad, 1963 in *Boa canina* (Reptilia) feces.
140. *E. bohemica* Ryšavý, 1957 in *Arvicola terrestris* (Mammalia) feces.
141. *E. bombayansis* Rao and Hiregaudar, 1954 in *Bos indicus* (Mammalia) feces.
142. *E. bonasae* Allen, 1934 in *Tetrastes bonasia* and possibly *Lagopus lagopus, Canachites canadensis,* and *Pediocetes phasianellus campestris* (Aves) cecum.
143. *E. borealis* Hair and Mahrt, 1970 in *Ursus americanus* (Mammalia) feces.
144. *E. boschadis* Waldén, 1961 in *Anas p. platyrhynchos* (Aves) kidneys.
145. *E. botelhoi* Carini, 1932 in *Sciurus (Guerlinguetus) ingrami* (Mammalia) small intestine.
146. *E. bothropis* Lainson, 1968 in *Bothrops atrox* (Reptilia) feces.
147. *E. boveroi* Carini and Pinto, 1930 in *Hemidactylus mabuia* (Reptilia) small intestine.
148. *E. bovis* (Züblin, 1908) Fiebiger, 1912 (syns., *Coccidium bovis* Züblin, 1908; *E. canadensis* Bruce, 1921 in part; *E. smithi* Yakimoff and Galouzo, 1927; *E. [Globidium] bovis* [Züblin, 1908] Reichenow, 1953; *E. aareyi* Rao and Bhatavdekar, 1959; *G. fusiformis* [?] Hassan, 1935) in *Bos taurus, B. indicus, B. gaurus, Bubalus bubalis, Bison bonasus, B. bison,* and *Bibos banteng* (Mammalia) small intestine.
149. *E. branchyphila* Dyková, Lom, and Grupcheva, 1983 in *Rutilus rutilus* (Osteichthys) gill filaments.
150. *E. brantae* Levine, 1953 in *Branta canadensis parvipes* and *B. c. hutchinsi* (Aves) feces.
151. *E. brasiliensis* Torres and Ramos, 1939 (syns., *E. helenae* Donçiu, 1961; *E. boehmi* Supperer, 1952; *E. orlovi* Basanova, 1952) in *Bos taurus, B. indicus,* and *Bubalus bubalis* (Mammalia) feces.
152. *E. brevoortiana* Hardcastle, 1944 in *Brevoortia tyrannus* (Osteichthys) testes.
153. *E. brinkmanni* Levine, 1953 in *Lagopus mutus rupestris* (Aves) feces.
154. *E. brodeni* Cerruti, 1930 in *Testudo graeca* (Reptilia) intestine.
155. *E. brunetti* P. P. Levine, 1942 in *Gallus gallus* (Aves) posterior ileum and colon.
156. *E. bubonis* Cawthorn and Stockdale, 1981 in *Bubo virginianus* and *B. bubo* (Aves) feces.
157. *E. bucephalae* Christiansen and Madsen, 1948 in *Bucephala c. clangula* (Aves) small intestine.
158. *E. bukidnonensis* Tubangui, 1931 in *Bos indicus, B. taurus,* and also probably *Bubalus bubalis* and *Bibos banteng* (Mammalia) intestine.
159. *E. bychowskyi* Musaev and Veisov, 1965 in *Rattus norvegicus* (Mammalia) feces.
160. *E. cabassi* Carini, 1933 emend. (syn., *E. cabassusi* Carini, 1933) in *Cabassus* sp. (Mammalia) feces.
161. *E. cagandzeeri* Sugár, 1983 in *Procapra gutturosa* (Mammalia) feces.
162. *E. calentinei* Duszynski and Brunson, 1973 in *Ochotona princeps, O. collaris,* and *O. hyperborea yesoensis* (Mammalia) feces.
163. *E. callosciuri* Colley, 1971 in *Callosciurus prevosti, C. notatus, C. caniceps,* and *C. nigrovitattus* (Mammalia) anterior small intestine.

164. *E. callospermophili* Henry, 1932 (syn., *E. callosphermophili [sic]* Henry of Svanbaev [1962] in *Spermophilus maximus*) in *S. lateralis, S. fulvus, S. spilosoma, S. maximus, S. columbianus, S. franklinii, S. richardsoni, S. beecheyi, S. tridecemlineatus, S. armatus,* and *Cyanomys leucurus* (Mammalia) intestine.

165. *E. calomyscus* Glebezdin, 1971 in *Calomyscus bailwardi* (Mammalia) feces.

166. *E. cameli* (Henry and Masson, 1932) Reichenow, 1952 (syns., *Globidium cameli* Henry and Masson, 1932; *E. [G.] cameli* [Henry and Masson, 1932] Reichenow, 1952 of Abdussalam and Rauf [1957]; *E. kazachstanica* Tsygankov, 1950; *E. casahstanica [sic]* Zigankoff, 1950 of Pellérdy [1965]; *E. noelleri* [Henry and Masson, 1932] Pellérdy, 1956 of Dubey and Pande [1965]) in *Camelus dromedarius* and *C. bactrianus* (Mammalia) small intestine.

167. *E. campania* Carvalho, 1943 (syns., *E. irresidua* form *campanius* Carvalho, 1943; *E. semisculpta* Madsen, 1938 of Pellérdy [1974] in part; *E. irresidua* Kessel and Jankiewicz, 1931 of Gill and Ray [1960] in *Lepus townsendii campanius, L. europaeus,* and *L. ruficaudatus* (Mammalia) presumably intestine.

168. *E. canadensis* Bruce, 1921 (syn., *E. zurnabadensis* Yakimoff, 1931) in *Bos taurus, B. indicus,* and also *Bison bonasus, Bibos banteng,* and *Bubalus bubalis* (Mammalia) feces.

169. *E. canaliculata* Lavier, 1935 in *Triturus alpestris, T. cristatus, T. helveticus,* and *T. vulgaris* (Amphibia) intestine.

170. *E. canna* Triffitt, 1924 in *Taurotragus oryx* (Mammalia) feces.

171. *E. capibarae* Carini, 1937 in *Hydrochoerus hydrochoerus* (syn., *H. capybara*) (Mammalia) feces.

172. *E. capoetobramae* Allamuratov and Iskov, 1970 (syn., *E. sareotobramae* Allamuratov and Iskov, 1970 of Dyková and Lom [1983]) in *Capoetobrama kuschakowitschi* (Osteichthys) kidney and urinary duct.

173. *E. capra* Musaev, 1970 (syn., *E. faurei* [Moussu and Marotel, 1902] Martin, 1909 of Svanbaev [1958] in *Capra sibirica*) in *Capra sibirica* (Mammalia) feces.

174. *E. capreoli* Galli-Valerio, 1927 in *Capreolus capreolus* (Mammalia) feces.

175. *E. caprina* Lima, 1979 in *Capra hircus* (Mammalia) feces.

176. *E. capromydis* Ryšavý, 1967 in *Capromys pilorides* (Mammalia) intestinal contents.

177. *E. caprovina* Lima, 1980 in *Capra hircus* and experimentally in *Ovis aries* (Mammalia) feces.

178. *E. carassii* Yakimoff and Gousseff, 1935 in *Carassius carassius* (Osteichthys) feces.

179. *E. carassiusaurati* Romero Rodriguez, 1978 in *Carassius auratus* (Osteichthys) small intestine.

180. *E. caripensis* Arcay de Peraza, 1964 in *Proechimys guyanensis* (Mammalia) small intestine.

181. *E. carolinensis* Von Zellen, 1959 in *Peromyscus leucopus* (Mammalia) intestinal contents.

182. *E. carri* Ernst and Forrester, 1973 in *Terrapene carolina* (Reptilia) feces.

183. *E. cascabeli* Vetterling and Widmer, 1968 in *Crotalus viridis* (Reptilia) gall bladder and bile ducts.

184. *E. catalana* Lom and Dyková, 1981 in *Crenilabrus mediterraneus* (Osteichthys) middle intestine.

185. *E. catostomi* Molnar and Hanek, 1974 in *Catostomus commersoni* and *Hypentelium nigricans* (Osteichthys) anterior gut.

186. *E. catubrina* Mantovani, Borelli, and Ricci Bitti, 1970 (syn., *E. cotubrina* Mantovani, Borrelli, and Ricci Bitti of Arnastauskene, 1974 *lapsus calami*) in *Capreolus capreolus* (Mammalia) feces.

187. *E. caucasica* Yakimoff and Buewitsch, 1932 (syn., *E. coturnicis* Chakravarty and

Kar, 1947 of Lizcano and Romero [1972]) in *Alectoris graeca, A. chukar,* and *A. rufa* (Aves) feces.

188. *E. causeyi* Ernst, Cooper, and Frydendall, 1970 in *Castor canadensis* (Mammalia) feces.

189. *E. caviae* Sheather, 1924 in *Cavia porcellus* and *C. aperea* (Mammalia) colon and sometimes cecum.

190. *E. celebii* Sayin, 1981 in *Spalax ehrenbergi* (Mammalia) mixed intestinal contents.

191. *E. centrocerci* Simon, 1939 in *Centrocercus urophasianus* (Aves) ceca.

192. *E. cephalophi* Pampiglione, Ricci-Bitti, and Kabla, 1973 in *Cephalophus monticola* (Mammalia) feces.

193. *E. cerastis* (Chatton, 1912) Phisalix, 1921 (syn., *Coccidium cerastis* Chatton, 1912) in *Cerastes cerastes* and *C. cornutus* (Reptilia) gall bladder.

194. *E. cernae* Levine and Ivens, 1965 (syn., *E. schueffneri* Yakimoff and Gousseff, 1938 of Černá [1962]) in *Clethrionomys glareolus* and *C. rutilis* (Mammalia) cecum.

195. *E. (?) certhiidarum* Gottschalk, 1972 in *Certhia brachydactyla* (Aves) feces. (*Remarks.* The oocysts did not sporulate, but Gottschalk, 1972 nevertheless assigned this species to *Eimeria.*)

196. *E. cervi* Galli-Valerio, 1927 in *Cervus elephaus* (Mammalia) feces.

197. *E. cervis* Mandal and Choudhury, 1982 in *Axis axis* (Mammalia) feces.

198. *E. chagasi* Yakimoff and Gousseff, 1935 in *Sorex araneus* and *S. arcticus* (Mammalia) intestine.

199. *E. charadrii* Mandal, 1965 in *Charadrius asiaticus* (Aves) small intestine.

200. *E. chatangae* Arnastauskene, 1980 in *Lemmus sibiricus* (Mammalia) feces.

201. *E. chaus* Ryšavý, 1955 in *Felis (Chaus) chaus* (Mammalia) intestine.

202. *E. chausinghi* Pande et al., 1970 in *Tetracerus quadricornis* (Mammalia) feces.

203. *E. cheetali* Bhatia, 1968 in *Antilope cervicapra* and *Axis axis* (Mammalia) feces.

204. *E. cheissini* Schulman and Zaika, 1962 in Bykhovskaya-Pavlovskaya et al., 1962 in *Gobio gobio, Hemiculter leucisculus,* and *Henibarbus labeo* (Osteichthys) peritoneum, intestine, swim bladder, and gall bladder.

205. *E. chelydrae* Ernst et al., 1969 in *Chelydra serpentia* (Reptilia) feces.

206. *E. cheni* Schulman and Zaika, 1962 in Bykhovskaya-Pavlovskaya et al., 1962 (syn., *E. intestinalis* Chen, 1956) in *Mylopharyngodon piceus* (Osteichthys) anterior intestine.

207. *E. chetae* Arnastauskene, 1980 in *Microtus middendorfi* (Mammalia) feces.

208. *E. chihuahuaensis* Short, Mayberry, and Bristol, 1981 in *Dipodomys merriami* (Mammalia) feces.

209. *E. chinchillae* De Vos and van der Westhuizen, 1968 in *Chinchilla laniger* and *Rhabdomys pumilio* (Mammalia) cecum.

210. *E. chinkari* Pande et al., 1970 in *Gazella gazella* (Mammalia) feces.

211. *E. chobotari* Ernst, Oaks, and Sampson, 1970 in *Dipodomys merriami, D. microps, D. agilis,* and *D. ordi* (Mammalia) feces.

212. *E. choloepi* Lainson and Shaw, 1982 emend. (syn., *E. choloepusi* Lainson and Shaw, 1968) in *Choloepus didactylus* (Mammalia) ileum.

213. *E. choudari* Bhatia et al., 1972 in *Streptopelia decaocto* and *Columba* spp. (Aves) feces.

214. *E. christenseni* Levine, Ivens, and Fritz, 1962 (syns., *E. tirupatiensis* Sivanarayana and Venkataratnam, 1969; *E. tuniensis* Musaev and Mamedova, 1981) in *Capra hircus* (Mammalia) small intestine.

215. *E. christianseni* Waldén, 1961 in *Cygnus olor* (Aves) kidney.

216. *E. chrysemydis* Deeds and Jahn, 1939 in *Chrysemys picta belli* (Reptilia) intestine.

217. *E. chudatica* Musaev, Veisov, and Alieva, 1963 in *Microtus socialis* (Mammalia) feces.

218. *E. cicaki* Else and Colley, 1975 in *Gehyra mutilata* and also *Hemidactylus frenatus* (Reptilia) small intestine.
219. *E. circumborealis* Hobbs and Samuel, 1974 in *Ochontona collaris, O. princeps,* and *O. hyperborea yesoensis* (Mammalia) feces.
220. *E. citelli* Kartchner and Becker, 1930 in *Spermophilus tridecemlineatus, S. citellus, S. fulvus, S. maximus, S. pygmaeus,* and *S. undulatus* (Mammalia) cecum, colon, and small intestine.
221. *E. citriformis* Dogel', 1948 in *Tilesina gibbosa* (Osteichthys) pyloric ceca.
222. *E. clarkei* Hanson, Levine, and Ivens, 1957 in *Anser c. caerulescens, Branta canadensis leucoporeia,* and *B. c. interior* (Aves) feces.
223. *E. clethrionomyis* Straneva and Kelley, 1979 in *Clethrionomys gapperi* (Mammalia) feces.
224. *E. clini* Fantham, 1932 in *Clinus superciliosus* (Osteichthys) intestine.
225. *E. cnemidophori* Carini, 1941 in *Cnemidophorus l. lemniscatus* (Reptilia) intestine.
226. *E. coahulliensis* Vance and Duszynski, 1985 in *Microtus mexicanus subsimus* (Mammalia) feces.
227. *E. cobitis* Stankovitch, 1924 in *Cobitis taenia* (Osteichthys) liver.
228. *E. coecicola* Kheisin, 1947 (syn., *E. oryctolagi* Ray and Banik, 1965) in *Oryctolagus cuniculus* (Mammalia) posterior ileum and cecum.
229. *E. coelopeltis* (Galli-Valerio, 1926) Hoare, 1933 (syn., *Isospora coelopeltis* Galli-Valerio, 1926) in *Malpolon monspessulanus* (Reptilia) feces.
230. *E. colchici* Norton, 1967 (syn., *E. colchica* Gottschalk, 1972 *lapsus calami*) in *Phasianus colchicus* (Aves) cecum and small intestine.
231. *E. colini* Fisher and Kelley, 1977 in *Colinus virginianus* (Aves) feces.
232. *E. collanuli* Wacha and Christiansen, 1974 in *Diadophis punctatus arnyi* (Reptilia) bile ducts (?).
233. *E. columbae* Mitra and Das-Gupta, 1937 in *Columba livia intermedia* (Aves) small intestine and ceca.
234. *E. conevi* Glebezdin, 1969 in *Rhombomys opimus* (Mammalia) feces.
235. *E. confusa* Joseph, 1969 in *Sciurus carolinensis* and *S. niger* (Mammalia) small intestine villi.
236. *E. congolensis* Ricci-Bitti, Pampiglione, and Kabala, 1973 in *Kobus defassa* (Mammalia) feces.
237. *E. connochaetei* Levine and Ivens, 1970 (syn., *E. ellipsoidalis* Becker and Frye, 1929 of Prasad [1960] in *Connochaetes taurinus* (Mammalia) feces.
238. *E. coracias* Rakhmatullina and Svanbaev, 1971 in *Coracias garrulus* (Aves) feces.
239. *E. correptionis* Veisov, 1962 in *Microtus majori* (Mammalia) intestine.
240. *E. cotiae* Carini, 1935 in *Dasyprocta aguti* (Mammalia) feces.
241. *E. cotti* Gauthier, 1921 (syn., *E. votti* Gauthier, 1921 *lapsus calami*) in *Cottus gobio* and *C. cognatus* (Osteichthys) intestine and pyloric ceca.
242. *E. coturnicis* Chakravarty and Kar, 1947 in *Coturnix c. coturnix* (Aves) feces.
243. *E. coucangi* Patnaik and Acharjyo, 1970 in *Nycticebus coucang* (Mammalia) intestine.
244. *E. couesii* Kruidenier, Levine, and Ivens, 1960 in *Oryzomys c. couesi* (Mammalia) feces.
245. *E. coypi* Obitz and Wadowski, 1937 in *Myocastor coypus* (Mammalia) feces.
246. *E. crandallis* Honess, 1942 in *Ovis canadensis, O. aries, O. musimon, O. ammon,* and possibly *Capra hircus, C. sibirica,* and *C. aegagrus* (Mammalia) small intestine.
247. *E. crassa* Farr, 1963 in *Branta canadensis* (Aves) feces.
248. *E. creutzi* Creutz and Gottschalk, 1969 in *Larus ridibundus* (Aves) esophagus.
249. *E. criceti* Nöller, 1920 emend. Pellérdy, 1956 (syn., *E. falciformis* var. *criceti* Nöller, 1956) in *Cricetus cricetus* (Mammalia) cecum and colon.

250. *E. cricetomysi* Prasad, 1960 in *Cricetomys gambianus* (Mammalia) intestinal contents.
251. *E. cricetuli* Musaev and Veisov, 1961 in *Cricetulus migratorius* (Mammalia) intestinal contents.
252. *E. crocidurae* Galli-Valerio, 1933 in *Sorex araneus* (Mammalia) apparently intestine.
253. *E. crocodyli* Lainson, 1968 in *Crocodylus acutus* (Reptilia) feces.
254. *E. crotali* Phisalix, 1919 emend. Pellérdy, 1965 (syn., *E. crotalae* Phisalix, 1919) in *Bitis gabonica* and *Crotalus terrificus* (Reptilia) liver or bile ducts.
255. *E. crotalviridis* Duszynski et al., 1977 in *Crotalus v. viridis* (Reptilia) feces.
256. *E. crusti* Duszynski and Gutiérrez, 1981 in *Oreortyx pictus* (Aves) feces.
257. *E. cryptobarretti* Duszynski and Brunson, 1973 in *Ochotona princeps* (Mammalia) feces.
258. *E. cubinica* Musaev, Veisov, and Alieva, 1963 in *Microtus socialis* (Mammalia) feces (?).
259. *E. culteri* Lee and Chen, 1964 in *Culter erythropterus* (Osteichthys) intestine. (See Dyková and Lom, 1983.)
260. *E. cusarica* Musaev, Veisov, and Alieva, 1963 in *Microtus socialis* (Mammalia) feces ?).
261. *E. cyanophlyctis* Chakravarty and Kar, 1944 in *Rana cyanophlyctis* (Amphibia) intestine.
262. *E. cyclopis* Lainson and Shaw, 1982 emend. (syn., *E. cyclopesi* Lainson and Shaw, 1980) in *Cyclopes didactylus* (Mammalia) ileum.
263. *E. cylindrica* Wilson, 1931 in *Box taurus, B. indicus,* and *Bubalus bubalis* (Mammalia) feces.
264. *E. cylindrospora* Stankovitch, 1921 in *Alburnus alburnus* (Osteichthys) intestine.
265. *E. cynomysis* Andrews, 1928 (syn., *E. cynaomysi* Andrews, 1928 *lapsus calami*) in *Cynomys ludovicianus* (Mammalia) feces.
266. *E. cyprini* Plehn, 1924 in *Carassius carassius* and *Cyprinus carpio* (Osteichthys) intestine.
267. *E. cyprinorum* Stankovitch, 1921 in *Abramis brama, Barbus barbus, Rutilus rutilus, Phoxinus phoxinus,* and *Scardinius eryuthrophthalmus* (Osteichthys) intestinal epithelial cells.
268. *E. cystisfelleae* Debaisieux, 1914 in *Natrix natrix* (Reptilia) gall bladder and bile ducts.
269. *E. dacunhai* Levine, 1984 (syn., *E. tatusi* Carini, 1933) in *Cabassous* sp. (Mammalia) feces.
270. *E. dalli* Clark and Colwell, 1974 in *Ovis dalli* (Mammalia) feces.
271. *E. damirchinica* Musaev and Veisov, 1965 (syn., *E. caucasica* Musaev and Veisov, 1963) in *Allactaga elater* (Mammalia) large intestine contents.
272. *E. danailovae* Gräfner, Graubmann, and Betke, 1965 in *Anas p. platyrhynchos* and experimentally *Anser anser* (Aves) small intestine.
273. *E. danielle* Dida, 1970 in *Ovis aries* (Mammalia) location unknown.
274. *E. darjeelingensis* Sinha and Sinha, 1980 in *Suncus murinus soccatus* (Mammalia) intestinal epithelium and subepithelium.
275. *E. dasymysis* Levine et al., 1959 in *Dasymys incomptus rufulus* (Mammalia) jejunum.
276. *E. dathei* Tscherner, 1976 in *Kobus ellipsiprymnus* (Mammalia) feces.
277. *E. dauki* Bhatia and Pande, 1965 in *Amaurornis phoenicurus* (Aves) small intestine.
278. *E. daurica* Machul'skii, 1947 in *Ochotona daurica* (Mammalia) feces (?).
279. *E. davisi* Ivens, Kruidenier, and Levine, 1959 in *Neotoma albigula* (Mammalia) intestinal contents.
280. *E. dawari* Bhatia et al., 1973 in *Muntiacus muntijaka* (Mammalia) feces.
281. *E. debliecki* Douwes, 1921 (syns., *Coccidium suis* Jaeger, 1921; *E. brumpti* Cauchemez, 1921 in part; *E. jalina* Krediet, 1921; *E. scrofae* Galli-Valerio, 1935; *E. polita* Pellérdy, 1949 in part) in *Sus scrofa* (Mammalia) small intestine.

282. *E. delagei* (Labbé, 1893) Reichenow, 1921 (syn., *Coccidium delagei* Labbé, 1893) in *Emys orbicularis* (Reptilia) intestine.
283. *E. delicata* Levine and Ivens, 1960 in *Peromyscus maniculatus* (Mammalia) intestinal contents.
284. *E. dendrocopi* Levine, 1953 (syn., *E. lyruri* Galli-Valerio, 1927 in part) in *Dendrocopus major pinetorum* (Aves) intestine.
285. *E. dendrohyracis* van den Berghe and Chardome, 1953 in *Dendrohyrax arboreus adolphifriederici* (Mammalia) intestinal contents.
286. *E. depuytoraci* Černá, 1976 in *Sylvia curruca* (Aves) feces.
287. *E. derenica* Veisov, 1963 in *Microtus arvalis* (Mammalia) feces (?).
288. *E. dericksoni* Roudabush, 1937 in *Trionyx spinifera* (Reptilia) intestine.
289. *E. deserticola* Davronov, 1973 in *Spermophilus fulvus* (Mammalia) feces (?).
290. *E. dhamini* Mandal and Mukherjee, 1977 in *Ptyas mucosus* (Reptilia) feces.
291. *E. dicrostonicis* Levine, 1952 in *Dicrostonyx groenlandicus richardsoni, D. torquatus,* and *Lemmus sibiricus* (Mammalia) feces.
292. *E. didelphidis* Carini, 1936 emend. Pellérdy, 1974 (syn., *E. didelphydis* Carini, 1936) in *Didelphis aurita* (Mammalia) feces.
293. *E. dingleyi* Davies, 1978 in *Blennius pholis* (Osteichthys) epithelial cells throughout intestine.
294. *E. dipodomysis* Levine, Ivens, and Kruidenier, 1958 in *Dipodomys merriami, D. phillipsi,* and *D. ordi* (Mammalia) intestinal contents.
295. *E. disaensis* Musaev and Veisov, 1960 in *Meriones persicus* (Mammalia) intestinal contents.
296. *E. dispersa* Tyzzer, 1929 in *Colinus virginianus, Phasianus colchicus, Meleagris gallopavo, M. mexicana, Gallus gallus, and also possibly Coturnix coturnix, Pediocetes phasianus campestris,* and *Tetrastes bonasia* (Aves) small intestine.
297. *E. dissanaikei* Fernando and Remmler, 1973 in *Gallus lafayettei* (Aves) feces.
298. *E. dissimilis* Yakimoff and Gousseff, 1935 in *Sorex araneus* and *S. minutus* (Mammalia) feces (?).
299. *E. distorta* Saxe, 1955 in *Ambystoma tigrinum* (Amphibia) intestine.
300. *E. divichinica* Musaev and Veisov, 1963 in *Apodemus sylvaticus* (Mammalia) feces.
301. *E. dogeli* Musaev and Veisov, 1965 (syns., *E. zemphirica* Veisov, 1965 of Musaev in Pellérdy [1974]; *E. dogieli Veisov, 1964 of Pellérdy [1974]*) in *Meriones persicus* (Mammalia) feces (?).
302. *E. dolichotis* Morini, Boero, and Rodriguez, 1955 in *Dolichotis p. patagonum* (Mammalia) small intestine.
303. *E. dorcadis* Mantovani, 1966 in *Gazella dorcas* (Mammalia) feces.
304. *E. dorneyi* Levine and Ivens, 1965 in *Glaucomys sabrinus* (Mammalia) cecal contents.
305. *E. dorsalis* van den Berghe and Chardome, 1962 in *Dendrohyrax dorsalis emini* (Mammalia) feces.
306. *E. dromedarii* Yakimoff and Matschoulsky, 1939 (syns., *E. cameli* Nöller, 1939 of Iwanoff-Gobzem [1934] in part; *E. cameli* Iwanoff-Gobzem of Tsygankov [1950] in part) in *Camelus dromedarius* and *C. bactrianus* (Mammalia) feces.
307. *E. dubeyi* Pande et al., 1970 in *Gallus gallus* (Aves) feces.
308. *E. duculae* Varghese, 1980 in *Ducula spilorrhoa* (Aves) feces.
309. *E. dukei* Lavier, 1927 in *Tadarida pumilus* (Mammalia) intestine.
310. *E. duncani* Varghese, 1977 in *Halcyon sancta* (Aves) feces.
311. *E. dunsingi* Farr, 1960 (syn., *Eimeria* sp. Brada, 1966) in *Melopsittacus undulatus* (Aves) small intestine.
312. *E. duodenalis* Norton, 1967 in *Phasianus colchicus* and probably also *Chrysolophus pictus, Syrmaticus reevesi, Crossoptilon tibetanum,* and *Lophura nycthemera* (Aves) small intestine.

313. *E. dusii* Wheat and Ernst, 1974 in *Neotoma floridana* (Mammalia) feces.
314. *E. duszynskii* Conder, Oberndorfer, and Heckman, 1981 in *Cottus bairdi* (Osteichthys) intestinal mucosa.
315. *E. dyromidis* Zolotarev, 1935 in *Dyromys nitedula* (Mammalia) intestinal contents.
316. *E. dzhahriana* Musaev and Veisov, 1960 in *Meriones shawi* (syn., *M. tristrami*) (Mammalia) large intestine contents.
317. *E. dzhulfaensis* Musaev and Veisov, 1959 in *Microtus socialis* (Mammalia) intestinal contents.
318. *E. echidnae* Barker, Beveridge, and Munday, 1985 in *Tachyglossus aculeatus* (Mammalia) small intestine.
319. *E. edkysios* Triffitt, 1928 in *Tachypodoiulus niger* (Diplopodasida) intestine.
320. *E. edwardsi* Colley and Mullin, 1971 in *Rattus edwardsi* (Mammalia) feces.
321. *E. egerniae* Cannon, 1967 in *Egernia whitii* (Reptilia) gall bladder epithelium.
322. *E. egypti* Prasad, 1960 in *Meriones s. shawi* (Mammalia) feces.
323. *E. elaphi* Jansen and van Haaften, 1966 in *Cervus elaphus* (Mammalia) feces.
324. *E. elater* Musaev and Veisov, 1963 in *Allactaga elater* (Mammalia) large intestine contents.
325. *E. elegans* Yakimoff, Gousseff, and Rastegaieff, 1932 in *Gazella subgutturosa* (Mammalia) feces.
326. *E. elerybeckeri* Levine, 1984 (syns., *E. levinei* Krishnamurthy and Kshirsagar, 1980; *E. lavinei* Krishnamurthy and Kshirsagar, 1980 *lapsus calami*) in *Rattus rattus* (Mammalia) intestinal contents.
327. *E. ellipsoidalis* Becker and Frye, 1929 in *Bos taurus, B. indicus, probably Bubalus bubalis,* and perhaps *Bison bonasus, Bibos banteng,* and *Bos gaurus* (Mammalia) small intestine.
328. *E. elliptica* Sayin, Dincer, and Meric, 1977 in *Spalax leucodon* (Mammalia) intestinal contents.
329. *E. ellobii* Svanbaev, 1956 in *Ellobius talpinus* (Mammalia) feces.
330. *E. elongata* Marotel and Guilhon, 1941 (syn., *E. neoleporis* Carvalho, 1942 [?]) in *Oryctolagus cuniculus* and possibly *Sylvilagus* floridanus (Mammalia) presumably intestine.
331. *E. environ* Honess, 1939 in *Sylvilagus nuttallii grangeri, S. floridanus mearnsii,* and *S. audubonii* (Mammalia) small intestine.
332. *E. (?) epidermica* Léger and Duboscq, 1917 in *Glossobalanus minutus* (Hemichordata) proboscis epithelium.
333. *E. eremiasica* Davronov, 1985 in *Eremias velox* (Reptilia) feces.
334. *E. eremici* Levine, Ivens, and Kruidenier, 1957 in *Peromyscus eremicus* (Mammalia) intestinal contents.
335. *E. ernsti* Todd and O'Gara, 1968 in *Oreamnos americanus* (Mammalia) feces.
336. *E. erschovi* Machul'skii, 1949 in *Ochotona daurica* and *O. pricei* (Mammalia) intestinal contents.
337. *E. erythrourica* Musaev and Alieva, 1961 in *Meriones libyicus* (syn., *M. erythrourus*) (Mammalia) cecum and colon.
338. *E. (?) escomeli* (Rastegaieff, 1930) Pellérdy, 1974 (syn., [gen.?] *escomeli* Rastegaieff, 1930) in *Myrmecophaga tridactyla* (Mammalia) feces.
339. *E. esoci* Schulman and Zaika, 1962 in *Esox lucius* (Osteichthys) intestine or wall of urinary bladder and *Lepomis gibbosus* (Osteichthys) cecum, gall bladder, gills, heart, kidney, liver, muscle, spleen, and swim bladder. (See Li and Desser, 1985.)
340. *E. etheostomae* Molnar and Hanek, 1974 in *Etheostoma exile* and *E. nigrum* (Osteichthys) anterior intestine epithelium.
341. *E. etrumei* Dogel', 1948 in *Etrumeus micropus* (Osteichthys) testes.

342. *E. eumopos* Marinkelle, 1968 in *Molossus (Eumops) trumbulli* (Mammalia) small intestine.
343. *E. europaea* Pellérdy, 1956 (syn., *E. belorussica* Litvenkova, 1969) in *Lepus europaeus* and *L. granatensis* (Mammalia) feces.
344. *E. eutamiae* Levine, Ivens, and Kruidenier, 1957 in *Eutamias dorsalis* (Mammalia) intestinal contents.
345. *E. evaginata* Dogel', 1948 in *Sebastodes taczanowskii* and *Myoxocephalus stelleri* (Osteichthys) pyloric ceca.
346. *E. exigua* Yakimoff, 1934 in *Oryctolagus cuniculus* (Mammalia) feces.
347. *E. falciformis* (Eimer, 1870) Schneider, 1875 (syns., *Gregarina falciformis* Eimer, 1870; *Coccidium falciforme* [Eimer, 1870] Schuberg, 1896; *G. muris* Rivolta, 1878; *Pfeifferia schubergi* Labbé, 1896; *Pfeifferinella schubergi* [Labbé, 1896] Labbé, 1899; *Pfeifferi nelle schubergi* [Labbé, 1896] Labbé, 1899; *Pfeifferella schubergi* [Labbé, 1896] Labbé, 1899; *E. schubergi* [Labbé, 1896] Doflein, 1916; *E. pragensis* Černá and Sénaud, 1969; *E. contorta* Haberkorn, 1971 in part) (TYPE SPECIES) in *Mus musculus* and very dubiously *Apodemus sylvaticus* (Mammalia) cecum and colon.
348. *E. fanthami* Levine, 1953 in *Lagopus mutus rupestris* (Aves) feces.
349. *E. farrae* Hanson, Levine, and Ivens, 1957 emend. Pellérdy, 1974 (syn., *E. farri* Hanson, Levine, and Ivens, 1957) in *Anser albifrons frontalis* (Aves) feces.
350. *E. faurei* (Moussu and Marotel, 1902) Martin, 1909 (syns., *Coccidium faurei* Moussu and Marotel, 1902; *C. ovis* Jaeger, 1921; *E. aemula* Yakimoff, 1931 in part) in *Ovis aries, O. canadensis, O. ammon polii (?), O. a. sewerzewi (?), O. musimon, O. orientalis, Ammotragus lervia*, and also reported, probably erroneously, from *Rupicapra rupicapra, Capreolus capreolus, Dama dama*, and *Ovibos moschatus* (Mammalia) small intestine.
351. *E. fausti* Yakimoff and Matschoulsky, 1936 (syn., *E. cunnamullensis* Mykytowycz, 1964) in *Macropus giganteus* (Mammalia) feces.
352. *E. fernandoae* Molnar and Hanek, 1974 in *Catostomus commersoni* and *Hypentelium nigricans* (Osteichthys) anterior gut.
353. *E. ferrisi* Levine and Ivens, 1965 in *Mus musculus* (Mammalia) cecum and colon.
354. *E. ferruginea* Colley, 1970 in *Tupaia glis ferruginea* (Mammalia) intestinal contents.
355. *E. fibrilosa* Mandal, 1976 in *Enhydris enhydris* (Reptilia) intestine.
356. *E. filamentifera* Wacha and Christiansen, 1979 in *Chelydra serpentina* (Reptilia) intestine.
357. *E. firestonei* Bray, 1958 in *Crocidura schweitzeri* (Mammalia) ileum.
358. *E. fitzgeraldi* Todd and Tryon, 1970 in *Thomomys talpoides* (Mammalia) feces.
359. *E. flavescens* Marotel and Guilhon, 1941 (syns., *E. pellerdyi* Coudert, 1977; *E. hakei* Coudert, 1978; *E. irresidua* Kessel and Jankiewicz, 1931 of Francalanci and Manfredini [1970]) in *Oryctolagus cuniculus* (Mammalia) small intestine, cecum, and colon.
360. *E. flexilis* Golemansky, 1978 in *Talpa europaea* (Mammalia) intestinal contents.
361. *E. fluviatilis* Lewis and Ball, 1984 in *Myocastor coypus* (Mammalia) feces.
362. *E. forresteri* Upton, Ernst, Clubb & Current, 1984 in *Ramphastos toco* (Aves) feces.
363. *E. franchinii* Brunelli, 1935 in *Sciurus vulgaris* (Mammalia) feces.
364. *E. francolini* Swarup and Chauhan, 1976 in *Francolinus francolinus* (Aves) feces.
365. *E. franklinii* Hall and Knipling, 1935 in *Spermophilus frankli* (Mammalia) intestine.
366. *E. freemani* Molnar and Fernando, 1974 in *Notropis cornutus* (Osteichthys) kidneys.
367. *E. fulica* Machul'skii, 1941 in *Fulica a. atra* and perhaps also *Rallus aquaticus* and *Porzana porzana* (Aves) apparently intestine.
368. *E. fulva* Farr, 1953 in *Anser c. caerulescens, Branta canadensis*, and experimentally *Anser anser* (Aves) intestine.
369. *E. furonis* Hoare, 1927 in *Mustela putorius furo* and *M. vison* (Mammalia) intestine.

370. *E. galago* Poelma, 1966 in *Galago senegalensis* (Mammalia) feces.
371. *E. galateai* Varghese, 1977 in *Tanysiptera galatea* (Aves) feces.
372. *E. gallatii* Straneva and Kelley, 1979 in *Clethrionomys gapperi* (Mammalia) feces.
373. *E. gallinagoi* Mandal, 1965 in *Gallinago gallinago* (Aves) intestine.
374. *E. gallivalerioi* Rastegaieff, 1930 (syns., *E. galli-valeriano* Rastegaieff, 1930 *lapsus calami; E. gallivalerici* Rastegaieff, 1930 *lapsus calami*) in *Cervus elaphus* (Mammalia) feces.
375. *E. gallopavonis* Hawkins, 1952 in *Meleagris g. gallopavo* (Aves) ileum, rectum, and to a lesser extent cecum.
376. *E. gambai* Carini, 1938 in *Didelphis aurita* (Mammalia) small intestine.
377. *E. gandobica* Musaev and Veisov, 1965 in *Apodemus agrarius* (Mammalia) feces.
378. *E. garnhami* McMillan, 1958 in *Xerus (Euxerus) erythropus* (Mammalia) small intestine.
379. *E. garridoi* Ryšavý, 1967 in *Capromys pilorides* (Mammalia) intestine.
380. *E. garzettae* Golemansky and Kuldjieva, 1980 in *Egretta garzetta* (Aves) small intestine.
381. *E. gasterostei* (Thélohan, 1890) Doflein, 1909 (syn., *Coccidium gasterostei* Thélohan, 1890) in *Gasterosteus.aculeatus* and *G. clupeatus* (Osteichthys) liver.
382. *E. gaviae* Montgomery, Novilla, and Shillinger, 1978 in *Gavia immer* (Aves) kidneys.
383. *E. gazella* Musaev, 1970 (syn., *E. ninakohlyakimovae* Yakimoff and Rastegaieff, 1930 of Svanbaev [1958] in *Gazella subgutturosa*) in *Gazella subgutturosa* (Mammalia) feces.
384. *E. gehyrae* Cannon, 1967 in *Gehyra variegata* (Reptilia) gall bladder.
385. *E. gekkonis* Tanabe, 1928 in *Gekko japonicus* (Reptilia) intestine.
386. *E. genettae* Agostinucci and Bronzini, 1953 in *Genetta dongolona* (Mammalia) intestinal contents.
387. *E. gennaeuscus* Ray and Hiregoudar, 1959 in *Gennaeuscus horsfieldi* (sp.?) (Aves) feces.
388. *E. geomydis* Skidmore, 1929 in *Geomys bursarius* (Mammalia) small and large intestines.
389. *E. (?) gigantea* (Labbé, 1896) Reichenow, 1921 emend. Levine, 1985 (syns., *Pfeifferia* sp. Labbé, 1894 in part; *Coccidium [?] giganteum* [Labbé, 1896] Labbé, 1896; *E. gigantea* [Labbé, 1896] Reichenow, 1921) in *Lamna cornubica* (Chondrichthys) spiral intestine.
390. *E. gilruthi* (Chatton, 1910) Reichenow and Carini, 1937 (syns., *Gastrocystis gilruthi* Chatton, 1910; *Globidium gilruthi* [Chatton, 1910] Nöller, 1920) in *Ovis aries, Capra hircus, C. sibirica,* and *Pseudois nahoor* (Mammalia) abomasum.
391. *E. glauceae* Wheat and Ernst, 1974 in *Neotoma floridana* (Mammalia) feces.
392. *E. glaucomydis* Roudabush, 1937 in *Glaucomys volans* (Mammalia) intestine.
393. *E. glenorensis* Molnar and Fernando, 1974 in *Morone americana* (Osteichthys) gut.
394. *E. gliris* Musaev and Veisov, 1961 in *Myoxus glis* (Mammalia) large intestine contents.
395. *E. glossogobii* Mukherjee and Haldar, 1980 in *Glossogobius giuris* (Osteichthys) intestinal contents.
396. *E. gobii* Fantham, 1932 in *Gobius nudiceps* (Osteichthys) intestine.
397. *E. gokaki* Rao and Bhatavdekar, 1959 (syn., *E. brasiliensis* Torres and Ramos, 1939 of Patnaik [1965]) in *Bubalus bubalis* (Mammalia) feces.
398. *E. golemanskii* Levine, 1985 (syn., *Eimeria* sp. Golemanski and Yankova, 1973) in *Apodemus sylvaticus* and *A. flavicollis* (Mammalia) feces.
399. *E. gomarchaica* Veisov, 1963 in *Microtus arvalis* (Mammalia) intestine.
400. *E. gomurica* Musaev and Veisov, 1963 in *Apodemus sylvaticus* (Mammalia) feces.
401. *E. gonzalezi* Bazalar and Guerrero, 1970 (syn., *Eimeria* sp. Patyk, 1965) in *Ovis aries* (Mammalia) feces.
402. *E. gonzalezcastroi* Lizcano and Romero, 1975 (syn., *E. gonzalezi* Lizcano and Romero, 1972) in *Alectoris rufa* (Aves) large intestine and posterior small intestine.

403. *E. gorakhpuri* Bhatia and Pande, 1967 in *Numida meleagris* (Aves) feces.
404. *E. gorgonis* Prasad, 1960 in *Connochaetes taurinus* (Mammalia) feces.
405. *E. gouriae* Varghese, 1980 emend. (syn., *E. gourai* Varghese, 1980) in *Gouria victoria* (Aves) feces.
406. *E. gousseffi* Glebezdin, 1978 in *Nesokia indica* (Mammalia) feces.
407. *E. goussevi* Yakimoff, 1935 in *Talpa europaea* (Mammalia) feces.
408. *E. granulosa* Christensen, 1938 in *Ovis aries, O. canadensis*, possibly *Capra hircus*, and dubiously *Ovibos moschatus* (Mammalia) feces.
409. *E. graptemydos* Wacha and Christiansen, 1976 in *Graptemys geographica* and also *G. pseudogeographica* and *Chrysemys picta belli* (Reptilia) intestine.
410. *E. grenieri* Yvoré and Aycardi, 1967 in *Numida meleagris* (Aves) large and small intestines.
411. *E. grobbeni* Rudovsky, 1925 in *Salamandra atra* (Amphibia) intestine.
412. *E. groenlandica* Madsen, 1938 emend. Levine and Ivens, 1972 (syn., *E. perforans* var. *groenlandica* Madsen, 1938 in part) in *Lepus arcticus groenlandicus* (Mammalia) intestine.
413. *E. gruis* Yakimoff and Matschoulsky, 1935 (syn., *E. kazanskii* Zolotarev, 1938) in *Anthropoides virgo, Grus canadensis*, and *G. americana* (Aves) feces.
414. *E. guentherii* Golemansky, 1978 in *Microtus guentheri* (Mammalia) intestinal contents.
415. *E. guevarai* Romero and Lizacano, 1971 in *Sus scrofa* (Mammalia) feces.
416. *E. gumbaschica* Musaev and Veisov, 1963 in *Apodemus sylvaticus* (Mammalia) feces.
417. *E. gundi* Mishra and Gonzalez, 1978 in *Ctenodactylus gundi* (Mammalia) ileum mucosa.
418. *E. gungahlinensis* Mykytowycz, 1964 in *Macropus giganteus* (Mammalia) feces.
419. *E. gupti* Bhatia, 1938 (syn., *E. cylindrica* Ray and Das Gupta, 1936) in *Natrix piscator* (Reptilia) rectum.
420. *E. haberfeldi* Carini, 1937 in *Caluromys philander* (Mammalia) posterior small intestine.
421. *E. hadrutica* Musaev, Veisov, and Alieva, 1963 in *Microtus socialis* (Mammalia) feces.
422. *E. haematodi* Varghese, 1977 in *Trichoglossus haematodus* (Aves) feces.
423. *E. hagani* P. P. Levine, 1938 in *Gallus gallus* (Aves) duodenum and jejunum.
424. *E. hagenmuelleri* (Léger, 1898) Levine and Becker, 1933 (syn., *Coccidium hagenmuelleri* Léger, 1898) in *Stigmatogaster gracilis* (Diplopodasida) intestine.
425. *E. haichengensis* Chen, 1962 in *Cyprinus carpio* (Osteichthys) intestine. (See Dyková and Lom, 1983.)
426. *E. hammondi* Dubey and Pande, 1963 in *Felis (Chaus) chaus* (Mammalia) feces.
427. *E. haneki* Molnar and Fernando, 1974 in *Culaea inconstans* (Osteichthys) gut.
428. *E. hansonorum* Levine and Ivens, 1965 in *Mus musculus* (Mammalia) intestinal contents.
429. *E. haranica* Sayin, 1981 in *Spalax ehrenbergi* (Mammalia) mixed intestinal contents.
430. *E. harbelensis* Levine et al., 1959 in *Lophuromys s. sikapusi* (Mammalia) intestinal contents.
431. *E. harpodoni* Setna and Bana, 1935 in *Harpodon nehereus* (Osteichthys) alimentary canal.
432. *E. (?) hartmanni* Rastegaieff, 1930 in *Leo tigris* (Mammalia) feces.
433. *E. hasei* Yakimoff and Gousseff, 1936 in *Rattus rattus* (Mammalia) intestine.
434. *E. hegneri* Rastegaieff, 1930 in *Cervus canadensis* (Mammalia) feces.
435. *E. heissini* Svanbaev, 1956 in *Vulpes corsac* (Mammalia) intestine.
436. *E. helenlevineae* n. nom. Bray (syn., *E. helenae* Bray, 1984) in *Hemidactylus brookei* (Reptilia) feces.
437. *E. helmisophis* Wacha and Christiansen, 1974 in *Carphophis amoenus vermis* (Reptilia) feces.
438. *E. hemiculterii* Chen and Hsieh, 1964 in *Hemiculter leucisculus* (Osteichthys) intestine.

439. *E. hemidactyli* Knowles and Das Gupta, 1935 in *Hemidactylus flaviviridis* (Reptilia) intestine.
440. *E. hermani* Farr, 1953 in *Branta canadensis, Anser c. coerulescens,* and experimentally *A. anser* (Aves) intestine.
441. *E. hessei* Lavier, 1924 in *Rhinolophus hipposideros* (Mammalia) small intestine.
442. *E. hestermani* Mykytowycz, 1964 in *Macropus giganteus* (Mammalia) feces.
442a. *E. heterocapita* Duszynski, 1985 in *Neurotrichus gibbsii* (Mammalia) feces.
443. *E. heterocephali* Levine and Ivens, 1965 (syn., *E. muris* Galli-Valerio, 1932 of Porter [1957]) in *Heterocephalus glaber* (Mammalia) cecum.
444. *E. hexagona* Lom and Dyková, 1981 in *Onos tricirratus* (Osteichthys) pyloric ceca and middle intestine.
445. *E. hiepei* Gräfner, Graubmann, and Dobbriner, 1967 in *Mustela lutreola vison* (Mammalia) bile ducts.
446. *E. himalayani* Ray and Misra, 1942 emend. (syns., *E. himalayanum* Ray and Misra, 1942; *E. himalayana* Ray and Misra, 1942 emend. Pellérdy, 1974) in *Bufo himalayanus* (Amphibia) intestine.
447. *E. hindlei* Yakimoff and Gousseff, 1938 in *Mus musculus* (Mammalia) feces.
448. *E. hippuri* Colley and Mullin, 1971 in *Sundasciurus hippurus* (Mammalia) intestinal contents.
449. *E. hirci* Chevalier, 1966 in *Capra hircus* (Mammalia) feces.
450. *E. (?) hirsuta* Schneider, 1886 in gyrinid beetle larvae (Coleopterorida) intestine.
451. *E. hispidi* Bastardo de San Jose, 1974 in *Tropidurus hispidus* (Reptilia) epithelial and submucosal cells of small intestine.
452. *E. hoffmani* Molnar and Hanek, 1974 in *Umbra limi* (Osteichthys) anterior gut.
453. *E. hoffmeisteri* Levine, Ivens, and Kruidenier, 1958 in *Spermophilus s. spilosoma* (Mammalia) intestinal contents.
454. *E. holmesi* Samoil and Samuel, 1977 in *Lepus americanus* (Mammalia) feces.
455. *E. honessi* Carvalho, 1943 emend. Levine and Ivens, 1972 (syn., *E. media* form *honessi* Carvalho, 1943) in *Sylvilagus floridanus mearnsii, S. audubonii,* and *S. nuttallii grangeri* (Mammalia) feces.
456. *E. hudsonii* Duszynski, Eastham, and Yates, 1982 in *Zapus hudsonius luteus* (Mammalia) feces.
457. *E. hungarica* Pellérdy, 1956 (syns., *E. exigua* Yakimoff, 1934 of Pellérdy [1954] in part; *E. exigua* var. *septentrionalis* Madsen, 1938 in part; *E. minima* Yakimoff, 1934 of Gill and Ray [1960]; *E. orbiculata* Lucas, Laroche, and Durand, 1959) in *Lepus europaeus* and *L. ruficaudatus* (Mammalia) probably small intestine.
458. *E. hungaryensis* Levine and Ivens, 1965 (syn., *E. muris* Galli-Valerio, 1932 of Pellérdy [1954] and Černá [1962]) in *Apodemus flavicollis, A. sylvaticus,* and *A. agrarius* (Mammalia) small intestine.
459. *E. hupehensis* Chen and Hsieh, 1964 in *Carassius auratus* (Osteichthys) intestine.
460. *E. hybognathi* Molnar and Fernando, 1974 in *Hybognathus hankinsoni* (Osteichthys) gut.
461. *E. hydrobatidis* Gottschalk, 1969 in *Hydrobates pelagicus* (Aves) feces.
462. *E. hydrochaeri* Carini, 1937 emend. Levine and Ivens, 1987 (syn. *E. hydrochoeri* Carini, 1937) in *Hydrochoerus hydrochoerus* (Mammalia) feces.
463. *E. hydrophis* Wacha and Christiansen, 1974 in *Natrix s. sipedon* and *N. r. rhombifera* (Reptilia) feces.
464. *E. hylopetis* Colley and Mullin, 1971 in *Hylopetes spadiceus* (Mammalia) intestinal contents.
465. *E. hypophthalmichthys* Akhmerov in Dogel' and Akhmerov, 1959 in *Hypophthalmichthys molitrix* (Osteichthys) kidneys.
466. *E. ibicis* Colombo, 1958 in *Capra i. ibex* (Mammalia) feces.

467. *E. ibrahimovae* Musaev, 1970 (syns., *E. scabra* Henry, 1931 of Svanbaev [1958] in *Sus scrofa*; *E. ibragimovae* Musaev, 1970 of Svanbaev [1979]) in *Sus scrofa* (Mammalia) feces.

468. *E. ictaluri* Molnar and Fernando, 1974 in *Ictalurus nebulosus* (Osteichthys) gut.

469. *E. ictidea* Hoare, 1927 in *Mustela putorius* (Mammalia) feces.

470. *E. illinoisensis* Levine and Ivens, 1967 in *Bos taurus* (Mammalia) feces.

471. *E. imantauica* Nukerbaeva and Svanbaev, 1973 in *Alopex lagopus* (Mammalia) intestine.

472. *E. immodulata* Musaev and Veisov, 1961 in *Cricetulus migratorius* (Mammalia) large intestine contents.

473. *E. impalae* Prasad and Narayan, 1963 in *Aepyceros melampus* (Mammalia) small intestine.

474. *E. indianensis* Joseph, 1974 in *Didelphis virginiana* (Mammalia) feces.

475. *E. innocua* Moore and Brown, 1952 in *Meleagris gallopavo* (Aves) small intestine.

476. *E. innominata* Kar, 1944 in *Lissemys punctata* (Reptilia) liver.

477. *E. insignis* Lom and Dyková, 1982 in *Scorpaena notata* (Osteichthys) pyloric ceca.

478. *E. intermedia* Ruiz, 1959 in *Anolis intermedius* (Reptilia) intestine and feces.

479. *E. intestinalis* Kheisin, 1948 (syns., *E. piriformis* Gvelisiani and Nadiradze, 1945; *E. piriformis* Kheisin, 1948 of Pellérdy [1953]; *E. agnosta* Pellérdy, 1954) in *Oryctolagus cuniculus* (Mammalia) lower ileum.

480. *E. intricata* Spiegl, 1925 in *Ovis aries, O. canadensis, O. musimon*, possibly *Capra hircus*, and perhaps *Capreolus capreolus* and *Dama dama* (Mammalia) small and large intestine.

481. *E. iowaensis* Wacha and Christiansen, 1974 in *Thamnophis sirtalis parietalis* (Reptilia) feces.

482. *E. iradiensis* Veisov, 1963 in *Microtus arvalis* (Mammalia) feces.

483. *E. iraqiensis* Mirza, 1970 in *Camelus dromedarius* (Mammalia) feces.

484. *E. irara* Carini and da Fonseca, 1938 in *Eira barbara* (Mammalia) feces.

485. *E. irregularis* Kar, 1944 in *Lissemys punctata* (Reptilia) intestine.

486. *E. irresidua* Kessel and Jankiewicz, 1931 in *Oryctolagus cuniculus* and experimentally *Sylvilagus floridanus mearnsii* (Mammalia) small intestine.

487. *E. ismailovi* Musaev, 1970 (syn., *E. ninakohlyakimovae* Yakimoff and Rastegaieff, 1930 of Svanbaev [1958] in *Saiga tatarica*) in *Saiga tatarica* (Mammalia) feces.

488. *E. iturina* Pampiglione, Ricci-Bitti, and Kabala, 1973 in *Cephalophus monticola* (Mammalia) feces.

489. *E. ivanae* Lom and Dyková, 1981 in *Serranus cabrilla* (Osteichthys) pyloric ceca.

490. *E. ivensae* Todd and O'Gara, 1970 in *Odocoileus h. hemionus* (Mammalia) feces.

491. *E. iwanoffi* Veisov, 1963 (syn., *E. ivanovi* Vejsov, 1963 of Musaev and Veisov [1965] in *Microtus arvalis* (Mammalia) feces.

492. *E. jaboti* Carini, 1942 in *Testudo tabulata* (Reptilia) feces.

493. *E. jaegeri* Carini, 1933 in *Liophis jaegeri* (Reptilia) intestine.

494. *E. jakunini* Dzerzhinskii, 1981 in *Sicista tianschanica* (Mammalia) feces.

495. *E. jalpaiguriensis* Bandyopadhyay, 1982 in *Herpestes edwardsi* (Mammalia) feces.

496. *E. japaluris* Bovee, 1971 in *Japalura polygonata* (Reptilia) gall bladder and bile duct epithelial cells.

497. *E. japonicus* Bovee, 1971 in *Gekko japonicus* (Reptilia) gall bladder and bile duct epithelial cells.

498. *E. jardimlinica* Musaev and Veisov, 1961 in *Cricetulus migratorius* (Mammalia) large intestine contents.

499. *E. jerfinica* Musaev and Veisov, 1963 in *Apodemus sylvaticus* (Mammalia) feces.

500. *E. jersenica* Davronov, 1973 in *Meriones libycus* (syn., *M. erythrourus*) (Mammalia) feces (?).

501. *E. jirovecki* Ryšavý, 1967 in *Capromys pilorides* (Mammalia) feces.
502. *E. jolchijevi* Musaev, 1970 (syn., *E. granulosa* Christensen, 1938 in *Capra hircus*) in *Capra hircus* (Mammalia) feces.
503. *E. joyeuxi* Yakimoff and Gousseff, 1936 in *Allactaga major* (Mammalia) feces.
504. *E. juniataensis* Pluto and Rothenbacher, 1976 in *Graptemys geographica* (Reptilia) feces.
505. *E. jurschaensis* Veisov, 1961 in *Meriones vinogradovi* (Mammalia) large intestine contents.
506. *E. kanchili* Mullin and Colley, 1971 in *Tragulus javanicus* (Mammalia) intestinal contents.
507. *E. kapotei* Chatterjee and Ray, 1969 in *Columba livia intermedia* (Aves) posterior intestine.
508. *E. karatauica* Utebaeva, 1973 in *Alectoris kakelik* (Aves) feces.
509. *E. karschinica* Davronov, 1973 in *Meriones meridianus* (Mammalia) feces.
510. *E. Kassaii* Molnar, 1978 in *Umbra krameri* Osteichthys) intestinal epithelium.
511. *E. katangensis* Ricci-Bitti, Pampiglione, and Kabla, 1973 in *Kobus defassa* (Mammalia) feces.
512. *E. kaunensis* Arnastauskene, Kazlauskas, and Mal'dzhyunaite, 1978 in *Apodemus agrarius* (Mammalia) feces.
513. *E. kazakhstanensis* Levine and Ivens, 1965 (syn., *E. volgensis* Sassuchin and Rauschenbach, 1932 of Svanbaev [1956]) in *Ellobius talpinus* (Mammalia) feces.
514. *E. keilini* Yakimoff and Gousseff, 1938 in *Mus musculus* (Mammalia) feces.
515. *E. keithi* Samoil and Samuels, 1977 in *Lepus americanus* (Mammalia) feces.
516. *E. kermorganti* (Simond, 1901) Braun, 1908 (syn., *Coccidium kermorganti* Simond, 1901) in *Gavialis gangeticus* (Reptilia) spleen.
517. *E. kingi* Saxe, 1955 in *Ambystoma tigrinum* (Amphibia) feces.
518. *E. kinsellai* Barnard, Ernst, and Roper, 1971 in *Oryzomys palustris* (Mammalia) feces.
519. *E. klondikensis* Hobbs and Samuel, 1974 in *Ochotona collaris, O. hyperborea yesomensis,* and *O. princeps* (Mammalia) feces.
520. *E. kniplingi* Levine and Ivens, 1965 in *Sciurus niger rufiventer* (Mammalia) cecum and large intestine.
521. *E. knowlesi* Bhatia, 1936 (syn., *Eimeria* "species A" Knowles and Das Gupta, 1935) in *Hemidactylus flaviviridis* (Reptilia) gut.
522. *E. kobi* Ricci-Bitti, Pampiglione, and Kabala, 1973 in *Kobus defassa* (Mammalia) feces.
523. *E. kocharii* Musaev, 1970 (syns., *E. intricata* Spiegl, 1925 in *Capra hircus; E. nazijrovi* Musaev, 1970) in *C. hircus, C. ibex,* and *C. sibirica* (Mammalia) feces.
524. *E. kofoidi* Yakimoff and Matikaschwili, 1936 in *Alectoris graeca chukar* and *A. rufa* (Aves) posterior small intestine and large intestine.
525. *E. koganae* Svanbaev and Rakhmatullina, 1967 in *Anas acuta, A. crecca, A. platyrhynchos, A. querquedula, A. strepera,* and possibly *Netta rufina* and *Aythya ferina* (Aves) feces.
526. *E. kogoni* Mykytowycz, 1964 in *Macropus giganteus* (Mammalia) feces.
527. *E. koidzumii* Matubayasi, 1941 in *Gekko japonicus* (Reptilia) intestine.
528. *E. kolabski* Musaev and Veisov, 1965 in *Microtus socialis* (Mammalia) feces.
529. *E. kolanica* Veisov, 1963 in *Microtus arvalis* (Mammalia) feces.
530. *E. komareki* Černá and Daniel, 1956 in *Sorex araneus, S. arcticus,* and *S. minutus* (Mammalia) small intestine.
531. *E. koormae* Das Gupta, 1938 in *Lissemys punctata* (Reptilia) small intestine.
532. *E. korros* Van Peenen, Ryan, and McIntyre, 1967 in *Ptyas korros* (Reptilia) gall bladder.

533. *E. kostencovi* Davronov, 1973 in *Meriones meridianus* (Mammalia) feces.
534. *E. kosti* Elibihari and Hussein, 1974 in "cow" (Mammalia) abomasum.
535. *E. kotlani* Gräfner and Graubmann, 1964 in *Anser anser* (Aves) large intestine.
536. *E. kotuji* Arnastauskene, 1980 in *Microtus middendorfi* (Mammalia) feces.
537. *E. krijgsmanni* Yakimoff and Gousseff, 1938 (syn., *E. kriygsmanni* Yakimoff and Gousseff, 1938 from *Mus musculus* of Svanbaev [1956] and Veisov [1973] *lapsus calami*) in *M. musculus* (Mammalia) lower small intestine.
538. *E. krilovi* Musaev and Veisov, 1965 in *Meriones persicus* (Mammalia) feces.
539. *E. kruidenieri* Levine et al., 1959 in *Lophuromys s. sikapusi* (Mammalia) jejunum.
540. *E. krylovi* Svanbaev and Rakhmatullina, 1967 (syns., *E. kustanaica* Rakhmatullina and Svanbaev, 1969; *E. kustanaica* [Svanbaev and Rakhmatullina, 1967] Rakhmatullina and Svanbaev, 1969) in *Anas clypeata, A. crecca, A. penelope, A. querquedula, A. strepera,* and also possibly *Aythya ferina* (Aves) feces.
541. *E. kwangtungensis* Chen, 1960 in *Channa argus* and *C. maculata* (Osteichthys) intestine.
542. *E. labbeana* Pinto, 1928 (syns., *Coccidium pfeifferi* Labbé, 1896; *E. pfeifferi* [Labbé, 1896] von Wasielewski, 1904; *E. columbarum* Nieschulz, 1935) in *Columba livia, C. aenas, C. palumbus, Streptopelia orientalis meena, S. decaocto, Columbigallina p. passerina* (Aves) small intestine.
543. *E. (?) labbei* Hardcastle, 1943 emend. Levine, 1985 (syns., *Pfeifferia tritonis* Labbé, 1896; *Pfeifferella tritonis* [Labbé, 1896] Labbé, 1899; *E. tritonis* [Labbé, 1896] Levine and Becker, 1933; *E. labbei* Hardcastle, 1943) in *Triturus cristatus* (Amphibia) intestine.
543a. *E. lachrymalis* Reduker, Hertel, and Duszynski, 1985 in *Peromyscus e. eremicus* (Mammalia) feces.
544. *E. ladronensis* Reduker and Duszynski, 1985 in *Neotoma albigula* (Mammalia) feces.
545. *E. lafayettei* Fernando and Remmler, 1973 in *Gallus lafayettei* (Aves) feces.
546. *E. lagopodi* Galli-Valerio, 1929 in *Lagopus mutus* (Aves) feces (?).
547. *E. lairdi* Lom and Dyková, 1981 in *Myoxocephalus scorpius* (Osteichthys) pyloric ceca.
548. *E. lalahanensis* Sayin, Dincer, and Meric, 1977 in *Spalax leucodon* (Mammalia) intestinal contents.
549. *E. lamae* Guerrero, 1967 in *Lama pacos* (Mammalia) feces.
550. *E. laminata* Ray, 1935 in *Bufo melanosticus* (Amphibia) intestine.
551. *E. lampropeltis* Anderson, Duszynski, and Marquardt, 1968 in *Lampropeltis c. calligaster* (Reptilia) feces.
552. *E. lampropholidus* Cannon, 1967 in *Lampropholis guichenoti* (Reptilia) middle and upper small intestine.
553. *E. lancasterensis* Joseph, 1969 in *Sciurus carolinensis* and *S. niger rufiventer* (Mammalia) small intestine villi.
554. *E. landersi* Colley, 1971 in *Trichys lipura* (Mammalia) feces.
555. *E. langebarteli* Ivens, Kruidenier, and Levine, 1959 in *Peromyscus boylii, P. leucopus tormillo,* and *P. t. truei* (Mammalia) intestinal contents.
556. *E. langeroni* Yakimoff and Matschoulsky, 1937 in *Phasianus colchicus gordius* and *P. c. turkestanicus* (Aves) feces (?).
557. *E. lari* Schwalbach, 1959 in *Larus argentatus* and *L. ridibundus* (Aves) small intestine.
558. *E. larimerensis* Vetterling, 1964 in *Cynomys l. ludovicianus, C. leucurus, Spermophilus armatus, S. variegatus, S. tridecemlineatus, S. lateralis,* and *S. beecheyi* (Mammalia) jejunum and ileum.
559. *E. lateralis* Levine, Ivens, and Kruidenier, 1957 in *Spermophilus lateralis, S. columbianus,* and *S. richardsoni* (Mammalia) intestinal contents.
560. *E. lavieri* Yakimoff and Gousseff, 1936 in *Allactaga major* (Mammalia) intestinal contents.

561. *E. legeri* (Simond, 1901) Stankovitch, 1920 (syn., *Coccidium legeri* Simond, 1901) in *Emyda granosa* (Reptilia) liver.
562. *E. leimadophi* Lainson and Shaw, 1973 in *Leimadophis poecilogyrus* (Reptilia) small intestine.
563. *E. leiolopismatis* Cannon, 1967 in *Leiolopisma challengeri* (Reptilia) small intestine.
564. *E. lemniscomysis* Levine et al., 1959 in *Lemniscomys s. striatus* (Mammalia) duodenum and jejunum.
565. *E. lemuris* Poelma, 1966 in *Galago senegalensis* (Mammalia) jejunum and colon.
566. *E. leninogorica* Utebaeva, 1973 in *Tetrastes bonasis* (Aves) feces.
567. *E. leporis* Nieschulz, 1923 (syn., *E. leporina* Nieschulz [*sic*] of Murai, Meszaris, and Sugar [1978]) in *Lepus europaeus, L. tolai, L. capensis, L. timidus, L. americanus,* and *L. ruficaudatus* (Mammalia) small intestine.
568. *E. leptodactyli* Carini, 1931 in *Leptodactylus ocellatus* (Amphibia) feces.
569. *E. lerikaensis* Musaev and Veisov, 1960 in *Meriones persicus* (Mammalia) large intestine contents.
569a. *E. lettyae* Ruff, 1985 in *Colinus virginianus* (Aves) duodenum and also ileum and ceca.
570. *E. leucisci* Schulman and Zaika, 1964 in *Leuciscus leuciscus baicalensis, Abramis brama, Alburnus alburnus, Rutilus rutilus, Scardinus erythropthalmus,* and *Blicca bjoerkna* (Osteichthys) kidneys (and also gall bladder [?]).
571. *E. leuckarti* (Flesch, 1883) Reichenow, 1940 (syn., *Globidium leuckarti* Flesch, 1883) in *Equus caballus* and *E. asinus* (Mammalia) small intestine.
572. *E. leucodonica* Veisov, 1975 in *Spalax leucodon* (Mammalia) feces.
573. *E. leucodontis* Musaev and Veisov, 1961 in *Crocidura leucodon* and *C. suavolens* (Mammalia) feces.
574. *E. leucopi* Von Zellen, 1961 in *Peromyscus leucopus* (Mammalia) small intestine.
575. *E. leucuri* Haskins, 1969 (syn., *E. leucuri* Stabler et al., 1979) in *Lagopus leucurus* (Aves) feces.
576. *E. levinei* Bray, 1958 in *Tadarida bemmelini* (Mammalia) ileum.
577. *E. li* Golemansky, 1975 in *Vulpes vulpes* (Mammalia) large intestine contents.
578. *E. liberiensis* Levine et al., 1959 in *Lophuromys s. sikapusi* (Mammalia) intestinal contents.
579. *E. liomysis* Levine, Ivens, and Kruidenier, 1958 in *Liomys pictus* and *L. irroratus* (Mammalia) intestinal contents.
580. *E. liophi* Lainson and Shaw, 1973 in *Liophis cobella* (Reptilia) feces.
581. *E. lipura* Colley, 1971 in *Trichys lipura* (Mammalia) feces.
582. *E. lomarii* Dubey, 1963 in *Vulpes bengalensis* (Mammalia) feces.
583. *E. longaspora* Barrow and Hoy, 1960 in *Notophthalmus viridescens* (Amphibia) feces.
584. *E. lophortygis* Liburd and Mahrt, 1970 in *Lophortyx californicus* and also *Oreortyx pictus* (Aves) feces.
585. *E. lophurae* Chauhan, Paliwal, and Swarup, 1976 in *Lophura leucomelana* (Aves) feces.
586. *E. lophuromysis* Levine et al., 1959 in *Lophuromys s. sikapusi* (Mammalia) intestinal contents.
587. *E. lucknowensis* Misra, 1947 in *Motacilla alba* (Aves) small intestine.
588. *E. ludoviciani* Vetterling, 1964 in *Cynomys l. ludovicianus* and *C. leucurus* (Mammalia) intestinal contents.
589. *E. luisieri* (Galli-Valerio, 1935) Reichenow, 1953 (syns., *Jarrina luisieri* Galli-Valerio, 1935; *E. [J.] luisieri* [Galli-Valerio] Reichenow, 1953) in *Sciurus vulgaris* var. *alpina* (Mammalia) intestinal contents.
590. *E. luteola* Arnastauskene, 1980 in *Microtus middendorfi* (Mammalia) feces.

591. *E. lutescensis* Musaev and Veisov, 1963 emend. (syn., *E. lutescenae* Musaev and Veisov, 1963) in *Ellobius lutescens* (Mammalia) intestinal contents.
592. *E. lutotestudinis* Wacha and Christiansen, 1976 in *Kinosternon flavescens spooneri* (Reptilia) intestinal contents.
593. *E. lyncis* Anpilogova and Sokov, 1973 in *Lynx lynx isabellina* (Mammalia) feces.
594. *E. lyruri* Galli-Valerio, 1927 in *Lyrurus t. tetrix* and possibly *Tetrao urogallus* (Aves) intestinal contents.
595. *E. mabouia* Carini, 1938 in *Mabouia mabouia* (Reptilia) small intestine.
596. *E. macieli* Yakimoff and Matchulski, 1938 in *Kobus ellipsiprymnus* (Mammalia) feces.
597. *E. macropodis* Wenyon and Scott, 1925 in *Wallabia rufogrisea frutica* (Mammalia) small intestine.
598. *E. macrosculpta* Sugar, 1979 in *Lepus europaeus* (Mammalia) large intestine.
599. *E. macrotis* Mayberry et al., 1980 in *Vulpes macrotis neomexicanus* (Mammalia) feces.
600. *E. macusaniensis* Guerrero et al., 1971 (syn., *Eimeria* sp. Guerrero, Hernandez, and Alva, 1967) in *Lama pacos* and *L. guanicoe* (Mammalia) feces.
601. *E. macyi* Wheat, 1975 in *Pipistrellus subflavus* (Mammalia) cecal contents.
602. *E. madagascariensis* Uilenberg, 1967 in *Setifer setosus* and *Hemicentetes semispinosus* (Mammalia) intestine (?).
603. *E. madisonensis* Anderson and Samuel, 1969 in *Odocoileus virginianus* and *O. hemionus* (Mammalia) feces.
604. *E. maggieae* Lom and Dyková, 1981 in *Pagellus erythrinus* (Osteichthys) middle intestine.
605. *E. magna* Pérard, 1925 in *Oryctolagus cuniculus* (Mammalia) jejunum and ileum.
606. *E. magnalabia* Levine, 1951 (syn., *E. striata* Farr, 1953) in *Branta canadensis, Anser albifrons frontalis, A. c. coerulescens,* and experimentally *A. anser* (Aves) feces.
607. *E. maior* Honess, 1939 in *Sylvilagus nuttallii grangeri, S. floridanus mearnsii,* and *S. audubonii* (Mammalia) intestine.
608. *E. majorici* Veisov, 1962 (syn., *E. arvalis* of Veisov [1963] in *Microtus arvalis*) in *M. majori* and *M. arvalis* (Mammalia) feces.
609. *E. malabaricas* Chowattukunnel, 1978 in *Funambulus tristriatus* (Mammalia) feces.
610. *E. malaccae* Chakravarty and Kar, 1944 in *Munia m. malacca* (Aves) feces.
611. *E. malayensis* Colley and Mullin, 1971 in *Petaurista elegans* and *P. petaurista* (Mammalia) intestinal contents.
612. *E. manafovae* Musaev, 1970 (syn., *E. elegans* Yakimoff, Gousseff, and Rastegaieff, 1933 of Svanbaev [1956] in *Saiga tatarica*) in *Saiga tatarica* (Mammalia) feces.
613. *E. mandali* Banik and Ray, 1964 in *Pavo cristatus* (Aves) small intestine.
614. *E. maralikiensis* Veisov, 1975 in *Spalax leucodon* (Mammalia) feces.
615. *E. marasensis* Sayin, 1981 in *Spalax ehrenbergi* (Mammalia) mixed intestinal contents.
616. *E. marconii* Straneva and Kelley, 1979 in *Clethrionomys gapperi* (Mammalia) feces.
617. *E. marginata* Deeds and Jahn, 1939 emend. Pellérdy, 1974 (syn., *E. delagei* var. *marginata* Deeds and Jahn, 1939) in *Chrysemys picta marginata* and also *Graptemys geographica* and *G. pseudogeographica* (Reptilia) intestine.
618. *E. markovi* Svanbaev, 1956 in *Meriones tamariscinus* (Mammalia) feces.
619. *E. marmotae* Galli-Valerio, 1923 in *Marmota marmota* (Mammalia) intestine.
620. *E. marsica* Restani, 1971 in *Ovis aries* (Mammalia) feces.
621. *E. marsupialium* Yakimoff and Matschoulsky, 1936 in *Macropus giganteus* (Mammalia) feces.
622. *E. martunica* Musaev and Alieva, 1961 in *Meriones libycus* (syn., *M. erythrourus*) (Mammalia) lower small intestine.
623. *E. mascoutini* Wacha and Christiansen, 1976 in *Trionyx spiniferus* (Reptilia) intestine.

624. *E. mastomyis* De Vos and Dobson, 1970 in *Rattus (Praomys) natalensis* (Mammalia) feces.
625. *E. mathurai* Dubey and Pande, 1963 in *Felis chaus* (Mammalia) feces.
626. *E. matschoulskyi* Pellérdy, 1974 in *Cricetulus barabensis* (Mammalia) feces (?).
627. *E. matskasii* Molnar, 1978 in *Umbra krameri* (Osteichthys) intestinal epithelium.
628. *E. matsubayashii* Tsunoda, 1952 in *Oryctolagus cuniculus* (Mammalia) primarily ileum.
629. *E. maxima* Tyzzer, 1929 (syn., *E. tyzzeri* Yakimoff and Rastegaieff, 1911) in *Gallus gallus* (Aves) middle and posterior small intestine.
630. *E. maxima* var. *indentata* (Fernando and Remmler, 1973) Long, 1974 (syn., *E. indentata* Fernando and Remmler, 1973) in *Gallus lafayettei* (Aves) feces.
631. *E. mayeri* Yakimoff, Sokoloff, and Matschoulsky, 1936 (syn., *Eimeria* sp. Yakimoff and Sokoloff, 1935) in *Rangifer tarandus* (Mammalia) feces.
632. *E. mayurai* Bhatia and Pande, 1966 in *Pavo cristatus* and *P. muticus* (Aves) small intestine, especially duodenum.
633. *E. mazzai* Yakimoff and Gousseff, 1934 in *Bufo bufo* (Amphibia) feces.
634. *E. mccordocki* Honess, 1941 in *Odocoileus h. hemionus* and *O. virginianus* (Mammalia) feces.
635. *E. mecistophori* Narasimhamurti, 1976 in *Mecistocephalus punctifrons* (Chilopodasida) intestine.
636. *E. media* Kessel, 1929 in *Oryctolagus cuniculus* (Mammalia) small intestine.
637. *E. megalostiedae* Wacha and Christiansen, 1974 in *Clemmys insculpta* (Reptilia) intestine.
638. *E. megalostomata* Ormsbee, 1939 in *Phasianus colchicus* and also possibly *Chrysolophus pictus, Syrmaticus reevesi, Crossoptilon tibetanum,* and *Lophura nycthemera* (Aves) feces.
639. *E. megaresidua* Barrow and Hoy, 1960 in *Notophthalmus viridescens* (Amphibia) feces.
640. *E. mehelyi* Musaev and Gauzer, 1971 in *Rhinolophus mehelyi* (Mammalia) feces.
641. *E. meleagridis* Tyzzer, 1927 in *Meleagris gallopavo* (Aves) small intestine and cecum.
642. *E. meleagrimitis* Tyzzer, 1929 in *Meleagris gallopavo* (Aves) small intestine.
643. *E. melis* Kotlan and Pospesch, 1933 in *Meles taxus* (Mammalia) feces.
644. *E. mellumi* Schwalbach, 1959 in *Larus argentatus* (Aves) feces.
645. *E. menzbieri* Svanbaev, 1963 in *Marmota marmota menzbieri* (Mammalia) feces.
646. *E. mephitidis* Andrews, 1928 in *Mephitis mephitis* (Mammalia) feces.
647. *E. meridiana* Veisov, 1964 in *Meriones meridianus* (Mammalia) feces.
648. *E. merionis* Machul'skii, 1949 (syn., *E. marionis* Svanbaev, 1956) in *Meriones unguiculatus* and *M. tamariscinus* (Mammalia) feces.
649. *E. merlangi* Zaika, 1966 in *Odontogadus merlangus euxinus* (Osteichthys) intestinal wall and gall.
650. *E. merriami* Stout and Duszynski, 1983 in *Dipodomys merriami* (Mammalia) feces.
651. *E. meservei* Coatney, 1935 (syn., *E. meservey* of Schwalbach [1959] in *Sterna forsteri* (Aves) feces.
652. *E. mesnili* Rastegaieff, 1929 in *Alopex lagopus* (Mammalia) feces.
653. *E. meszarosi* Molnar, 1978 in *Umbra krameri* (Osteichthys) intestinal epithelium.
654. *E. metelkini* Machul'skii, 1949 in *Ochotona daurica* (Mammalia) feces.
655. *E. michikoa* Bovee, 1971 in *Gekko japonicus* (Reptilia) intestine.
656. *E. microcapi* Duszynski et al., 1972 in *Ambystoma tigrinum* (Amphibia) feces.
657. *E. micromydis* Golemansky, 1978 in *Micromys minutus* (Mammalia) intestinal contents.
658. *E. micropiliana* Musaev, Veisov, and Alieva, 1963 in *Microtus socialis* (Mammalia) feces.
659. *E. micropteri* Molar and Hanek, 1974 in *Micropterus dolomieui* and *M. salmoides* (Osteichthys) anterior gut.

660. *E. microtina* Musaev and Veisov, 1959 in *Microtus socialis* (Mammalia) intestinal contents.
661. *E. micruri* Lainson and Shaw, 1973 in *Micrurus filiformis* (Reptilia) feces.
662. *E. middendorfi* Arnastauskene, 1977 in *Microtus middendorfi* (Mammalia) feces.
663. *E. migratorii* Musaev and Veisov, 1961 emend. (syn., *E. migratoria* Musaev and Veisov, 1961) in *Cricetulus migratorius* (Mammalia) large intestine contents.
664. *E. mikanii* Carini, 1933 in *Sybinomorphus mikanii* (Reptilia) intestine.
665. *E. milleri* Bray, 1958 in *Crocidura schweitzeri* (Mammalia) feces.
666. *E. minetti* Ray, Raghavachari, and Sapre, 1942 in *Mabuia* sp. (Reptilia) small intestine.
667. *E. minima* Carvalho, 1943 (syn., *E. exigua* Yakimoff, 1934 of Morgan and Waller [1940]) in *Sylvilagus floridanus mearnsii* (Mammalia) small intestine.
668. *E. mira* Pellérdy, 1954 (syn., *E. piriformis* Lubimov, 1934) in *Sciurus vulgaris* (Mammalia) small intestine.
669. *E. mirabilis Yakimoff, 1936 in Ophisaurus apodus* (Reptilia) intestine.
670. *E. misgurni* Stankovitch, 1924 in *Misgurnus fossilis* and *Cobitis taenia* (Osteichthys) intestine.
671. *E. mitis* Tyzzer, 1929 (syn., *E. beachi* Yakimoff and Rastegaieff, 1931) in *Gallus gallus* (Aves) small intestine.
672. *E. mitraria* (Laveran and Mesnil, 1902) Doflein, 1909 (syns., *Coccidium mitrarium* Laveran and Mesnil, 1902; *Mitrocystis mitraria* [Laveran and Mesnil, 1902] Pinto, 1927) in *Chinemys reevesii, Chrysemys picta, Graptemys geographica, G. pseudogeographica,* and *Kinosternon flavescens spooneri* (Reptilia) intestine.
673. *E. mivati* Edgar and Seibold, 1964 (syn., *E. acervulina* var. *mivati* [Edgar and Seibold, 1964]) in *Gallus gallus* (Aves) small intestine. (A *nomen dubium*; possibly a mixture of *E. mitis* and *E. acervulina*. [See Shirley, Jeffers, and Long, 1983.])
674. *E. miyairii* Ohira, 1912 (syn., *E. carinii* Pinto, 1928) in *Rattus norvegicus, R. whiteheadi,* and *R. tiomanicus* (Mammalia) small intestine.
675. *E. modesta* Van Peenen, Ryan, and McIntyre, 1967 in *Tupaia glis modesta* (Mammalia) small intestine.
676. *E. moelleri* Levine and Ivens, 1965 (syn., *E. sciurorum* Galli-Valerio of Möller [1923]) in *Sciurus carolinensis* (Mammalia) small intestine.
677. *E. mohavensis* Doran and Jahn, 1949 in *Dipodomys panamintinus mohavensis* and experimentally *D. merriami, D. nitratoides, D. heermanni, D. deserti,* and *D. agilis* (Mammalia) small intestine and cecum.
678. *E. molochis* Bovee and Telford, 1965 in *Moloch horridus* (Reptilia) feces.
679. *E. monacis* Fish, 1930 (syn., *E. dura* Crouch and Becker, 1931) in *Marmota monax, M. bobak,* and *M. sibirica* (Mammalia) intestinal contents.
680. *E. mongolica* Machul'skii, 1941 in *Fulica a. atra* (Aves) feces.
681. *E. monocrustae* Veisov, 1963 in *Microtus arvalis* (Mammalia) feces.
682. *E. montanaensis* Todd and O'Gara, 1968 in *Oreamnos americanus* (Mammalia) feces.
683. *E. montgomeryae* Lewis and Ball, 1983 in *Apodemus sylvaticus* (Mammalia) feces.
684. *E. morainensis* Torbett, Marquardt, and Carey, 1982 in *Spermophilus lateralis* (Mammalia) feces.
685. *E. moreli* Vassiliades, 1966 in *Cricetomys gambianus* (Mammalia) feces.
686. *E. moronei* Molnar and Fernando, 1974 in *Morone americana* (Osteichthys) intestine.
687. *E. moschati* Duszynski, Samuel, and Gray, 1977 in *Ovibos moschatus* (Mammalia) feces.
688. *E. mrigai* Pande et al., 1972 in *Antilope cervicapra* (Mammalia) jejunum.
689. *E. (?) mucosa* (Blanchard, 1885) Levine, 1979 (syns., *Balbiania mucosa* Blanchard, 1885; *Sarcocystis mucosa* [Blanchard, 1885] Labbé, 1899; *Globidium mucosum* [Blanchard, 1885] Nöller, 1920; *G. mucosae* [Blanchard, 1885] Wenyon, 1926 of Mackerras [1958]) in *Petrogale penicillata* (Mammalia) colon submucosa.

690. *E. muehlensi* Yakimoff, Sokoloff, and Matschoulsky, 1936 in *Rangifer tarandus* (Mammalia) feces.
691. *E. mundaragi* Hiregaudar, 1956 in *Bos indicus* (Mammalia) feces.
692. *E. muraiae* Molnar, 1978 in *Misgurnus fossilis* (Osteichthys) kidney tubules.
693. *E. muris* Galli-Valerio, 1932 in *Apodemus sylvaticus* (Mammalia) intestine.
694. *E. musajevi* Veisov, 1961 in *Meriones vinogradovi* (Mammalia) large intestine contents.
695. *E. musculi* Yakimoff and Gousseff, 1938 in *Mus musculus* (Mammalia) feces.
696. *E. musculoidei* Levine et al., 1959 in *Mus musculoides* (Mammalia) upper ileum.
697. *E. musophagi* Kutzer, 1963 in *Tauraco porphyreolophus* (Aves) small intestine.
698. *E. mustelae* Iwanoff-Gobzem, 1934 in *Mustela nivalis* (Mammalia) small intestine.
699. *E. mutum* Grechi, 1939 in *Crax fasciolata* (Aves) feces (?).
700. *E. muuli* Mullin, Colley, and Stevens, 1972 in *Chiropodomys gliroides* (Mammalia) intestinal contents.
701. *E. mylopharyngodoni* Chen, 1956 in *Mylopharyngodon piceus* (Osteichthys) anterior small intestine, sometimes liver and kidney.
702. *E. myocastori* Prasad, 1960 in *Myocastor coypus* (Mammalia) feces.
703. *E. myocastoris* Ringuelet and Coscaron, 1961 in *Myocastor coypus bonariensis* (Mammalia) feces (?).
704. *E. myopotami* Yakimoff, 1933 in *Myocastor coypus* (Mammalia) feces.
705. *E. myotis* Gottschalk, 1969 in *Myotis myotis* (Mammalia) small and large intestines.
706. *E. myoxi* Galli-Valerio, 1940 in *Eliomys quercinus* (Mammalia) presumably intestine.
707. *E. myoxocephali* Fitzgerald, 1975 in *Myoxocephalus polyacanthocephalus* (Osteichthys) upper intestine.
708. *E. nachitschevanica* Musaev and Veisov, 1959 in *Dryomys nitedula* (Mammalia) intestinal contents.
709. *E. nadsoni* Yakimoff and Gousseff, 1936 in *Lyrurus tetrix* and *Tetrao urogallus* (Aves) feces.
710. *E. nagpurensis* Gill and Ray, 1961 in *Oryctolagus cuniculus* (Mammalia) feces.
711. *E. najae* Ray and Das Gupta, 1936 in *Naja naja* (Reptilia) small intestine.
712. *E. nasuae* Carini and Grechi, 1938 in *Nasau nasica* (Mammalia) intestinal contents.
713. *E. nativa* Arnastauskene, 1980 in *Lemmus sibiricus* (Mammalia) feces.
714. *E. natricis* Wacha and Christiansen, 1975 in *Natrix s. sipedon* (Reptilia) feces.
715. *E. (?) navillei* (Harant and Cazal, 1934) Levine, 1984 (syn., *Globidium navillei* Harant and Cazal, 1934) in *Tropidonotus viperinus* (Reptilia) small intestine.
716. *E. naye* Galli-Valerio, 1940 in *Apodemus sylvaticus* (Mammalia) intestine.
717. *E. nazijrovi* Svanbaev, 1979 (syn., *E. intricata* Spiegl, 1925 of Svanbaev [1969] in *Capra sibirica*) in *Capra sibirica* (Mammalia) feces.
718. *E. nebulosi* Colley and Else, 1975 emend. (syn., *E. nebulosa* Colley and Else, 1975) in *Varanus nebulosus* (Reptilia) small intestine.
719. *E. necatrix* Johnson, 1930 in *Gallus gallus* (Aves) small intestine and then ceca.
720. *E. neglecta* Nöller, 1920 in *Rana esculenta* and *R. temporaria* (Amphibia) intestine.
721. *E. nehramaensis* Musaev and Veisov, 1960 in *Meriones shawi* (syn., *M. tristrami*) (Mammalia) large intestine contents.
722. *E. neitzi* McCully et al., 1970 in *Aepyceros melampus* (Mammalia) uterus.
723. *E. nemethi* Molnar, 1978 in *Alburnus alburnus* (Osteichthys) spleen, liver, and kidney parenchyma.
724. *E. neodebliecki* Vetterling, 1965 in *Sus scrofa* (Mammalia) feces.
725. *E. neoirresidua* Duszynski and Marquardt, 1969 in *Sylvilagus audubonii* and *S. floridanus* (Mammalia) feces.
726. *E. neoleporis* Carvalho, 1942 (syns., *E. leporis* Nieschulz of Morgan and Waller [1940] and Waller and Morgan [1941]; *E. heoleporis* Carvalho, 1943 of Gill and Ray

[1960] *lapsus calami*) in *Sylvilagus floridanus mearnsii* and *Oryctolagus cuniculus* (Mammalia) posterior small intestine and large intestine.

727. *E. neomyi* Golemansky, 1978 (syn., *E. komareki* Černá and Daniel of Černá [1961] in part) in *Neomys anomalus* and *N. fodiens* (Mammalia) intestinal contents.

728. *E. neosciuri* Prasad, 1960 (syn., *E. sciurorum* Galli-Valerio of Ryšavý [1954]) in *Sciurus (Neosciurus) carolinensis* and *S. vulgaris* (Mammalia) small intestine.

729. *E. neotomae* Henry, 1932 in *Neotoma fuscipes* (Mammalia) intestinal contents.

730. *E. nereensis* Glebezdin, 1973 in *Apodemus sylvaticus* (Mammalia) feces.

731. *E. nesokiae* Mirza, 1975 emend. (syn., *E. nesokiai* Mirza, 1975) in *Nesokia indica* (Mammalia) intestinal contents.

731a. *E. neurotrichi* Duszynski, 1985 in *Neurotrichus gibbsii* (Mammalia) feces.

732. *E. newalai* Dubey and Pande, 1963 in *Herpestes edwardsi* (Mammalia) feces.

733. *E. nicollei* Yakimoff and Gousseff, 1935 in *Carassius carassius* (Osteichthys) feces (?).

734. *E. nieschulzi* Dieben, 1924 (syns., *E. falciformis* [Eimer] Schneider of *auctores* in part; *E. miyairii* Ohira of *auctores* in part; *E. halli* Yakimoff, 1935; *E. contorta* Haberkorn, 1971 in part) in *Rattus norvegicus*, *R. rattus*, *R. hawaiiensis*, *R. surifer*, *R. fulvescens*, and *R. muelleri* (Mammalia) small intestine.

735. *E. nilgai* Pande et al., 1970 in *Boselaphus tragocamelus* (Mammalia) feces.

736. *E. ninakohlyakimovae* Yakimoff and Rastegaieff, 1930 emend. Levine, 1961 (syns., *E. nina-kohl-yakimovi* Yakimoff and Rastegaieff, 1930; *E. galouzoi* Yakimoff and Rastegaieff, 1930 in part; *E. nina* Bjelica, 1964 *lapsus calami*) in *Capra hircus, C. ibex*, and other members of the genus *Capra* (Mammalia) small intestine and also cecum and colon.

737. *E. nishin* Fujita, 1934 in *Clupea harengus* (Osteichthys) testis.

738. *E. nitedulae* Musaev and Veisov, 1965 in *Dryomys nitedula* (Mammalia) feces.

739. *E. nocens* Kotlán, 1933 in *Anser anser, A. erythropus*, and *A. caerulescens* (Aves) usually small intestine.

740. *E. nochti* Yakimoff and Gousseff, 1936 in *Rattus rattus* and perhaps *R. norvegicus* (Mammalia) intestine.

741. *E. noctisauris* Bovee and Telford, 1965 (syn., *E. cystifelleae* var. *americana* of Amrein [1952] in part) in *Klauberina riversiana* (Reptilia) gall bladder.

742. *E. (?) noelleri* (Rastegaieff, 1930) Becker, 1956 (syn., [gen.?] *nolleri* Rastegaieff, 1930) in *Agouti paca* (Mammalia) feces.

743. *E. nonbrumpti* Levine, 1953 (syns., *E. yakisevi* Hardcastle, 1943 in part; *E. brumpti* Yakimoff and Gousseff, 1936) in *Dendrocopus major pinctorum* (Aves) intestine.

744. *E. noraschenica* Musaev and Veisov, 1960 in *Meriones persicus* (Mammalia) large intestine contents.

745. *E. normanlevinei* Ryšavý, 1967 in *Capromys pilorides* (Mammalia) small intestine.

746. *E. notati* Colley and Mullin, 1971 in *Callosciurus notatus* (Mammalia) intestinal contents.

747. *E. notopteri* Chakravarty and Kar, 1944 in *Notopterus notopterus* and *N. chitala* (Osteichthys) intestinal contents.

748. *E. novowenyoni* Rastegaieff, 1929 in *Leo (Tigris) tigris* (Mammalia) feces.

749. *E. nucleocola* Lom and Dyková, 1981 in *Myoxocephalus scorpius* (Osteichthys) pyloric ceca.

750. *E. numenii* Mandal, 1965 in *Numenius arquata* (Aves) small intestine.

751. *E. numidae* Pellérdy, 1962 in *Numida meleagris* (Aves) small and large intestine.

752. *E. nutriae* Prasad, 1960 in *Myocastor coypus* (Mammalia) feces.

753. *E. nuttalli* Yakimoff and Matikaschwili, 1933 in *Procyon lotor* (Mammalia) feces.

754. *E. nyctali* Gottschalk, 1974 in *Nyctalus noctula* (Mammalia) feces.

755. *E. nycticebi* Patnaik and Acharjyo, 1970 in *Nycticebus coucang* (Mammalia) intestine.
756. *E. nyroca* Svanbaev and Rakhmatullina, 1967 in *Aythya nyroca* and several other duck genera (Aves) feces.
757. *E. nyschanica* Davronov, 1985 in *Agama lehmanni* (Reptilia) feces.
758. *E. ochetobii* Lee and Chen, 1964 emend. (syn., *E. ochetobiusi* Lee and Chen, 1964) in *Ochetobius elongatus* (Osteichthys) intestine.
759. *E. ochotona* Machul'skii, 1949 in *Ochotona daurica* (Mammalia) feces.
760. *E. ochrogasteri* Ballard, 1970 in *Microtus ochrogaster, M. mexicanus,* and *M. pennsylvanicus* (Mammalia) feces.
761. *E. odocoilei* Levine, Ivens, and Senger, 1967 in *Odocoileus h. hemionus* and *O. virginianus* (Mammalia) feces.
762. *E. ojastii* Arcay-de-Peraza, 1964 in *Oryzomys albigularis* (Mammalia) feces.
763. *E. ojibwana* Molnar and Fernando, 1974 in *Cottus bairdi* (Osteichthys) gut epithelium.
764. *E. okanaganensis* Liburd and Mahrt, 1970 in *Lophortyx californicus* and also *Oreortyx pictus* (Aves) feces.
765. *E. ondatraezibethicae* Martin, 1930 in *Ondatra zibethica* (Mammalia) jejunum (and liver [?]).
766. *E. ondinae* Carini, 1939 in *Drymobius bifossatus* (Reptilia) feces.
767. *E. ontarioensis* Lee and Dorney, 1971 in *Sciurus carolinensis* and *S. niger rufiventer* (Mammalia) intestinal contents.
768. *E. onychomysis* Levine, Ivens, and Kruidenier, 1957 in *Onychomys leucogaster* (Mammalia) intestinal contents.
769. *E. oomingmakensis* Duszynski, Samuel, and Gray, 1977 in *Ovibos moschatus* (Mammalia) feces.
770. *E. operculata* Levine, Ivens, and Kruidenier, 1957 in *Neotoma stephensi* (Mammalia) intestinal contents.
771. *E. ophiocephali* Chen and Hsieh, 1960 emend. (syn., *E. ophiocephalae* Chen and Hsieh, 1960) in *Ophiocephalus argus* and *O. maculatus* (Osteichthys) intestine and pyloric ceca.
772. *E. ordubadica* Musaev and Veisov, 1965 in *Meriones persicus* (Mammalia) feces.
773. *E. oreamni* Shah and Levine, 1964 in *Oreamnos americanus* (Mammalia) feces.
774. *E. oreoecetes* Haskins, 1969 in *Lagopus leucurus* and *Dendragapus obscurus* (Aves) feces.
775. *E. oreortygis* Duszynski and Gutiérrez, 1981 in *Oreortyx pictus* and *Lophortyx californicus* (Aves) feces.
776. *E. orthogeomydos* Lainson, 1968 in *Orthogeomys grandis scalops* (Mammalia) feces (?).
777. *E. oryzomysi* Carini, 1937 in *Oryzomys* sp. (Mammalia) small intestine.
778. *E. os* Crouch and Becker, 1931 in *Marmota monax, M. menzbieri, M. sibirica, M. bobak, M. baibacina,* and *M. caudata* (Mammalia) feces.
779. *E. osmeri* Molnar and Fernando, 1974 in *Osmerus mordax* (Osteichthys) gut epithelium.
780. *E. ostertagi* Yakimoff and Gousseff, 1936 in *Erinaceus europaeus* (Mammalia) feces.
781. *E. otolicni* Poelma, 1966 in *Galago senegalensis* (Mammalia) intestinal contents.
782. *E. otomyis* De Vos and Dobson, 1970 in *Otomys irroratus* (Mammalia) feces (?).
783. *E. ottojiroveci* Dyková and Lom, 1983 (syn., *E. jiroveci* Lom and Dyková, 1981) in *Raja clavata* (Chondrichthys) intestinal epithelium and spiral valve.
784. *E. ovata* Duncan, 1968 in *Tamias striatus* (Mammalia) intestinal contents.
785. *E. ovibovis* Duszynski, Samuel, and Gray, 1977 in *Ovibos moschatus* (Mammalia) feces.
786. *E. ovina* Krylov, 1961 emend. (syns., *E. arloingi* forma *ovina* Krylov, 1961; *E. arloingi auctores* from sheep; *E. bakuensis* Musaev, 1970; *E. hawkinsi* Ray, 1952 in

part; *E. ovina* Levine and Ivens, 1970) in *Ovis aries* and also *O. canadensis, O. ammon,* and *O. musimon*; reported, probably erroneously, from *Ovibos moschatus* (Mammalia) small intestine.

787. *E. ovinoidalis McDougald, 1979 (syn., E. ninakohlyakimovae* Yakimoff and Raste-gaieff, 1930 in *Ovis*) in *O. aries* and also *O. musimon, O. canadensis,* and probably other members of the genus *Ovis* (Mammalia) small intestine and also cecum and colon.

788. *E. ovoidalis* Ray and Mandal, 1962 in *Bubalus bubalis* (Mammalia) feces.

789. *E. oytuni* Sayin, 1981 in *Spalax ehrenbergi* (Mammalia) mixed intestinal contents.

790. *E. pachylepyron* Colley and Mullin, 1972 in *Nycticebus coucang* (Mammalia) feces.

791. *E. pacifica* Ormsbee, 1939 in *Phasianus colchicus, Chrysolophus pictus, Syrmaticus reevesi, Crossoptilon tibetanum,* and *Lophura nycthemera* (Aves) anterior small intestine.

792. *E. padulensis* Romero and Lizcano 1974 (syn., *E. phasiani* Tyzzer, 1929 of Lizcano and Romero [1972]) in *Alectoris rufa* (Aves) posterior small intestine.

793. *E. pahangi* Colley and Mullin, 1971 in *Callosciurus notatus* (Mammalia) intestinal contents.

794. *E. pallasi* Lepp, Todd, and Samuel, 1972 (syn., *E. kriygsmanni* of Svanbaev [1958] in *Ochotona pallasi*) in *O. pricei* (syn., *O. pallasi*) (Mammalia) feces.

795. *E. pallida Christensen, 1938 in Ovis aries* and also possibly *Capra hircus* and *C. ibex* (Mammalia) small intestine.

796. *E. paludosa* (Léger and Hesse, 1922) Hoare, 1933 (syn., *Jarrina paludosa* Léger and Hesse, 1922) in *Fulica a. atra, F. americana,* and also perhaps *Gallinula c. chloropus* (Aves) small intestine.

797. *E. palustris* Barnard, Ernst, and Stevens, 1971 in *Oryzomys palustris* (Mammalia) small intestine.

798. *E. panda* Supperer and Kutzer, 1961 in *Capreolus capreolus* (Mammalia) feces.

799. *E. pandei* Patnaik and Ray, 1965 emend. Patnaik and Ray, 1966 (syn., *E. pandeii* Patnaik and Ray, 1965) in *Herpestes edwardsii* (Mammalia) intestine.

800. *E. papillata* Ernst, Chobotar, and Hammond, 1971 in *Mus musculus* (Mammalia) feces.

801. *E. paradisaeai* Varghese, 1977 in *Paradiseae raggiana* (Aves) feces.

802. *E. paraensis* Carini, 1935 in *Dasyprocta aguti* (Mammalia) feces.

803. *E. paragachaica* Musaev and Veisov, 1965 in *Mus musculus* (Mammalia) feces.

804. *E. parahi* Pande et al., 1970 in *Axis porcinus* (Mammalia) feces.

805. *E. parasciurorum* Bond and Bovee, 1957 (syn., *E. sciurorum* Galli-Valerio of Rou-dabush [1937]) in *Glaucomys volens* (Mammalia) feces.

806. *E. parasiluri* Lee and Chen, 1964 in *Parasilurus asotus* (Osteichthys) gall bladder.

806a. *E. parastiedica* Duszynski, 1985 in *Neurotrichus gibbsii* (Mammalia) feces.

807. *E. parva* Kotlán, Mócsy, and Vajda, 1929 (syns., *E. nana* Yakimoff, 1933; *E. galouzoi* Yakimoff and Rastegaieff, 1930 in part) in *Ovis aries* and also *O. orientalis, O. ammon, O. musimon, O. canadensis,* and possibly *Capra hircus, C. ibex,* and *Am-motragus lervia* (Mammalia); meronts in small intestine and sexual stages in cecum, colon, and to a lesser extent small intestine; this species also has been reported from *Capreolus capreolus, Dama dama, Cervus elaphus,* and *Rupicapra rupicapra* (Mam-malia), but there is considerable doubt that it occurs in them.

808. *E. patavina* Mantovani, Borrelli, and Ricci Bitti, 1970 in *Capreolus capreolus* (Mam-malia) feces.

809. *E. patnaiki* Ray, 1966 (syns., *E. pavonis* Patnaik, 1965; *E. indica* Patnaik, 1966) in *Pavo cristatus* (Aves) feces.

810. *E. pattersoni* Honess and Post, 1955 in *Centrocercus urophasianus* and *Lyrurus tetrix* (Aves) feces.

811. *E. paulistana* da Fonseca, 1933 in *Sylvilagus brasiliensis minensis* (Mammalia) feces.
812. *E. pavlovskyi* Machul'skii, 1949 in *Allactaga saltator mongolica* (Mammalia) feces.
813. *E. pavonina* Banik and Ray, 1961 in *Pavo cristatus* (Aves) feces.
814. *E. pavonis* Mandal, 1965 (syn., *E. cristata* Patnaik, 1965) in *Pavo cristatus* (Aves) small intestine.
815. *E. paynei* Ernst, Fincher, and Stewart, 1971 in *Gopherus polyphemus* (Reptilia) feces.
816. *E. pelandoki* Mullin and Colley, 1971 in *Tragulus javanicus* (Mammalia) intestinal contents.
817. *E. pelecani* Courtney and Ernst, 1975 in *Pelecanus occidentalis carolinensis* (Aves) feces.
818. *E. pellerdyi* Prasad, 1960 in *Camelus bactrianus* (Mammalia) feces.
819. *E. pellita* Supperer, *1952 in Bos taurus* and *B. indicus* (Mammalia) feces.
820. *E. pellopleuris* Bovee, 1971 in *Lygosoma pellopleurum* (Reptilia) gall bladder and bile duct epithelium.
821. *E. pellucida* Yakimoff, 1933 (syn., *E. pellicuda* Yakimoff, 1936 of Nukerbaeva and Svanbaev [1973] *lapsus calami*) in *Myocastor coypus* (Mammalia) ileum and other parts of small intestine.
822. *E. penicillati* Ivens, Kruidenier, and Levine, 1959 in *Perognathus penicillatus* and *P. flavus* (Mammalia) intestinal contents.
823. *E. perardi* Yakimoff and Gousseff, 1936 in *Erinaceus europaeus* and *Hemiechinus auritus* (Mammalia) presumably feces.
824. *E. perazae* Levine, 1981 (syn., *E. flaviviridis americana* Arcay de Peraza, 1963) in *Cnemidophorus l. lemniscatus* (Reptilia) gall bladder epithelium.
825. *E. percae* (Dujarric de la Rivière, 1914) Reichenow, 1921 (syns., *Coccidium percae* Dujarric de la Rivière, 1914; *E. rivierei* Yakimoff, 1929) in *Perca fluviatilis* (Osteichthys) stomach wall (and liver [?]).
826. *E. perforans* (Leuckart, 1879) Sluiter and Swellengrebel, 1912 (syns., *Coccidium perforans* Leuckart, 1879; *Pfeifferia princeps* Labbé, 1896; *Pfeifferella princeps* [Labbé, 1896] Labbé, 1899; *E. nana* Marotel and Guilhon, 1941; *E. lugdunumensis* Marotel and Guilhon, 1942) in *Oryctolagus cuniculus* (Mammalia) small intestine, cecum (?).
827. *E. perforoides* Crouch and Becker, 1931 in *Marmota monax* (Mammalia) feces.
828. *E. (?) perichaetae* (Beddard, 1888) Levine and Becker, 1933 (syn., *Coccidium perichaetae* Beddard, 1888) in *Perichaeta armata* and *P. novozelandiae* (Oligochaetasida) intestine (?).
829. *E. perminuta* Henry, 1931 in *Sus scrofa* (Mammalia) feces.
830. *E. perognathi* Levine, Ivens, and Kruidenier, 1957 in *Perognathus intermedius* (Mammalia) intestinal contents.
831. *E. peromysci* Levine, Ivens, and Kruidenier, 1957 in *Peromyscus truei* and *P. maniculatus rufinus* (Mammalia) intestinal contents.
832. *E. persica* (Phisalix, 1925) Levine and Becker, 1933 (syn., *Coccidium persicum* Phisalix, 1925) in *Natrix natrix* var. *persa* (Reptilia) bile duct epithelium.
833. *E. peruviana* Yakimoff, 1934 in *Lama glama* (Mammalia) feces.
834. *E. peschankae* Levine and Ivens, 1965 (syn., *E. kriygsmanni* Yakimoff and Gousseff of Svanbaev [1962]) in *Meriones tamariscinus* (Mammalia) feces.
835. *E. petauristae* Ray and Singh, 1950 in *Petaurista petaurista* (Mammalia) feces.
836. *E. pfeifferi* Labbé, 1896 in *Geophilus ferruginosus* (Chilopodasida) intestine.
837. *E. phasiani* Tyzzer, 1929 (syn., *E. phasiana* Tyzzer, 1929 *lapsus calami*) in *Phasianus colchicus, Chrysolophus pictus, Syrmaticus reevesi, Crossoptilon tibetanum,* and *Lophura nycthemera* (Aves) small intestine.
838. *E. phisalixae* Vetterling and Widmer, 1968 emend. Pellérdy, 1974 (syns., *E. cerastis [Chatton, 1912] Phisalix, 1921 in part; E. phisalixi* Vetterling and Widmer, 1968) in *Cerastes cerastes* (Reptilia) gall bladder and bile duct epithelium.

839. *E. phocae* Hsu, Melby, and Altman, 1974 in *Phoca vitulina concolor* (Mammalia) colon and ileum.

840. *E. phyllotis* Gonzales-Mugaburu, 1942 in *Phyllotis a. amicus* (Mammalia) epithelial cells of cecum.

841. *E. picti* Levine, Ivens, and Kruidenier, 1958 in *Liomys pictus* (Mammalia) intestinal contents.

842. *E. pictus* Bhatia, 1968 (syn., *E. picta* Bhatia, 1968 emend. Pellérdy, 1974) in *Chrysolophus pictus* and *Lophura nycthemera* (Aves) feces.

843. *E. pigra* Léger and Bory, 1932 in *Scardinius erythrophthalmus* (Osteichthys) intestine.

844. *E. pileata* Straneva and Kelley, 1979 in *Clethrionomys gapperi* (Mammalia) feces.

845. *E. pintoensis* da Fonseca, 1932 in *Sylvilagus brasiliensis minensis* (Mammalia) feces.

846. *E. pintoi* Carini, 1932 in *Caiman* sp. (Reptilia) intestinal contents.

847. *E. piraudi* Gauthier, 1921 in *Cottus gobio* (Osteichthys) intestine.

848. *E. piriformis* Kotlan and Pospesch, 1934 (syns., *E. piriformis* Marotel and Guilhon, 1941; *E. pyriformis* Kotlan and Pospesch, 1934 *lapsus calami*) in *Oryctolagus cuniculus* (Mammalia) small and large intestines.

849. *E. piscatori* Ray and Das Gupta, 1936 in *Natrix piscator* (Reptilia) rectum.

850. *E. pitymydis* Golemanski and Yankova, 1973 in *Pitymys subterraneus* (Mammalia) feces.

851. *E. plecoti* Gottschalk, 1969 in *Plecotus auritus* (Mammalia) rectum.

852. *E. pluvialina* Mandal, 1965 emend. Pellérdy, 1974 (syn., *E. roscoviensis pluvialina* Mandal, 1965) in *Pluvialis apricaria* (Aves) intestine.

853. *E. pneumatophori* Dogel', 1948 in *Scomber* (syn., *Pneumatophorus*) *japonicus* (Osteichthys) liver.

854. *E. poecilogyri* Carini, 1933 (syn., *E. poecilogyri* Carini, 1933 *lapsus calami*) in *Leimadophis poecilogyrus* (Reptilia) small intestine.

855. *E. (?) polaris* Yakimoff and Sokoloff, 1935 in *Rangifer tarandus* (Mammalia) feces.

856. *E. polita* Pellérdy, 1949 (syn., *E. cerdonis* Vetterling, 1965) in *Sus scrofa* (Mammalia) small intestine.

857. *E. poljanskii* Veisov, 1961 emend. Levine and Ivens, 1965 (syn., *E. poljanski* Veisov, 1961) in *Meriones vinogradovi* (Mammalia) small intestine.

858. *E. polycephali* Yakimoff and Matschoulsky, 1939 in *Porphyrio coeruleus* (Aves) feces.

859. *E. polydaedali* Borst, Peters, and Weijerman, 1975 in *Varanus niloticus* (Reptilia) intestinal contents.

860. *E. ponderosa* Wetzel, 1942 in *Capreolus capreolus* (Mammalia) feces.

861. *E. popovi* Machul'skii, 1949 in *Allactaga saltator mongolica* (Mammalia) feces.

862. *E. porci* Vetterling, 1965 in *Sus scrofa* (Mammalia) jejunum and ileum.

863. *E. poti* Lainson, 1968 in *Potos flavus* (Mammalia) feces.

864. *E. poudrei* Duszynski and Marquardt, 1969 in *Sylvilagus audubonii* (Mammalia) feces.

865. *E. praecox* Johnson, 1930 in *Gallus gallus* (Aves) upper third of small intestine.

865a. *E. praecox* var. *ceylonensis* Long, Fernando, and Remmler, 1974 in *Gallus lafayettei* (Aves) mostly duodenum just proximal to bile duct, but also ileum in heavy infections.

866. *E. praomysis* Levine et al., 1959 in *Praomys* (syn., *Rattus) tullbergi rostratus* (Mammalia) intestinal contents.

867. *E. prasadi* Levine and Ivens, 1965 (syn., *E. hindlei* Yakimoff and Gousseff of Svanbaev [1956]) in *Apodemus sylvaticus* and *A. flavicollis* (Mammalia) feces.

868. *E. pretoriensis* De Vos and Dobson, 1970 in *Rhabdomys pumilio* (Mammalia) small intestine (most often ileum).

869. *E. prevoti* (Laveran and Mesnil, 1902) Doflein, 1909 (syns., *Paracoccidium prevoti* Laveran and Mesnil, 1902; *E. prevunti* Laveran and Mesnil, 1902 *lapsus calami*) in *Rana esculenta* (Amphibia) small intestine epithelium.

870. *E. primbelica* Veisov, 1963 in *Microtus arvalis* (Mammalia) feces.
871. *E. princeps* Duszynski and Brunson, 1973 in *Ochotona princeps* (Mammalia) intestine.
872. *E. procera* Haase, 1939 in *Perdix p. perdix* and perhaps *Tetrao urogallus* and *Alectoris rufa* (Aves) intestine (?).
873. *E. procyonis* Inabnit, Chobotar, and Ernst, 1972 in *Procyon lotor* (Mammalia) feces.
874. *E. proechimyi* Arcay-de-Peraza, 1964 in *Proechimys guyanensis* (Mammalia) small intestine.
875. *E. propria* (Schneider, 1881) Doflein, 1909 (syns., *Orthospora propria* Schneider, 1881; *Coccidium proprium* [Schneider, 1881] Schneider, 1887) in *Salamandra atra, Triturus alpestris, T. cristatus, T. helveticus, T. marmoratus,* and *T. vulgaris* (Amphibia) intestine.
876. *E. pseudemydis* Lainson, 1968 in *Pseudemys ornata* and *P. scripta elegans* (Reptilia) feces.
877. *E. pseudogeographica* Wacha and Christiansen, 1976 in *Graptemys pseudogeographica* and *Chrysemys picta belli* (Reptilia) intestine.
878. *E. psittacina* Gottschalk, 1972 in *Melopsittacus undulatus* (Aves) intestine.
879. *E. pternistis* Agostinucci and Bronzini, 1956 in *Pternistis leucoscepus* (Aves) feces.
880. *E. ptilocerci* Mullin, Colley, and Stevens, 1972 in *Ptilocercus lowii* (Mammalia) intestinal contents.
881. *E. ptyas* Prasad, 1963 in *Ptyas mucosus* (Reptilia) feces.
882. *E. pulchella* Farr, 1963 in *Branta canadensis* (Aves) feces.
883. *E. pumilioi* De Vos and Dobson, 1970 in *Rhabdomys pumilio* (Mammalia) cecum and colon.
884. *E. punctata* Landers, 1955 (syn., *E. honessi* Landers, 1952) in *Ovis aries* and *Capra hircus* (Mammalia) feces.
885. *E. pungitii* Molnar and Hanek, 1974 in *Pungitius pungitius* (Osteichthys) anterior gut.
886. *E. punjabensis* Gill and Ray, 1961 in *Lepus ruficaudatus* (Mammalia) feces.
887. *E. punoensis* Guerrero, 1967 in *Lama pacos* (Mammalia) feces.
888. *E. purchasei* Mykytowycz, 1964 in *Macropus giganteus* (Mammalia) feces.
889. *E. putevelata* Bray, 1958 in *Lemniscomys s. striatus* (Mammalia) epithelial cells of posterior third of ileum.
890. *E. pythonis* Triffitt, 1925 in *Python molurus* and *P. sebae* (Reptilia) feces.
891. *E. quentini* Boulard, 1977 in *Aetobatis narinari* (Chondrichthys) nuclei of peritoneum.
892. *E. quiyarum* Ringuelet and Coscaron, 1961 in *Myocastor coypus bonariensis* (Mammalia) feces.
893. *E. rabiti* Mandal and Nair, 1973 in *Gymnodactylus rabitus* (Reptilia) small intestine.
894. *E. rachmatullinae* Svanbaev, 1979 (syn., *E. arloingi* [Marotel, 1905] Martin, 1909 of Svanbaev [1969] in *Ovis ammon*) in *Ovis ammon* (Mammalia) feces.
895. *E. raiarum* van den Berghe, 1937 in *Raja batis* (Chondrichthys) epithelial cells of posterior intestine.
896. *E. raillieti* (Léger, 1899) Galli-Valerio, 1930 (syn., *Coccidium raillieti* Léger, 1899) in *Anguis fragilis* (Reptilia) small intestine.
897. *E. rajasthani* Dubey and Pande, 1963 in *Camelus dromedarius* (Mammalia) feces.
898. *E. ramgai* Pande et al., 1970 in *Tragulus meminna* (Mammalia) feces.
899. *E. ranae* Dobell, 1908 in *Rana esculenta* and *R. temporaria* (Amphibia) intestine.
900. *E. ranarum* (Labbé, 1894) Doflein, 1909 (syns., *Karyophagus ranarum* Labbé, 1894; *Caryophagus ranarum* Labbé, 1894 emend. Labbé, 1899; *Coccidium ranarum* [Labbé, 1894] Laveran and Mesnil, 1902; *Acystis parasitica* Labbé, 1894 in part) in *Rana esculenta, R. temporaria,* and possibly *Ambystoma opacum* (Amphibia) intestine.
901. *E. ratti* Yakimoff and Gousseff, 1936 in *Rattus rattus* (Mammalia) intestinal contents.
902. *E. rayii* Rao and Bhatavdekar, 1957 in *Canis familiaris* (Mammalia) feces (a dubious species).

903. *E. razgovica* Musaev and Veisov, 1962 in *Mesocricetus auratus* (Mammalia) feces.
904. *E. reedi* Ernst, Oaks, and Sampson, 1970 in *Perognathus formosus* (Mammalia) intestinal contents.
905. *E. reichenowi* Yakimoff and Matschoulsky, 1935 (syns., *E. nicolscii* Zolotarev, 1937; *E. grusi* Pande et al., 1970) in *Anthropoides virgo, Grus canadensis, G. antigone,* and *G. americana* (Aves) feces and viscera.
906. *E. renicola* Creutz and Gottschalk, 1969 in *Laurus ridibundus* (Aves) intestine.
907. *E. residua* Henry, 1932 in *Neotoma fuscipes* (Mammalia) intestinal contents.
908. *E. residualis* Martinez and Hernandez, 1973 in *Sus scrofa* (Mammalia) feces.
909. *E. rhynchonycteridis* Lainson, 1968 in *Rhynchonycteris naso* (Mammalia) small intestine.
910. *E. rhynchoti* Reis and Nobrega, 1936 in *Rhynchotus rufescens* (Aves) intestine.
911. *E. ridjakovi* Golemansky, 1976 in *Perdix p. perdix* (Aves) small intestine.
912. *E. riedmuelleri* Yakimoff and Matschoulsky, 1940 in *Rupicapra rupicapra* (Mammalia) feces.
913. *E. rissae* Soulsby and Jennings, 1957 in *Rissa t. tridactyla* (Aves) feces.
914. *E. robertsoni* Madsen, 1938 (syns., *E. magna* var. *robertsoni* Madsen, 1938 emend. Carvalho, 1943; *E. perforans* var. *groenlandica* Madsen, 1938 in part) in *Lepus arcticus groenlandicus, L. townsendi campanius, L. europaeus, L. americanus,* and *L. ruficaudatus* (Mammalia) feces.
915. *E. robusta* Supperer and Kutzer, 1961 in *Cervus elephus* (Mammalia) feces.
916. *E. rochalimai* Carini and Pinto, 1926 in *Hemidactylus mabuya* (Reptilia) gall bladder.
917. *E. rochesterensis* Samoil and Samuel, 1977 in *Lepus americanus* (Mammalia) feces.
918. *E. roperi* Barnard, Ernst, and Dixon, 1974 in *Sigmodon hispidus* (Mammalia) cecum and colon.
919. *E. ropotamae* Golemansky, 1978 in *Crocidura l. leucodon* (Mammalia) intestinal contents.
920. *E. roscoviensis* (Labbé, 1893) Labbé and von Wasielewski, 1904 (syn., *Coccidium roscoviense* Labbé, 1894) reported in various charadriiform and other birds, including *Charadrius cantianus, C. philippinus, Strepsilas interpres, Calidris arenaria, Pelidna torquata, Tringa alpina, Actitis hypoleucos,* and perhaps *Motacilla alba* (Aves) feces.
921. *E. rotunda* Pellérdy, 1955 in *Capreolus capreolus* (Mammalia) feces.
922. *E. roudabushi* Levine and Ivens, 1960 in *Peromyscus leucopus* (Mammalia) intestinal contents.
923. *E. rountreei* Bovee, 1971 in *Takydromus tachydromoides* (Reptilia) intestinal contents.
924. *E. roussillona* Lom and Dyková, 1981 in *Labrus turdus* (Osteichthys) middle intestine.
925. *E. rouxi* (Elmassian, 1909) Reichenow, 1921 (syn., *Coccidium rouxi* Elmassian, 1909) in *Tinca tinca* (Osteichthys) midgut.
926. *E. rowani* Samoil and Samuel, 1977 in *Lepus americanus* (Mammalia) feces.
927. *E. rufi* Prasad, 1960 emend. Levine, 1985 (syn., *E. rufusi* Prasad, 1960) in *Macropus rufus* (Mammalia) feces.
928. *E. ruficaudati* Gill and Ray, 1961 in *Lepus ruficaudatus* and *L. americanus* (Mammalia) feces.
929. *E. rugosa* Pellérdy, 1954 in *Apodemus flavicollis* (Mammalia) intestine.
930. *E. rupicaprae* Galli-Valerio, 1923 in *Rupicapra rupicapra* (Mammalia) feces.
931. *E. russiensis* Levine and Ivens, 1965 (syn., *E. musculi* Yakimoff and Gousseff of Svanbaev [1965] in *Apodemus sylvaticus*) in *Apodemus sylvaticus* (Mammalia) feces.
932. *E. rutili* Dogel' and Bykhovskii, 1939 in *Rutilus rutilus caspicus* (Osteichthys) kidneys.
933. *E. rysavyi* Levine and Ivens, 1965 (syns., *E. apodemi* Pellérdy of Ryšavý [1957]; *E. hindlei* Yakimoff and Gousseff of Ryšavý [1954] and Černá [1962] in *Clethrionomys glareolus*) in *C. glareolus* (Mammalia) small intestine.
934. *E. sabani* Colley and Mullin, 1971 in *Rattus sabanus* and *R. whiteheadi* (Mammalia) intestinal contents.

935. *E. sablii* Nukerbaeva, 1981 in *Martes zibellina* (Mammalia) feces.
936. *E. sadaraktica* Veisov, 1961 in *Meriones vinogradovi* (Mammalia) large intestine contents.
937. *E. saiga* Svanbaev, 1958 in *Saiga tatarica* (Mammalia) feces.
938. *E. saitamae* Inoue, 1967 in *Anas platyrhynchos domestica* (Aves) intestine.
939. *E. sajanica* Machul'skii, 1947 in *Saiga tatarica* (Mammalia) feces.
940. *E. salamandrae* (Steinhaus, 1889) Dobell, 1909 (syns., *Karyophagus salamandrae* Steinhaus, 1889; *Caryophagus salamandrae* Steinhaus, 1889 emend. Labbé, 1899; *Cytophagus tritonis* Steinhaus, 1891 in part; *Acystis parasitica* Labbé, 1894 in part; *K. tritonis* [Steinhaus, 1891] von Wasielewski, 1896; *Caryophagus tritonis* [Steinhaus, 1891] von Wasielewski, 1896 emend. Labbé, 1899; *Coccidium salamandrae* Steinhaus, 1889] Simond, 1897; *E. tritonis* [Steinhaus, 1891] Walton, 1941) in *Salamandra salamandra* and *Triton* sp. (Amphibia) intestine.
941. *E. salamandraeatrae* (Phisalix, 1927) Levine and Becker, 1933 (syn., *Coccidium salamandrae atrae* Phisalix, 1927) in *Salamandra atra* (Amphibia) intestinal epithelial cells.
942. *E. salasuzica* Musaev and Veisov, 1960 in *Meriones persicus* (Mammalia) large intestine contents.
943. *E. salvelini* Molnar and Hanek, 1974 in *Salvelinus fontinalis* (Osteichthys) anterior gut.
944. *E. sami* Bovee, 1971 in *Eumeces oshimensis* (Reptilia) intestine.
945. *E. samiae* Iskander and Tadros, 1980 in *Python reticulata* (Reptilia) intestine.
946. *E. sardari* Bhatia et al., 1973 in *Muntiacus muntijaka* (Mammalia) feces.
947. *E. sardinae* (Thélohan, 1890) Reichenow, 1921 (syns., *Coccidium sardinae* Thélohan, 1890; *E. oxyphila* Dobell, 1919 *lapus calami; E. oxyspora* Dobell, 1919; *E. snijdersi* Dobell, 1920) in *Clupea* (syn., *Sardina) pilcharda, C. harengus, Engraulis encrasicholus, Alosa fallax,* and *Sprattus sprattus* (Osteichthys) testes.
948. *E. saurogobii* Chen, 1964 in *Ctenopharyngodon idella* (Osteichthys) intestine.
949. *E. saxei* Vance and Duszynski, 1985 in *Microtus pennsylvanicus, M. californicus, M. mexicanus,* and *M. oregoni* (Mammalia) feces.
950. *E. scabra* Henry, 1931 (syns., *E. scarba* of Yakimoff and Matikaschwili [1932] *lapsus calami; E. romaniae* Donçiu, 1962) in *Sus scrofa* (Mammalia) small intestine.
951. *E. scapani* Henry, 1932 in *Scapanus latimanus* and also *Talpa micrura coreana* (Mammalia) intestinal contents.
952. *E. scardinii* Pellérdy and Molnár, 1968 in *Scardinius erythrophthalmus* (Osteichthys) kidneys.
953. *E. sceloporis* Bovee and Telford, 1965 (syn., *Eimeria* sp. Telford and Bovee, 1964) in *Sceloporus clarki boulangeri, S. occidentialis biseriatus, S. o. occidentalis,* and *S. magister* (Reptilia) small intestine.
954. *E. schachdagica* Musaev et al., 1966 in *Anas platyrhynchos* and several other ducks (Aves) feces.
955. *E. schachtachtiana* Musaev and Veisov, 1960 in *Meriones shawi* (syn., *M. tristrami*) (Mammalia) large intestine contents.
956. *E. schamchorica* Musaev and Alieva, 1961 in *Meriones libycus* (syn., *M. erythrourus*) (Mammalia) large intestine contents.
957. *E. schelkovnikovi* Musaev, 1967 in *Microtus (Pitymys) schelkovnikovi* (Mammalia) feces.
958. *E. schiwicki* Arnastauskene, 1977 in *Clethrionomys rutilus* (Mammalia) feces.
959. *E. schoenbuchi* Boch, 1963 in *Cervus elephus* (Mammalia) feces.
960. *E. schokpaki* Rakhmatullina and Svanbaev, 1971 in *Coracias garrulus* (Aves) feces.
961. *E. scholtysecki* Ernst, Frydendall, and Hammond, 1967 in *Dipodomys ordii, D. agilis, D. gravipes, D. panamintinus,* and *D. spectabilis* (Mammalia) feces.

962. *E. schoutedeni* van den Berghe and Chardome, 1957 in *Cricetomys dissimilis* (Mammalia) feces.
963. *E. schueffneri* Yakimoff and Gousseff, 1938 in *Mus musculus* (Mammalia) feces.
964. *E. schulmani* Kulemina, 1969 (syn., *E. achulmani* Kulemina, 1969 *lapsus calami*) in *Leuciscus idus* (Osteichthys) intestinal contents.
965. *E. scinci* (Phisalix, 1923) Levine and Becker, 1933 (syns., *Coccidium scinci* Phisalix, 1923; *E. sinci* [Phisalix, 1923] of Pellérdy [1974] *lapsus calami*) in *Scincus officinalis* and also possibly *Hemidactylus flaviviridis* (Reptilia) bile ducts and gall bladder.
966. *E. sciurorum* Galli-Valerio, 1922 (syns., *E. sciuri* Yakimoff and Terwinsky, 1931; *E. sciurarum* Litvenkova, 1969 *lapsus calami*) in *Sciurus vulgaris* and also presumably *S. carolinensis* and *S. aureogaster* (Mammalia) small intestine.
967. *E. scorpaenae* Zaika, 1966 in *Scorpaena porcus* (Osteichthys) intestine.
968. *E. scriptae* Sampson and Ernst, 1969 in *Pseudemys scripta elegans* (Reptilia) intestinal contents.
969. *E. sculpta* Madsen, 1938 in *Lepus arcticus groenlandicus* and *L. townsendii campanius* (Mammalia) intestine.
970. *E. (?) scyllii* (Drago, 1902) Levine and Becker, 1933 emend. Levine, 1985 (syns., *Coccidium scyllii* Drago, 1902; *E. scyllii* [Drago, 1902] Levine and Becker, 1933) in *Scyllium stellare* (Chondrichthys) spiral valve epithelium.
971. *E. seideli* Pellérdy, 1957 (syns., *E. fulva* Seidel, 1954; *E. [Globidium] fulva* Seidel, 1954; *E. [G.] perniciosa* Sprehn in Seidel [1954]) in *Myocastor coypus* (Mammalia) intestine.
972. *E. sejnijevi* Musaev, 1970 (syn., *E. intricata* Spiegl, 1925 of Svanbaev [1958] in *Ovis ammon*) in *Ovis ammon* (Mammalia) feces.
973. *E. semisculpta* Madsen, 1938 emend. Pellérdy, 1956 (syn., *E. magna* var. *robertsoni* forma *semisculpta* Madsen, 1938) in *Lepus arcticus groenlandicus* and *L. europaeus* (Mammalia) duodenum.
974. *E. semispinosi* Uilenberg, 1970 in *Hemicentetes semispinosus* (Mammalia) feces.
975. *E. separata* Becker and Hall, 1931 (syn., *E. separatica* Becker and Hall, 1931 of Litvenkova [1969] *lapsus calami*) in *Rattus norvegicus* and also *R. cremoriventer*, *R. defua*, *R. fulvescens*, *R. hawaiiensis*, *R. muelleri*, *R. tiomanicus*, probably *R. rattus*, and experimentally certain strains of *Mus musculus* (Mammalia) cecum and colon.
976. *E. septentrionalis* Yakimoff, Matschoulsky, and Spartansky, 1936 (syn., *E. exigua* var. *septentrionalis* Yakimoff, 1936 of Madsen [1938] in part) in *Lepus timidus*, *L. arcticus groenlandicus*, *L. europaeus*, and *L. townsendi campanius* (Mammalia) duodenum.
977. *E. serbica* Pop-Cenitch and Bordjochki, 1957 (syns., *E. serbca* Pop-Cenitch and Bordjochki, 1957 *lapsus calami*; *E. sciurorum* Galli-Valerio, 1922 of Jirovec [1942] and Ryšavý [1954]) in *Sciurus vulgaris* (Mammalia) feces.
978. *E. sericei* n. sp. (syn., *E. rhodei* Kandilov, 1963 *nomen nudum*) in fish *Rhodeus sericeus* (Osteichthys [?]) intestine. (*Remarks*. Kandilov (1963) called this species *E. rhodei* n. sp., but he gave no description, so this name is a *nomen nudum* and I am calling it *E. sericei* n. sp. It was described [poorly] by Mikhailov, 1975 and discussed by Dyková and Lom, 1983.
979. *E. serventyi* Pellérdy, 1974 in *Puffinus tenuirostris* (Aves) kidneys.
980. *E. setosi* Uilenberg, 1967 in *Setifer setosus* (Mammalia) feces.
981. *E. shubini* Lepp, Todd, and Samuel, 1972 (syn., *E. musculi* Yakimoff and Gousseff, 1938 of Svanbaev [1958] in *Ochotona pallasi*) in *O. pricei* and *O. pallasi* (Mammalia) feces.
982. *E. sibirica* Yakimoff and Terwinsky, 1931 in *Martes zibellina* and *M. foina* (Mammalia) intestinal contents.

983. *E. sicistae* Levine and Ivens, 1987 (syn., *E. tianschanica* Dzerzhinskii and Svanbaev, 1980 *nomen nudum*) in *Sicista tianschanica* (Mammalia).
984. *E. sigmodontis* Barnard, Ernst, and Dixon, 1974 in *Sigmodon hispidus* (Mammalia) feces.
985. *E. sikapusii* Levine et al., 1959 in *Lophuromys s. sikapusi* (Mammalia) intestinal contents.
986. *E. silvana* Pellérdy, 1954 in *Sciurus vulgaris* (Mammalia) small intestine.
987. *E. simondi* (Léger, 1898) Levine and Becker, 1933 (syn., *Coccidium simondi* Léger, 1898) in *Himantarium gabrielis*, *Lithobius castaneus*, and *L. pilicornis* (Chilopodasida) intestine.
988. *E. sinensis* Chen, 1956 in *Hypophthalmichthys molitrix* and *Aristichthys nobilis* (Osteichthys) intestine.
989. *E. siniffi* Levine and Ivens, 1963 in *Peromyscus maniculatus* (Mammalia) feces.
990. *E. sipedon* Wacha and Christiansen, 1975 in *Natrix s. sipedon* (Reptilia) feces.
991. *E. skrjabini* Dashnyam, 1961 in *Capra hircus* (Mammalia) feces (?).
992. *E. solipedum* Gousseff, 1935 in *Equus caballus* and *E. asinus* (Mammalia) feces.
993. *E. somateriae* Christiansen, 1952 in *Somateria m. mollissima*, also probably *Clangula hyemalis*, and possibly *Aythya marila* (Aves) kidneys.
994. *E. sordida* Supperer and Kutzer, 1961 in *Cervus elaphus* (Mammalia) feces.
995. *E. soricinae* Galli-Valerio, 1927 in *Sorex araneus* (Mammalia) feces.
996. *E. soricis* Henry, 1932 in *Sorex californicus* and also *S. araneus* (Mammalia) intestine.
997. *E. soufiae* Stankovitch, 1921 (syns., *E. sontlae* Stankovitch, 1921 *lapsus calami; E. souflae* Stankovitch, 1921 of Becker [1956] *lapsus calami*) in *Leuciscus soufia agassizi* (Osteichthys) midgut.
998. *E. southwelli* Halawani, 1930 in *Aetobatis narinari* and possibly also *Scoliodon sorrakowah* (Chondrichthys) embryo spiral valve.
999. *E. spalacis* Sayin, Dincer, and Meric, 1977 in *Spalax leucodon* (Mammalia) intestinal contents.
1000. *E. speotytoi* Carini, 1939 in *Speotyto funicularia grallaria* (Aves) intestine.
1001. *E. spermophili* Hilton and Mahrt, 1971 in *Spermophilus richardsoni* and *S. franklinii* (Mammalia) intestinal contents.
1002. *E. sphaerica* Dogel', 1948 (syns., *E. dogieli* [Dogiel, 1948] Pellérdy, 1963 of Pellérdy [1974]; *E. dogieli* Pellérdy, 1963) in *Opisthocentrus ocellatus* (Osteichthys) kidneys.
1003. *E. sphenocerci* Ray, 1952 emend. (syn., *E. sphenocercae* Ray, 1952) in *Sphenocercus sphenurus* (Aves) feces.
1004. *E. spherica* (Schneider, 1887) Levine and Becker, 1933 (syns., *Coccidium sphericum* Schneider, 1887; *Cytophagus tritonis* Steinhaus, 1891 in part) in *Triturus alpestris*, *T. cristatus*, *T. helveticus*, and *T. vulgaris* (Amphibia) feces (?).
1005. *E. spinosa* Henry, 1931 in *Sus scrofa* (Mammalia) small intestine.
1006. *E. sporadica* Plaan, 1951 in *Gallus gallus* (Aves) feces (?).
1007. *E. sprehni* Yakimoff, 1934 in *Castor canadensis* and *C. fiber* (Mammalia) feces.
1008. *E. squali* Fitzgerald, 1975 in *Squalus acanthias* (Chondrichthys) spiral valve.
1009. *E. stefanskii* Pastuszko, 1961 in *Lepus europaeus* (Mammalia) small intestine.
1010. *E. stenocerci* Carini, 1940 (syn., *E. tropiduri* Carini, 1941) in *Tropidurus t. torquatus* (Reptilia) feces.
1011. *E. stercorarii* Galli-Valerio, 1940 emend. (syn., *E. stercorariae* Galli-Valerio, 1940) in *Stercorarius parasiticus* (Aves) feces.
1012. *E. stiedai* (Lindemann, 1895) Kisskalt and Hartmann, 1907 (syns., *Monocystis stiedae* Lindemann, 1896; *Psorospermium cuniculi* Rivolta, 1878; *Coccidium oviforme* Leuckart, 1879; *C. cuniculi* [Rivolta, 1878] Labbé, 1899; *E. oviformis* [Leuckart,

1879] Fantham, 1911; *E. cuniculi* [Rivolta, 1878] von Wasielewski, 1904; *E. stiedae* [Lindemann, 1896] *auctores; E. stiedae* var. *cuniculi* Graham, 1933) in *Oryctolagus cuniculus* and also *Sylvilagus floridanus, S. nuttallii,* and *Lepus europaeus* (Mammalia) bile duct epithelium; *S. floridanus* and *S. audubonii* have been infected experimentally.

1013. *E. stigmosa* Klimeš, 1963 in *Anser anser* (Aves) small intestine.

1014. *E. stolatae* Ray and Das Gupta, 1938 in *Natrix stolata* (Reptilia) small intestine.

1015. *E. strangfordensis* Straneva and Gallati, 1980 in *Neotoma floridana* (Mammalia) feces.

1016. *E. strelkovi* Schulman and Zaika, 1962 in *Pseudorasbora parva* (Osteichthys) presumably feces.

1017. *E. strigis* Kutzer, 1963 in *Strix aluco* (Aves) feces.

1018. *E. subrotunda* Moore, Brown, and Carter, 1954 in *Meleagris gallopavo* (Aves) duodenum, jejunum, and upper ileum.

1019. *E. subsimi* Vance and Duszyn-ski, 1985 in *Microtus mexicanus subsimus* (Mammalia) feces.

1020. *E. subspherica* Christensen, 1941 (syn., *E. subsfericalis* Casorso and Zaraza, 1963) in *Bos taurus, B. indicus,* and *Bubalus bubalis* (Mammalia) feces.

1021. *E. suis* Nöller, 1921 (syns., *E. brumpti* Cauchemez, 1921 in part; *E. debliecki* Douwes, 1921 in part) in *Sus scrofa* (Mammalia) feces.

1022. *E. sultanovi* Davronov, 1973 in *Rhombomys opimus* (Mammalia) feces.

1023. *E. sumgaitica* Musaev and Alieva, 1961 in *Meriones libycus* (syn., *M. erythrourus*) (Mammalia) large intestine contents.

1024. *E. suncus* Ahluwalia et al., 1979 in *Suncus m. murinus* (Mammalia) small intestine.

1025. *E. superba* Pellérdy, 1955 (syn., *E. auburnensis* Christensen and Porter, 1939 of Böhm and Supperer [1956]) in *Capreolus capreolus* (Mammalia) feces.

1026. *E. suppereri* Kutzer, 1964 in *Rupicapra rupicapra* (Mammalia) feces.

1027. *E. surifer* Colley and Mullin, 1971 in *Rattus surifer* (Mammalia) intestinal contents.

1028. *E. surkovae* Musaev, 1970 (syn., *E. ninakohlyakimovae* Yakimoff and Rastegaieff, 1930 of Svanbaev [1958] in *Ovis ammon*) in *Ovis ammon* (Mammalia) feces.

1029. *E. susliki* Levine and Ivens, 1965 (syn., *E. ussuriensis* Yakimoff and Sprinholtz-Schmidt of Svanbaev [1962]) in *Spermophilus maximus* and *S. fulvus* (Mammalia) feces.

1030. *E. svanbaevi* Levine and Ivens, 1965 (syns., *E. kriygsmanni [sic]* Yakimoff and Gousseff of Svanbaev [1956] in part; *E. krijgsmani [sic]* of Ryšavý [1954]) in *Apodemus sylvaticus* (Mammalia) feces.

1031. *E. sylvatici* Prasad, 1960 emend. (syn., *E. sylvatica* Prasad, 1960) in *Apodemus sylvaticus* (Mammalia) feces.

1032. *E. sylvilagi* Carini, 1940 in *Sylvilagus brasiliensis minensis* and *S. floridanus mearnsi* (Mammalia) jejunum and ileum.

1033. *E. symmetrica* Fernando and Remmler, 1973 in *Gallus lafayettei* (Aves) feces.

1034. *E. syngnathi* Yakimoff and Gousseff, 1936 in *Syngnathus nigrolineatus* (Osteichthys) intestinal contents.

1035. *E. tachyglossi* Barker, Beveridge, and Munday, 1984 in *Tachyglossus aculeatus* (Mammalia) small intestine.

1036. *E. tachyoryctis* van den Berghe and Chardome, 1956 in *Tachyoryctes ruandae* (Mammalia) probably intestine.

1037. *E. tadshikistanica* Veisov, 1964 in *Ellobius talpinus* (Mammalia) feces.

1038. *E. tadzhikistanica* Anpilogova and Sokov, 1973 in *Lynx lynx isabellina* (Mammalia) feces.

1039. *E. taimyrica* Arnastauskene, 1977 in *Microtus middendorfi* (Mammalia) feces.

1040. *E. talboti* Prasad and Narayan, 1963 in *Alcelaphus cokei* (Mammalia) feces.

1041. *E. taldykurganica* Svanbaev and Utebaeva, 1973 in *Coturnix coturnix* (Aves) cecum and small intestine.

1042. *E. talikiensis* Veisov, 1975 in *Spalax leucodon* (Mammalia) feces.

1043. *E. talischaensis* Musaev and Veisov, 1960 in *Arvicola terrestris* (Mammalia) large intestine contents.

1044. *E. talpini* Levine and Ivens, 1965 (syn., *E. beckeri* Yakimoff and Sokoloff of Svanbaev [1956] in *Ellobius talpinus*) in *Ellobius talpinus* (Mammalia) feces.

1045. *E. tamanduae* Lainson, 1968 in *Tamandua tetradactyla* (Mammalia) ileum.

1046. *E. tamariscini* Levine and Ivens, 1965 (syn., *E. musculi* Yakimoff and Gousseff of Svanbaev [1956] in *Meriones tamariscinus*) in *Meriones tamariscinus* (Mammalia) feces.

1047. *E. tamiasciuri* Levine, Ivens, and Kruidenier, 1957 in *Tamiasciurus hudsonicus* (Mammalia) intestinal contents and also presumably *Microtus montanus arizonensis* and *Eutamias* spp.

1048. *E. tanabei* Levine and Ivens, 1979 (syn., *Eimeria* sp. Tanabe, 1938) in *Talpa micrura coreana* (Mammalia) intestine.

1049. *E. tarandi* Yakimoff, Sokoloff, and Matschoulsky, 1936 emend. (syn., *E. tarandina* Yakimoff, Sokoloff, and Matschoulsky, 1936) in *Rangifer tarandus* (Mammalia) feces.

1050. *E. tarichae* Levine, 1981 (syn., *E. grobbeni* Rudovsky, 1925 of Doran [1953]) in *Taricha torosus* (Amphibia) feces.

1051. *E. tasakendica* Veisov, 1961 in *Meriones vinogradovi* (Mammalia) large intestine contents.

1052. *E. tatarica* Musaev, 1970 (syn., *E. faurei* [Moussu and Marotel, 1902] Martin, 1909 of Svanbaev [1958] in *Saiga tatarica*) in *Saiga tatarica* (Mammalia) feces.

1053. *E. taterae* Mirza and Al-Rawas, 1975 in *Tatera indica* (Mammalia) feces.

1054. *E. (?) tatusi* (da Cunha and Torres, 1924) Levine 1984 (syn., *Globidium tatusi* da Cunha and Torres, 1923 of da Cunha and Torres [1924]) in *Cabassus novemcinctus* (Mammalia) small intestine.

1055. *E. tedlai* Molnar and Fernando, 1974 in *Perca flavescens* (Osteichthys) gut.

1056. *E. teetari* Bhatia, Pandey, and Pande, 1966 in *Francolinus francolinus* and *F. pondicerianus* (Aves) anterior small intestine.

1057. *E. tekenovi* Svanbaev, 1979 (syn., *E. arloingi* [Marotel, 1905] Martin, 1909 of Svanbaev [1969] in *Saiga tatarica*) in *Saiga tatarica* (Mammalia) feces.

1058. *E. telfordi* Bovee, 1971 in *Gehyra mutilata* (Reptilia) intestinal contents.

1059. *E. tenella* (Railliet and Lucet, 1891) Fantham, 1909 (syns., *Psorospermium avium* Silvestrini and Rivolta, 1871 in part; *Coccidium avium* [Silvestrini and Rivolta, 1871] Leuckart, 1879 in part; *C. tenellum* Railliet and Lucet, 1891; *C. globosum* Labbé, 1893; *Pfeifferia avium* Labbé, 1894; *Pfeifferella avium* [Labbé, 1894] Labbé, 1899; *E. avium auctores; E. bracheti* Gérard, 1913; *E. globosa* [Labbé, 1893] Levine and Becker, 1933) in *Gallus gallus* (Aves) ceca.

1060. *E. tenggilingi* Else and Colley, 1976 in *Manis javanica* (Mammalia) feces.

1061. *E. terrestris* Musaev and Veisov, 1960 in *Arvicola terrestris* (Mammalia) large intestine contents.

1062. *E. tertia* Lavier, 1935 in *Triturus alpestris* and *T. cristatus* (Amphibia) intestine.

1063. *E. tetartooimia* Wacha, 1973 in *Phasianus colchicus* (Aves) feces.

1064. *E. tetradacrutata* Wacha and Christiansen, 1976 in *Chrysemys picta belli* (Reptilia) bile ducts.

1065. *E. tetricis* Haase, 1939 in *Lyrurus t. tetrix* (Aves) feces (?).

1066. *E. theileri* De Vos and Dobson, 1970 (syn., *Eimeria* sp. Fantham, 1926) in *Rattus natalensis* (Mammalia) presumably jejunum and ileum.

1067. *E. thianethi* Gwéléssiany, 1935 in *Bubalus bubalis* and also *Bos taurus* and *B. indicus* (Mammalia) feces.
1068. *E. thomomysis* Levine, Ivens, and Kruidenier, 1957 in *Thomomys bottae* and *T. talpoides* (Mammalia) intestinal contents.
1069. *E. tikusi* Colley and Mullin, 1971 in *Rattus edwardsi* (Mammalia) intestinal contents.
1070. *E. toddi* Dorney, 1962 in *Tamiasciurus hudsonicus* (Mammalia) intestinal contents.
1071. *E. toganmainensis* Mykytowycz, 1964 in *Macropus rufus* (Mammalia) feces.
1072. *E. tolucadensis* Vance and Duszynski, 1985 in *Microtus m. mexicanus* (Mammalia) feces.
1073. *E. tomopea* Duszynski and Barkley, 1985 in *Tomopeas ravus* (Mammalia) feces.
1074. *E. torosicum* Sayin, 1981 in *Spalax ehrenbergi* (Mammalia) mixed intestinal contents.
1075. *E. townsendi* Carvalho, 1943 emend. Pellérdy, 1956 (syn., *E. magna* form *townsendii* Carvalho, 1943) in *Lepus townsendi campanius, L. europaeus, L. timidus, L. americanus,* and *L. californicus* (Mammalia) feces.
1076. *E. tragocamelis* Bhatia, 1968 in *Boselaphus tragocamelus* (Mammalia) feces.
1077. *E. traguli* Mullin and Colley, 1971 in *Tragulus javanicus* (Mammalia) intestinal contents.
1078. *E. transcaucasica* Yakimoff and Gousseff, 1937 in *Bufo bufo* (Amphibia) feces (?).
1079. *E. travassosi* da Cunha and Muniz, 1928 (syns., *Globidium travassosi* da Cunha and Muniz, 1928 of *auctores; E. (?) travassosi* da Cunha and Muniz, 1928; *E. [G.] travassosi* [da Cunha and Muniz, 1928] Reichenow, 1953) in *Euphractus sexcinctus* and "*Muletia*" *hybridus* (presumably a hybrid armadillo of some sort) (Mammalia) small intestine.
1080. *E. triangularis* Chakravarty and Kar, 1943 in *Trionyx gangeticus* (Reptilia) rectal contents.
1081. *E. trichechi* Lainson et al., 1983 in *Trichechus inunguis* (Mammalia) feces.
1082. *E. triffittae* Yakimoff, 1934 emend. Levine and Ivens, 1970 (syn., *E. triffitt* Yakimoff, 1934) in *Taurotragus oryx* (Mammalia) feces.
1083. *E. trionyxae* Chakravarty and Kar, 1943 in *Trionyx gangeticus* (Reptilia) intestine.
1084. *E. tristrami* Musaev and Veisov, 1965 in *Meriones blackleri* (Mammalia) feces.
1085. *E. tropicalis* Malhotra and Ray, 1961 in *Columba livia intermedia* (Aves) duodenum and anterior jejunum.
1086. *E. tropidonoti* Guyenot, Naville, and Ponse, 1922 in *Natrix natrix* (Reptilia) intestine.
1087. *E. truncata* (Railliet and Lucet, 1891) von Wasielewski, 1904 (syn., *Coccidium truncatum* Railliet and Lucet, 1891) in *Anser anser, A. rossi, A. caerulescens,* and *Branta canadensis* (Aves) kidneys.
1088. *E. tsunodai* Tsutsumi, 1972 in *Coturnix coturnix japonica* (Aves) ceca.
1089. *E. tuberculata* Krishnamurthy and Kshirsagar, 1980 in *Rattus r. rattus* (Mammalia) intestine.
1090. *E. tupaiae* Mullin, Colley, and Stevens, 1972 in *Tupaia glis* and *T. minor* (Mammalia) feces.
1091. *E. turaco* Pellérdy, 1967 in *Leucoptes turaco* (Aves) feces.
1092. *E. turkestanica* Veisov, 1964 in *Rattus turkestanicus* (Mammalia) feces.
1093. *E. turkmenica* Sayin, Dincer, and Meric, 1977 in *Spalax leucodon* (Mammalia) intestinal contents.
1094. *E. turnbulli* Pampiglione, Ricci-Bitti, and Kabala, 1973 in *Cephalophus dorsalis, C. monticola,* and *C. nigrifrons* (Mammalia) feces.
1095. *E. turturi* Golemansky, 1976 in *Streptopelia t. turtur* (Aves) feces.
1096. *E. tuscarorensis* Dorney, 1965 in *Marmota monax* (Mammalia) feces.
1097. *E. tuskegeensis* Barnard, Ernst, and Dixon, 1974 in *Sigmodon hispidus* (Mammalia) small intestine.

1098. *E. tuzdili* Sayin, Dincer, and Meric, 1977 in *Spalax leucodon* (Mammalia) intestinal contents.
1099. *E. tyanshanensis* Levine and Ivens, 1964 (syn., *E. monacis* Fish, 1930 of Svanbaev [1963]) in *Marmota menzbieri* (Mammalia) feces.
1100. *E. typhlopisi* Ovezmukhamedov, 1968 in *Typhlops vermicularis* (Reptilia) intestine.
1101. *E. uilenbergi* Pellérdy, 1969 (syn., *E. madagascariensis* Uilenberg, 1967 in *Tenrec ecaudatus*) in *Tenrec ecaudatus* (Mammalia) small intestine.
1102. *E. umis* Bovee, 1969 in *Uma notata* (Reptilia) gall bladder epithelial cells.
1103. *E. uniungulata* Gousseff, 1935 in *Equus caballus* and *E. asinus* (Mammalia) feces.
1104. *E. uptoni* Lewis and Ball, 1983 in *Apodemus sylvaticus* and *A. flavicollis* (Mammalia) feces.
1105. *E. urfensis* Sayin, 1981 in *Spalax ehrenbergi* (Mammalia) mixed intestinal contents.
1106. *E. urnula* Hoare, 1933 (syn., *E. cernula* Hoare, 1933 *lapsus calami*) in *Phalacrocorax carbo lugubris* (Aves) small intestine.
1107. *E. urodela* Duszynski et al., 1972 in *Ambystoma tigrinum* (Amphibia) feces.
1108. *E. urosauris* Bovee, 1966 in *Urosaurus graciosus* (Reptilia) gall bladder epithelium.
1109. *E. ursi* Yakimoff and Machul'skii, 1935 in *Ursus arctos* (Mammalia) feces.
1110. *E. ursini* Supperer, 1957 in *Lasiorhinus latifrons* (Mammalia) feces.
1111. *E. ustkamenogorica* Utebaeva, 1973 in *Tetrastes bonasia* (Aves) feces.
1112. *E. utahensis* Ernst, Hammond, and Chobotar, 1968 in *Dipodomys ordii, D. agilis, D. merriami, D. californicus,* and *D. microps* (Mammalia) feces.
1113. *E. uzbekistanica* Davronov, 1973 in *Meriones meridianus* (Mammalia) feces.
1114. *E. uzura* Tsunoda and Muraki, 1971 in *Coturnix coturnix japonica* (Aves) duodenum and upper small intestine.
1115. *E. vahidovi* Musaev and Veisov, 1965 in *Meriones vinogradovi* (Mammalia) feces.
1116. *E. vanelli* Mandal, 1965 in *Vanellus malabaricus* (Aves) intestine.
1117. *E. vanmurghavi* Pande et al., 1970 in *Gallus gallus* and *G. sonneratti* (Aves) feces.
1118. *E. varani* Bhatia and Chauhan, 1972 in *Varanus monitor* (Reptilia) feces.
1119. *E. vejsovi* Černá, 1976 in *Nyctalus noctula* (Mammalia) intestinal epithelium.
1120. *E. ventriosa* Haase, 1939 (syn., *E. ventricosa* Haase of Reichenow [1949] *lapsus calami*) in *Tetrao u. urogallus* (Aves) feces.
1121. *E. vermiformis* Ernst, Chobotar, and Hammond, 1971 in *Mus musculus* (Mammalia) small intestine.
1122. *E. vesicostieda* Wacha and Christiansen, 1977 in *Trionyx spiniferus* (Reptilia) feces.
1123. *E. vespertilii* Musaev and Veisov, 1961 in *Vespertilio kuehli* (Mammalia) feces.
1124. *E. vilasi* Dorney, 1962 in *Tamias striatus* (Mammalia) posterior small intestine.
1125. *E. vinckei* Rodhain, 1954 in *Thamnomys s. surdaster* (Mammalia) cecum.
1126. *E. vinogradovi* Veisov, 1961 in *Meriones vinogradovi* (Mammalia) large intestine contents.
1127. *E. virginianus* Anderson and Samuel, 1969 in *Odocoileus virginianus* (Mammalia) feces.
1128. *E. viridis* (Labbé, 1893) Reichenow, 1921 (syn., *Coccidium viride* Labbé, 1893) in *Rhinolophus ferrumequinum* (Mammalia) intestine.
1129. *E. vison* Kingscote, 1935 (syn., *E. mustelae* Kingscote, 1934) in *Mustela vison* and *M. putorius furo* (Mammalia) intestine.
1130. *E. voinjamensis* Bray, 1964 in *Galago (Galagoides) demidovi* (Mammalia) feces.
1131. *E. volgensis* Zasukhin and Rauschenbach, 1932 in *Spermophilus pygmaeus, S. relictus,* and *S. undulatus* (Mammalia) intestine.
1132. *E. voronezhensis* Levine and Ivens, 1981 (syn., *E. mephitidis* Andrews, 1928 of Yakimoff and Matikaschwili [1932] from *Mephitis mephitis*) in *Mephitis mephitis* (Mammalia) feces.

1133. *E. vulpis* Galli-Valerio, 1929 in *Vulpes vulpes* (Mammalia) feces.

1134. *E. waiganiensis* Varghese, 1978 in *Chalcophaps indica* and *Otidiphaps nobilis* (Aves) feces.

1135. *E. walleri* Prasad, 1960 in *Litocranius walleri* (Mammalia) feces.

1136. *E. waltoni* Saxe, 1955 in *Ambystoma tigrinum* (Amphibia) feces.

1137. *E. wapiti* Honess, 1955 in *Cervus canadensis nelsoni* (Mammalia) feces.

1138. *E. wassilewskyi* Rastegaieff, 1930 (syns., *E. wassilewsky* Rastegaieff, 1930 *lapsus calami; E. wassielewskyi* Rastegaieff of Pellérdy [1963] *lapsus calami*) in *Axis axis* (Mammalia) feces.

1139. *E. webbae* Barnard, Ernst, and Dixon, 1974 in *Sigmodon hispidus* (Mammalia) feces.

1140. *E. weissi* Gonzalez-Mugaburu, 1946 in *Phyllotis a. amicus* (Mammalia) ileum and cecum.

1141. *E. wenrichi* Saxe, Levine, and Ivens, 1960 in *Microtus pennsylvanicus* and *M. breweri* (Mammalia) cecal contents and also *M. mexicanus* and *M. montanus* feces.

1142. *E. weybridgensis* norton, Joyner, and Catchpole, 1974 (syn., *E. arloingi* Type B of Pout, Norton, and Catchpole [1973]) in *Ovis aries* (Mammalia) small intestine.

1143. *E. wilcanniensis* Mykytowycz, 1964 in *Macropus rufus* (Mammalia) feces.

1144. *E. williamsi* Musaev and Veisov, 1965 in *Allactaga williamsi* (Mammalia) feces.

1145. *E. wisconsiensis* Dorney, 1962 in *Tamias striatus* (Mammalia) middle small intestine.

1146. *E. wombati* (Gilruth and Bull, 1912) Barker, Munday, and Presidente, 1979 (syns., *Ileocystis wombati* Gilruth and Bull, 1912; *E. [Globidium] tasmaniae* Supperer, 1957; *E. tasmaniae* Supperer, 1957 emend. Pellérdy, 1974) in *Lasiorhinus latifrons* (Mammalia) small intestine.

1147. *E. worleyi* Lepp, Todd, and Samuel, 1972 in *Ochotona princeps* (Mammalia) feces.

1148. *E. wyomingensis* Huizinga and Winger, 1942 in *Bos taurus, B. indicus*, and *Bubalus bubalis* (Mammalia) feces.

1149. *E. xeri* Vassiliades, 1967 in *Xerus (Euxerus) erythropus* (Mammalia) feces.

1150. *E. yakimoffmatschoulskyi* Supperer and Kutzer, 1961 (syns., *E. arloingi* Marotel, 1905 of Yakimoff and Matschoulsky [1940] in part; *E. böhmi* Supperer, 1952 of Böhm and Supperer [1956]) in *Rupicapra rupicapra* (Mammalia) feces.

1151. *E. yakimovi* Rastegaieff, 1929 in *Boselaphus tragocamelus* (Mammalia) feces.

1152. *E. yakisevi* Hardcastle, 1943 (syn., *E. brumpti* Yakimoff and Gousseff, 1936 in part) in *Tetrao urogallus* (Aves) feces.

1153. *E. yunnanensis* Zuo and Chen, 1984 in *Bos taurus* (Mammalia) feces.

1154. *E. yukonensis* Sampson, 1969 in *Spermophilus undulatus* (Mammalia) feces.

1155. *E. zakirica* Musaev, 1967 in *Vespertilio kuehlii* (Mammalia) large intestine.

1156. *E. zamenis* Phisalix, 1921 in *Coluber constrictor, C. c. flaviventris, Coluber* sp., *Lampropeltis c. calligaster, L. getulus holbrooki, L. t. triangulum*, and *Masticophis f. flagellum* (Reptilia) bile.

1157. *E. zapi* Gerard, Chobotar, and Ernst, 1977 in *Zapus hudsonius, Z. trinotatus*, and *Z. princeps* (Mammalia) feces.

1158. *E. zaurica* Musaev and Veisov, 1963 in *Apodemus sylvaticus* (Mammalia) feces.

1159. *E. zuernii* (Rivolta, 1878) Martin, 1909 (syns., *Cytospermium zurnii* Rivolta, 1878; *Coccidium zürni* [Rivolta, 1878] Railliet and Lucet, 1891; *E. bovis* [Züblin, 1908] Fiebiger, 1912 in part; *E. canadensis* Bruce, 1921 in part; *E. zuerni* [Rivolta, 1878] Martin, 1909 of Pellérdy [1965]) in *Bos taurus* and *Bubalus bubalis* (Mammalia) cecum, colon, rectum, and throughout small intestine. (Some authors have reported this species from the wisent, white-tailed deer, roe deer, and elk, but without descriptions; they were probably dealing with other species.)

1160. *E. zulfiaensis* Veisov, 1961 in *Meriones vinogradovi* (Mammalia) large intestine contents.

1161. *E. zuvandica* Veisov, 1963 in *Microtus arvalis* (Mammalia) large intestine contents.
1162. *E. zygaenae* Mandal and Chakravarty, 1965 in *Sphyrna* (syn., *Zygaena*) *blochii* (Chondrichthys) small intestine.

Genus *Epieimeria* Dyková and Lom, 1981. Oocysts with four sporocysts, each with two sporozoites; sporocysts with Stieda body; merogony and gamogony extracellular; sporogony intracellular; in fish. TYPE SPECIES *E. anguillae* (Léger and Hollande, 1922) Dyková and Lom, 1981.

1.   *E. anguillae* (Léger and Hollande, 1922) Dyková and Lom, 1981 (syn., *Eimeria anguillae* Léger and Hollande, 1922) (TYPE SPECIES) in *Anguilla anguilla, A. rostrata,* and probably *A. australis* and *A. dieffenbachi* (Osteichthys) intestine.
2.   *E. isabellae* Lom and Dyková, 1982 in *Conger conger* (Osteichthys) middle part of intestine.

Genus *Mantonella* Vincent, 1936. Oocysts with one sporocyst, each containing four sporozoites. TYPE SPECIES *M. peripati* Vincent, 1936.

1.   *M. hammondi* Wacha and Christiansen, 1976 in *Kinosternon flavescens spooneri* (Reptilia) intestinal contents.
2.   *M. meriones* Glebezdin, 1971 in *Meriones erythrourus* (Mammalia) feces.
3.   *M. peripati* Vincent, 1936 (TYPE SPECIES) in *Peripatopsis sedgwicki* and *P. moseleyi* (Onychophora) intestine.
4.   *M. podurae* Manier and Bouix, 1981 in springtail *Podura aquatica* (Collembolorida) intestinal epithelium.
5.   *M. potamobii* Gousseff, 1936 (syn., *Yakimovella potamobii* Gousseff, 1936) in *Astacus leptodactylus* (Crustaceasida) intestine.

Genus *Cyclospora* Schneider, 1881. Oocysts with two sporocysts, each with two sporozoites. TYPE SPECIES *C. glomericola* Schneider, 1881.

1.   *C. babaulti* Phisalix, 1924 in *Vipera berus* (Reptilia) intestine.
2.   *C. caryolytica* Schaudinn, 1902 in *Talpa europaea*, presumably *T. micrura coreana*, and possibly *Parascalops breweri* (Mammalia) small and large intestines.
3.   *C. glomericola* Schneider, 1881 (TYPE SPECIES) in *Glomeris* sp. (Diplopodasida) intestine.
4.   *C. niniae* Lainson, 1965 in *Ninia s. sebae* (Reptilia) small intestine.
5.   *C. scinci* Phisalix, 1924 in *Scincus officinalis* (Reptilia) intestine.
6.   *C. talpae* Pellérdy and Tanyi, 1968 in *Talpa europaea* (Mammalia) liver.
7.   *C. tropidonoti* Phisalix, 1924 in *Natrix natrix* and *N. stolata* (Reptilia) intestine.
8.   *C. viperae* Phisalix, 1923 in *Vipera aspis* and possibly also *Coluber scalaris, Coronella austriaca,* and *Natrix viperinus* (Reptilia) intestine.
9.   *C. zamenis* Phisalix, 1924 in *Coluber v. viridiflavus* (Reptilia) intestine.

Genus *Caryospora* Léger, 1904. Oocysts with one sporocyst containing eight sporozoites (a few species are known to have hypnozoites). TYPE SPECIES *C. simplex* Léger, 1904.

1.   *C. argentati* Schwalbach, 1959 in *Larus argentatus* (Aves) feces.
2.   *C. bengalensis* Mandal, 1976 in *Enhydris enhydris* (Reptilia) intestine.
3.   *C. bigenetica* Wacha and Christiansen, 1982 in *Crotalus horridus* and *Sistrurus catenatus* (Reptilia) duodenum and jejunum and also experimentally *Mus musculus* (Mammalia) connective tissue, dermis, and hypodermis.

4.   *C. brasiliensis* Carini, 1932 in *Philodryas aestivus* and also *P. olfersi, P. nattereri,* and *Leimadophis poecilogyrus schotti* (Reptilia) small intestine epithelial cells.
5.   *C. bubonis* Cawthorn and Stockdale, 1981 in *Bubo virginianus* (Aves) feces.
6.   *C. cheloniae* Leibovitz, Rebell, and Boucher, 1978 in *Chelonia m. mydas* (Reptilia) posterior small intestine and large intestine.
7.   *C. cobrae* Nandi, 1985 in *Naja naja* (Reptilia) feces.
8.   *C. colubris* Matuschka, 1984 in *Coluber viridiflavus* (Reptilia) feces.
9.   *C. corallae* Matuschka, 1984 in *Corallus caninus* (Reptilia) feces.
10. *C. demansiae* Cannon, 1967 in *Demansia psammophis* (Reptilia) intestine.
11. *C. dendrelaphis* Cannon and Rzepczyk, 1974 in *Dendrelaphis punctulatus* (Reptilia) duodenal epithelium.
12. *C. duszynskii* Upton, Current, and Barnard, 1984 in *Elaphe guttata* and *E. o. obsoleta* (Reptilia) feces.
13. *C. ernsti* Upton, Current, and Barnard, 1984 in *Anolis carolinensis* (Reptilia) anterior intestine.
14. *C. falconis* Wetzel and Enigk, 1937 in *Falco p. peregrinus* (Aves) feces.
15. *C. gekkonis* Chakravarty and Kar, 1947 in *Gekko gecko* (Reptilia) feces.
16. *C. gloriae* Pellérdy, 1967 in *Ptiloxena atroviolacea* (Aves) feces.
17. *C. henryae* (Yakimoff and Matikaschwili, 1932) Yakimoff and Machul'skii, 1936 emend. Pellérdy, 1974 (syns., *Isospora henryi* Yakimoff and Matikaschwili, 1932; *C. henryi* [Yakimoff and Matikaschwili, 1932] Yakimoff and Machul'skii, 1936) in *Bubo bubo* and possibly also *Tinnunculus tinnunculus, Falco subbuteo,* and *Milvus ater* (Aves) feces.
18. *C. hermae* Bray, 1960 in *Psammophis sibilans phillipsi* (Reptilia) intestine.
19. *C. japonicum* Matsubayashi, 1936 in *Natrix tigrina* (Reptilia) intestine.
20. *C. jararacae* Carini, 1939 (syn., *C. jaracae* Carini, 1939 *lapsus calami*) in *Bothrops jararaca* (Reptilia) intestine.
21. *C. jiroveci* Černá, 1976 in *Erithacus rubecula* (Aves) feces.
22. *C. kutzeri* Böer, 1982 in *Falco mexicanus, F. tinnunculus, F. biarmicus, F. jugger, F. cherrug, F. risticolus,* and *F. peregrinus* (Aves) feces.
23. *C. lampropeltis* Anderson, Duszynski, and Marquardt, 1968 in *Lampropeltis c. calligaster* (Reptilia) presumably feces.
24. *C. legeri* Hoare, 1933 in *Psammophis sibilans* (Reptilia) small intestine.
25. *C. microti* Saxe, Levine, and Ivens, 1960 (syn., *Caryospora* sp. Saxe, 1952) in *Microtus pennsylvanicus* (Mammalia) cecal contents.
26. *C. najae* Matuschka, 1982 in *Naja nigricollis pallida* (Reptilia) small intestine and liver.
27. *C. neofalconis* Böer, 1982 in *Falco mexicanus, F. subbuteo, F. biarmicus,* and *F. peregrinus* (Aves) feces.
28. *C. psammophi* Bray, 1960 in *Psammophis sibilans phillipsi* (Reptilia) small intestine.
29. *C. simplex* Léger, 1904 (TYPE SPECIES) in *Vipera aspis* and *V. xanthina* (Reptilia) posterior intestine and experimentally *Mus musculus* and *Sigmodon hispidus* (Mammalia) cheek, tongue, and nose.
30. *C. strigis* Gottschalk, 1972 in *Tyto alba* (Aves) small intestine.
31. *C. tremula* (Allen, 1933) Hoare, 1934 (syn., *Eumonospora tremula* Allen, 1933) in *Cathartes aura septentrionalis* (Aves) intestine.
32. *C. undata* Schwalbach, 1959 in *Laurus argentatus* and also *Uria a. aalge* (Aves) feces.
33. *C. weyerae* Bray, 1960 in *Psammophis sibilans phillipsi* (Reptilia) small intestine.
34. *C. zuckermanae* Bray, 1960 in *Coluber ravergieri nummifer* (Reptilia) intestine.

Genus *Isospora* Schneider, 1881. Oocysts with two sporocysts, each with four sporozoites; usually in vertebrates. TYPE SPECIES *I. rara* Schneider, 1881. (*Remarks*. The species names and synonymy of this genus will be uncertain until the complete life cycles of all its members are known. Some species may have been named *Haemogregarina* or *Isospora* because their blood forms and fecal oocysts were not thought to be related; they actually might be *Atoxoplasma* species.)

1. *I. ablephari* Cannon, 1967 in *Ablepharus boutonii* and also apparently *Heteronota bineoi* (Reptilia) middle to upper intestine.
2. *I. africana* Prasad, 1961 in *Poecilictis libyca alexandrae* (Mammalia) feces.
3. *I. aksaica* Bazanova, 1952 in *Bos taurus* (Mammalia) feces.
4. *I. almaataensis* Paichuk, 1955 emend. Orlov, 1956 (syn., *I. almataensis* Paichuk, 1950) in *Sus scrofa* (Mammalia) feces.
5. *I. alaudae* Gottschalk, 1972 in *Alauda arvensis* (Aves) intestine.
6. *I. altaica* Svanbaev and Rakhmatullina, 1971 in *Mustela altaica* (Mammalia) feces.
7. *I. ameivae* Carini, 1932 in *Ameiva ameiva* and *Cnemidophorus l. lemniscatus* (Reptilia) intestine.
8. *I. amphiboluri* Cannon, 1967 in *Amphibolurus barbatus* (Reptilia) duodenum and small intestine. (This species may be *Sarcocystis*.)
9. *I. ampullacea* Schwalbach, 1959 in *Sylvia borin* (Aves) feces.
10. *I. anatolicum* Sayin, Dincer, and Meric, 1977 in *Spalax leucodon* (Mammalia) intestinal contents.
11. *I. anseris* Skene, Remmler, and Fernando, 1981 in *Branta canadensis* and experimentally *Anser anser* and *A. caerulescens* (Aves) feces.
12. *I. anthi* Schwalbach, 1959 (syn., *Eimeria anthi* Schwalbach, 1959 of Pellérdy [1974] *lapsus calami*) in *Anthus pratensis* (Aves) feces.
13. *I. aranei* Golemansky, 1978 emend. (syn., *I. araneae* Golemansky, 1978) in *Sorex araneus* (Mammalia) intestinal contents.
14. *I. arctopitheci* Rodhain, 1933 in *Callithrix penicillata*, *Saguinus geoffroyi*, *Cebus capucinus*, and experimentally *Aotus trivirgatus*, *Ateles fuscipes*, *Alouatta vilosa*, *Saimiri sciureus*, *Galago senegalensis*, *Canis familiaris*, *Nasua nasua*, *Potos flavus*, *Eira barbara*, *Felis catus*, and *Didelphis marsupialis* (Mammalia) intestine.
15. *I. arvalis* Mikeladze, 1973 in *Microtus arvalis* (Mammalia) feces.
16. *I. ashkhabadensis* Ovezmukhamedov, 1968 in *Vipera lebetina* (Reptilia) hindgut contents.
17. *I. assensis* Svanbaev, 1979 (syn., *I. laguri* Iwanoff-Gobzem, 1935 of Svanbaev [1962] in *Spermophilus fulvus*) in *Spermophilus fulvus* (Mammalia) feces.
18. *I. aurangabadensis* Kshirsagar, 1980 in *Rattus r. rattus* (Mammalia) intestinal contents.
19. *I. balaericae* Banik and Ray, 1962 in *Balaerica* (syn., *Bellerica) regulorum* (Aves) feces.
20. *I. basilisci* Lainson, 1968 in *Basiliscus vittatus* (Reptilia) feces.
21. *I. batabatica* Musaev and Veisov, 1960 in *Arvicola terrestris* (Mammalia) large intestine.
22. *I. belli* Wenyon, 1923 (syn., *Lucetina belli* [Wenyon, 1923] Henry and Leblois, 1926) in *Homo sapiens* (Mammalia) upper small intestine.
23. *I. bengalensi* Patnaik and Acharjyo, 1971 (syn., *I. nandankanani* Patnaik and Acharjyo, 1977) in leopard cat (Mammalia) intestine.
24. *I. bisoni* Mandal and Choudhury, 1983 in *Bos gaurus* (Mammalia) feces.
25. *I. boxae* Grulet, Landau, and Baccam, 1982 in *Passer domesticus* (Aves) small intestine.
26. *I. brayi* Levine, Van Riper, and Van Riper, 1980 in *Zosterops japonicus* (Aves) feces.
27. *I. brumpti* Lavier, 1941 in *Bufo viridis* (Amphibia) small intestine.
28. *I. buriatica* Yakimoff and Machul'skii, 1940 in *Vulpes corsac*, *V. vulpes*, *Alopex lagopus*, and possibly *V. bengalensis* (Mammalia) feces.

29. *I. burrowsi* Trayser and Todd, 1977 in *Canis familiaris* (Mammalia) posterior small intestine and cecum; transport hosts (experimental) *Mus musculus* and *Rattus norvegicus* (Mammalia) liver, spleen, mesentery, and occasionally skeletal muscles.

30. *I. buteonis* Henry, 1932 in *Buteo jamaicensis* and probably *B. swainsoni* and other raptorial birds including *Accipiter nisus, A. gracilis, Athene noctua, Aquila chrysaetos, Asio otus, B. buteo, B. rufinus, Falco naumanni, F. tinnunculus, F. peregrinus, F. sparvarius, Strix aluco,* and *Tyto alba* (Aves) feces. (This species actually may belong to the genus *Frenkelia*.)

31. *I. californica* Davis, 1967 in *Peromyscus californicus, P. boylii, P. truei,* and *P. maniculatus* (Mammalia) cecal contents and feces.

32. *I. callimico* Hsu and Melby, 1974 in *Callimico goeldii* (Mammalia) feces.

33. *I. calomyscus* Musaev and Veisov, 1965 in *Calomyscus bailwardi* (Mammalia) feces.

34. *I. calotesi* Bhatia, 1938 in *Calotes versicolor* (Reptilia) intestinal epithelium.

35. *I. camillerii* (Hagenmüller, 1898) Sergent, 1902 (syn., *Diplospora camillerii* Hagenmüller, 1898) in *Chalcides ocellatus* (Reptilia) intestine.

36. *I. canaria* Box, 1975 (syn., *I. chloridis* of Romero Rodriguez [1973]) in *Serinus canarius* (Aves) small intestine epithelium.

37. *I. canis* Neméséri, 1959 (syns., *I. felis* Wenyon, 1923 in the dog; *Cystoisospora canis* [Neméséri, 1959] Frenkel, 1977; *Levinea canis* [Neméséri, 1959] Dubey, 1977) in *Canis familiaris* and experimentally *C. latrans*; transport host (experimental) *Mus musculus* (Mammalia) mostly small intestine, also liver, spleen, mesentery, and occasionally skeletal muscles.

38. *I. canivelocis* (Weidman, 1915) Wenyon, 1923 (syns., *Coccidium bigeminum* var. *canivelocis* Weidman, 1915; *I. bigemina* var. *canivelocis* [Weidman, 1915] Mesnil, 1916; *I. canivecolis [sic]* [Weidman, 1915] Wenyon, 1923 *lapsus calami; I. canivelocis* [Weidman, 1915] Wenyon, 1923 of *auctores; Lucetina canivelocis* [Weidman, 1915] Henry and Leblois, 1926) in *Vulpes velox* and also *V. vulpes* and experimentally *Alopex lagopus* (Mammalia) feces.

39. *I. cannabinae* Gottschalk, 1972 in *Carduelis cannabina* (Aves) intestine.

40. *I. capistrata* Sinha and Sinha, 1981 in *Heterophasia capistrata* (Aves) feces.

41. *I. capreoli* Svanbaev, 1958 in *Capreolus capreolus* (Mammalia) feces.

42. *I. cardinalis* Gottschalk, 1972 in *Lophospingus pusillus* (Aves) small intestine.

43. *I. carduelis* Gottschalk, 1969 in *Carduelis carduelis* (Aves) small intestine.

44. *I. caryophila* Rogier and Colley, 1976 in *Gonyocephalus grandis* (Reptilia) posterior small intestine.

45. I. cebi Marinkelle, 1969 in *Cebus albifrons* (Mammalia) feces.

46. *I. certhiae* Gottschalk, 1972 in *Certhia brachydactyla* (Aves) feces.

47. *I. ceylonensis* Sinha et al., 1978 (syn., *I. calochrysea* Sinha et al., 1978 of *Zool. Rec.* *[1978] lapsus calami*) in *Culicicapa ceylonensis calochrysea* (Aves) small intestine.

48. *I. cheeli* Bhatia, Pande, and Garg, 1970 in *Milvus migrans* (Aves) small intestine. (This species might be *Sarcocystis* or *Frenkelia*.)

49. *I. chloridis* Anwar, 1966 in *Carduelis chloris* and also possibly *Fringilla coelebs* and dubiously *Passer domesticus* (Aves) duodenum.

50. *I. chobotari* Levine and Ivens, 1981 (syn., *Isospora* sp. Inabnit, Chobotar, and Ernst, 1972) in *Procyon lotor* (Mammalia) feces.

51. *I. choudari* Bhatia et al., 1972 (syn., *I. choudrai* Bhatia et al., 1972 of Pellérdy [1974] *lapsus calami*) in *Gallus sonneratii* (Aves) feces.

52. *I. cincli* Gottschalk, 1972 in *Cinclus cinclus* (Aves) intestine.

53. *I. citelli* Levine, Ivens, and Kruidenier, 1957 in *Spermophilus variegatus utah* and *S. fulvus* (Mammalia) feces.

54. *I. clethrionomydis* Golemanski and Yankova, 1973 in *Clethrionomys glareolus* (Mammalia) small intestine.

55. *I. cnemidophori* Carini, 1942 in *Cnemidophorus lemniscatus, Ameiva undulata,* and *A. ameiva* (Reptilia) feces.
56. *I. concinnus* Sinha and Sinha, 1979 in *Aegithalos concinnus* (Aves) small intestine.
57. *I. corvi* Ray et al., 1952 emend. Levine, 1985 (syn., *I. corviae* Ray et al., 1952) in *Corvus macrorhynchus intermedius* (Aves) small intestine.
58. *I. crotali* (Triffitt, 1925) Hoare, 1933 (syn., *Cryptosporidium crotali* Triffitt, 1925) in *Crotalus confluentus, C. adamanteus,* and perhaps other snakes (Reptilia) feces. (This species might be *Sarcocystis.*)
59. *I. cruzi* Pinto and Vallim, 1926 in *Hyla crospedospila, H. nasica, H. fuscovaria,* and *H. rubra* (Amphibia) intestinal epithelium. (This species might be *Sarcocystis.*)
60. *I. dasguptai* Levine, Ivens, and Healy, 1975 (syns., *I. rivolta* (Grassi, 1879) of Knowles and Das Gupta [1931]; *I. garnhami* [small form] Bray, 1954 of Dubey and Pande [1963]) in *Herpestes auropunctatus* and *H. edwardsi* (Mammalia) feces.
60a. *I. dawadimiensis* Kasim and Al Shawa, 1985 in *Jaculus jaculus* (Mammalia) cecal and colon contents.
61. *I. decipiens* Lainson and Shaw, 1973 in *Philodryas olfersi* (Reptilia) small intestine.
62. *I. dilatata* Schwalbach, 1959 in *Anthus pratensis,* possibly *Sylvia borin, S. communis, S. atricapilla,* and perhaps *Sturnus vulgaris* and *S. contra* (Aves) feces.
63. *I. divitis* Pellérdy, 1967 in *Dives atraviolacea* (Aves) feces.
64. *I. dutoiti* Yakimoff, Matikaschwili, and Rastegaieff, 1933 (syn., *Eimeria dutoiti* Yakimoff, Matikaschwili, and Rastegaieff, 1933 *lapsus calami*) in *Canis aureus* (Mammalia) feces.
65. *I. dyromidis* Glebezdin, 1974 in *Dyromys nitedula* (Mammalia) feces.
66. *I. egerniae* Cannon, 1967 in *Egernia whitii* (Reptilia) mid-intestine epithelium. (This species might be *Sarcocystis.*)
67. *I. egypti* Prasad, 1960 in *Meriones shawi* (Mammalia) feces.
68. *I. emberizae* Mandal and Chakravarty, 1964 in *Emberiza bruniceps* (Aves) small intestine.
69. *I. endocallimici* Duszynski and File, 1974 in *Callimico goeldii* (Mammalia) feces.
70. *I. erinacei* Yakimoff and Gousseff, 1936 in *Erinaceus europaeus* (Mammalia) feces.
71. *I. erithaci* Anwar, 1972 in *Erithacus rubecula* (Aves) duodenum.
72. *I. erythrourica* Veisov, 1964 in *Meriones libycus* (syn., *M. erythrourus*) (Mammalia) feces.
73. *I. eversmanni* Svanbaev, 1956 in *Mustela eversmanni* and *M. vison* (Mammalia) feces.
74. *I. fatiguei* Grulet, Landau, and Baccam, 1982 in *Passer domesticus* (Aves) small intestine.
75. *I. felis* Wenyon, 1923 (syns., *Diplospora bigemina* of von Wasielewski [1904] in part; *I. bigemina* of Swellengrebel [1914]; *I. rivolta* Dobell and O'Connor, 1921; *I. cati* Marotel, 1921; *I. felis* var. *servalis* Mackinnon and Dibb, 1938; *Lucetina felis* [Wenyon, 1923] Henry and Leblois, 1926; *Cystoisospora felis* [Wenyon, 1923] Frenkel, 1977; *Levinea felis* [Wenyon, 1923] Dubey, 1977) in *Felis catus, F. sylvestris, F. serval, Leo tigris, L. leo, L. onca,* and *Lynx lynx;* transport hosts (all experimental) *Mus musculus, Rattus norvegicus, Mesocricetus auratus,* and *Bos taurus* (Mammalia) small intestine, sometimes cecum, occasionally colon, also (in transport hosts) liver, spleen, mesentery, and occasionally skeletal muscles.
76. *I. fennechi* Prasad, 1961 in *Fennecus zerda* (Mammalia) feces.
77. *I. ficedulae* Schwalbach, 1959 in *Ficedula hypoleuca* (Aves) feces.
78. *I. fonsecai* Yakimov and Machul'skii, 1940 in *Ursus arctos isabellinus* (Mammalia) feces.
79. *I. fragilis* (Léger, 1904) Lavier, 1941 (syn., *Diplospora fragilis* Léger, 1904) in *Vipera aspis* (Reptilia) intestine. (This species might be *Sarcocystis.*)

80. *I. freundi* Yakimoff and Gousseff, 1935 in *Cricetus cricetus* (Mammalia) feces.
81. *I. fringillae* Yakimoff and Gousseff, 1938 (syns., *I. rivoltai* [Labbé, 1893] Levine, 1982); *Diplospora rivoltae* Labbé, 1893) in *Fringilla coelebs* (Aves) feces (*nomen dubium*).
82. *I. gallicolumbae* Varghese, 1978 in *Gallicolumba beccarii* (Aves) feces.
83. *I. gallinae* Scholtyseck, 1954 (syn., *I. iallinae* Scholtyseck, 1954 of Bhatia et al. [1972] *lapsus calami*) in *Gallus gallus* (Aves) feces.
84. *I. gallinarum* Glebezdin, 1964 in *Gallus gallus* (Aves) feces.
85. *I. garnhami* Bray, 1954 in *Helogale undulata rufula* and *Crossarchus obscurus* (Mammalia) ileum and colon.
86. *I. garrulacis* Ray et al., 1952 emend. Levine, 1982 (syn., *I. garrulae* Ray et al., 1952) in *Garrulax l. lineatus* (Aves) small intestine.
87. *I. garruli* Ray et al., 1952 emend. Levine, 1982 (syn., *I. garrulusae* Ray et al., 1952) in *Garrulus glandarius bispecularis* (Aves) small intestine.
88. *I. ginginiani* Chakravarty and Kar, 1944 emend. Levine, 1982 (syn., *I. ginginiana* Chakravarty and Kar, 1944) in *Acridotheres ginginianus* (Aves) intestine.
89. *I. golemanskii* Levine, 1982 (syn., *Isospora* sp. Golemanski and Yankova, 1973) in *Apodemus flavicollis* and *A. sylvaticus* (Mammalia) feces.
90. *I. gonnetae* Grulet, Landau, and Baccam, 1982 in *Passer domesticus* (Aves) small intestine.
91. *I. graculae* Bhatia et al., 1973 emend. Levine, 1982 (syn., *I. graculai* Bhatia et al., 1973) in *Gracula religiosa* (Aves) feces.
92. *I. guersae* Yakimoff and Matschoulsky, 1937 (syn., *I. gursae* Yakimov and Machul'skii, 1940) in *Vipera lebentina* (Reptilia) feces.
93. *I. guzarica* Davronov, 1985 in *Coluber karelini* (Reptilia) feces.
94. *I. gymnodactyli* Ovezmukhammedov, 1972 in *Gymnodactylus fedtschenkoi* (Reptilia) feces.
95. *I. gypsi* Patnaik and Mohanti, 1969 in *Gyps bengalensis* (Aves) feces.
96. *I. hammondi* Barnard, Ernst, and Stevens, 1971 in *Oryzomys palustris* (Mammalia) intestine.
97. *I. hastingsi* Davis, 1967 in *Peromyscus truei* (Mammalia) cecal contents.
98. *I. heissini* Svanbaev, 1955 in *Meleagris gallopavo* (Aves) feces.
99. *I. hemidactyli* Carini, 1936 in *Hemidactylus mabuya* (Reptilia) small intestine.
100. *I. herpestei* Levine, Ivens, and Healy, 1975 in *Herpestes auropunctatus* (Mammalia) feces.
101. *I. hirundinis* Schwalbach, 1959 in *Hirundo rustica* and also possibly *Phoenicurus ochruros* and *P. phoenicurus* (Aves) feces.
102. *I. hoarei* Bray, 1954 in *Helogale undulata rufula* (Mammalia) epithelial cells of duodenum and jejunum.
103. *I. hoogstraali* Prasad, 1961 in *Poecilictis libyca alexandrae* (Mammalia) feces.
104. *I. hylae* Mesnil, 1907 in *Hyla arborea* (Amphibia) small intestine. (This species might be *Sarcocystis*.)
105. *I. hylomysis* Colley and Mullin, 1971 in *Hylomys suillus* (Mammalia) feces.
106. *I. iansmithi* Grulet, Landau, and Baccam, 1982 in *Passer domesticus* (Aves) small intestine.
107. *I. ichneumonis* Levine, Ivens, and Healy, 1975 (syn., *I. rivolta* [Grassi, 1879] of Balozet [1933]) in *Herpestes ichneumon* (Mammalia) feces.
108. *I. ivensae* Levine, Van Riper and Van Riper, 1980 in *Lonchura punctulata* (Aves) feces.
109. *I. jacarei* Carini and Biocca, 1940 in *Caiman latirostris* (Reptilia) intestine. (This species might be *Sarcocystis*.)

110. *I. jacobfrenkeli* Grulet, et al., 1985 (syn., *I. frenkeli* Grulet, Landau, and Baccam, 1982) in *Passer domesticus* (Aves) small intestine.
111. *I. jeffersonianum* Doran, 1953 in *Ambystoma jeffersonianum* (Amphibia) feces.
112. *I. kaschcadarinica* Davronov, 1985 in *Eremias lineolata* (Reptilia) feces.
113. *I. knowlesi* Ray and Das Gupta, 1937 in *Hemidactylus flaviviridis* (Reptilia) small intestine (intranuclear).
114. *I. koreana* Bhatia et al., 1972 in *Phasianus colchicus* and *P. versicolor* (Aves) feces.
115. *I. kouyatei* Grulet, Landau, and Baccam, 1982 in *Passer domesticus* (Aves) small intestine.
116. *I. krishnamurthyi* Kshirsagar, 1980 in *Rattus r. rattus* (Mammalia) intestinal contents.
117. *I. lacazei* Labbé, 1893 (syns., *Diplospora rivoltae* Labbé, 1893; *D. lacazii* Labbé, 1893; *D. lacazei* Labbé, 1893 of Labbé [1899]) in *Carduelis carduelis* (Aves) small intestine.
118. *I. laguri* Iwanoff-Gobzem, 1934 in *Lagurus lagurus* (Mammalia) intestine.
119. *I. laidlawi* Hoare, 1927 in *Mustela putorius* var. *furo* and *M. vison* (Mammalia) feces; transport host *Mus musculus* (Mammalia) liver, spleen, mesentery, and occasionally skeletal muscles.
120. *I. laverani* (Hagenmüller, 1898) Sergent, 1902 (syns., *Diplospora laverani* Hagenmüller, 1898; *Lucetina laverani* [Hagenmüller, 1898] Henry and Leblois, 1926) in *Malpolon monspessulanus* (Reptilia) intestine. (This species might be *Sarcocystis*.)
121. *I. lenti* Pinto, 1934 in *Bothrops jararaca* (Reptilia) intestine. (This species might be *Sarcocystis*.)
122. *I. leonina* Mandal and Ray, 1960 in *Leo leo* (Mammalia) feces.
123. *I. levinei* Dubey, 1963 in *Hyaena hyaena* (syn., *H. striata*) (Mammalia) feces.
124. *I. lickfeldi* Schwalbach, 1959 in *Sylvia borin, S. communis,* and *S. atricapilla* (Aves) feces.
125. *I. lieberkuehni* (Labbé, 1894) Laveran and Mesnil, 1902 (syns., *Klossia lieberkuehni* Labbé, 1894; *Hyaloklossia lieberkuehni* [Labbé, 1894] Labbé, 1986) in *Rana esculenta, R. pipiens,* and *Bombina variegata* (Amphibia) kidneys. (This species might be *Sarcocystis*.)
126. *I. lonchurae* Mandal and Chakravarty, 1964 in *Lonchura punctulata* and perhaps *Sturnus contra* (Aves) small intestine.
127. *I. loxopis* Levine, Van Riper, and Van Riper, 1980 in *Viridonia* (syns., *Loxopis, Loxops) virens* (Aves) feces.
128. *I. lusciniae* Golemansky, 1977 in *Erithacus* (syn., *Luscinia) megarhynchus* (Aves) small intestine.
129. *I. lutreolinae* Carini, 1939 in *Lutreolina crassicauda* (Mammalia) subepithelial tissues of small intestine villi. (This species might be *Sarcocystis*.)
130. *I. lyncis* Levine and Ivens, 1981 (syn., *I. felis* [von Wasielewski, 1904] Wenyon, 1923 from lynx of Triffitt [1927]) in *Lynx* sp. (Mammalia) feces.
131. *I. lyruri* Galli-Valerio, 1931 in *Lyrurus tetrix* and *Tetrao urogallus* (Aves) feces.
132. *I. malabaricae* Swarup and Chauhan, 1975 in *Sturnus m. malabarica* (Aves) intestine.
133. *I. mandari* Bhatia et al., 1972 in *Aix galericulata* (Aves) feces.
134. *I. marquardti* Duszynski and Brunson, 1972 in *Ochotona princeps* (Mammalia) feces.
135. *I. martes* Nukerbaeva, 1981 emend. (syns., *I. martesi* Nukerbaeva, 1981; *I. martessi* Nukerbaeva, 1981 *lapsus calami*) in *Martes zibellina* (Mammalia) feces.
136. *I. masoni* Upton et al., 1985 in *Sigmodon hispidus* (Mammalia) jejunum and ileum.
137. *I. mayuri* Patnaik, 1966 (syn., *I. pellerdyi* Patnaik, 1965) in *Pavo cristatus* and *P. muticus* (Aves) feces.
138. *I. mcdowelli* Saxe, Levine, and Ivens, 1960 (syns., *Isospora* sp. Saxe, 1952) in *Microtus pennsylvanicus* (Mammalia) cecal contents.

139. *I. megalaimae* Mandal and Chakravarty, 1964 in *Megalaima haemacephala* (Aves) intestine.
140. *I. melis* Pellérdy, 1955 (syns., *Lucetina* sp. Kotlan and Pospesch, 1933; *I. melis* [Kotlan and Pospesch, 1933] Pellérdy, 1955 of Pellérdy [1965]) in *Meles meles* (Mammalia) feces.
141. *I. melopsittaci* Bhatia et al., 1973 emend. Levine, 1985 (syn., *I. melopsittacusi* Bhatia et al., 1973) in *Melopsittacus undulatus* (Aves) feces.
142. *I. merionis* Musaev and Veisov, 1965 in *Meriones vinogradovi* (Mammalia) feces.
143. *I. mesnili* Sergent, 1902 in *Chamaeleon vulgaris* (Reptilia) in small intestine epithelium (intranuclear). (This species might be *Sarcocystis.*)
144. *I. mexicanasubsimi* Vance and Duszynski, 1985 in *Microtus mexicanus subsimus* (Mammalia) feces.
145. *I. michaelbakeri* Grulet, Landau, and Baccam, 1982 in *Passer domesticus* (Aves) feces.
146. *I. miltgeni* Grulet, Landau, and Baccam, 1982 in *Passer domesticus* (Aves) feces.
147. *I. mikei* Grulet, Landau, and Baccam, 1982 in *Passer domesticus* (Aves) small intestine.
148. *I. minuta* Mitra and Das-Gupta, 1937 in *Naja naja* (Reptilia) feces.
149. *I. monedulae* Yakimoff and Machul'skii, 1936 (syn., *Isospora* sp. Neméséri, 1949) in *Corvus* (syn., *Coloeus) monedula collaris* (Aves) feces.
150. *I. mungoi* Levine, Ivens, and Healy, 1975 (syn., large *I. garnhami* Bray, 1954 of Dubey and Pande [1963]) in *Herpestes edwardsi* (Mammalia) feces.
151. *I. muniae* Chakravarty and Kar, 1944 in *Lonchura* (syn., *Munia) m. malacca* (Aves) small intestine.
152. *I. naiae* Fantham, 1932 in *Naja nivea* (Reptilia) small intestine. (This species might be *Sarcocystis.*)
153. *I. nancyae* Grulet, Landau, Baccam, 1982 in *Passer domesticus* (Aves) feces.
154. *I. nankinovi* Golemansky, 1976 in *Garrulus glandarius graecus* (Aves) small intestine.
155. *I. natalensis* Elsdon-Dew, 1953 in *Homo sapiens* (Mammalia) feces.
156. *I. natricis* Yakimoff and Gousseff, 1935 in *Natrix natrix* (Reptilia) feces.
157. *I. neivai* Pinto and Maciel, 1929 in *Bothrops jararaca* (Reptilia) intestine.
158. *I. neomyi* Golemansky, 1978 in *Neomys anomalus* and *N. fodiens* (Mammalia) intestinal contents.
159. *I. neorivolta* Dubey and Mahrt, 1978 in *Canis familiaris* (Mammalia) posterior small and sometimes large intestine.
160. *I. neos* Yakimoff and Gousseff, 1936 in *Rana arvalis* (Amphibia) small intestine.
161. *I. neurotrichi* Duszynski, 1985 in *Neurotrichus gibbsii* (Mammalia) feces.
162. *I. neyrai* Romero Rodriguez and Lizcano Herrera, 1971 in *Sus scrofa* (Mammalia) feces.
163. *I. nucifragae* Galli-Valerio, 1933 in *Nucifraga caryocatactes* (Aves) feces.
164. *I. ohioensis* Dubey, 1975 (syns., *Cystoisospora ohioensis* [Dubey, 1975] Frenkel, 1977; *Levinea ohioensis* [Dubey, 1975] Dubey, 1977) in *Canis familiaris* and also *C. latrans, C. dingo (?), Vulpes vulpes (?)*, and perhaps *Nyctereutes procyonoides ussuriensis* (Mammalia); transport host (experimental) *Mus musculus* (Mammalia) generally posterior small intestine, also liver, spleen, mesentery, and occasionally skeletal muscles.
165. *I. ordubadica* Musaev and Veisov, 1960 in *Meriones persicus* (Mammalia) intestinal contents.
166. *I. orlovi* Tsygankov, 1950 in *Camelus* sp. (Mammalia) feces.
167. *I. papionis* McConnell et al., 1971 in *Papio ursinus* (Mammalia) muscles, jejunum, and ileum. (This species might be *Sarcocystis.*)
168. *I. pari* Ray et al., 1952 emend. Levine, 1982 (syns., *I. parusae* Ray et al., 1952; *I. lophophuriae* Ray et al., 1952) in *Parus dichrous* (Aves) feces.

169. *I. parvae* Chatterjee and Choudhury, 1979 in *Muscicapa parva* (Aves) feces.
170. *I. passeris* Levine, 1982 (syn., *I. lacazei auctores* in part) in *Passer domesticus* and possibly *P. montanus saturatus* (Aves) feces.
171. *I. passerum* Sjöbring, 1897 emend. Levine, 1982 (syn., *I. communis passerum* Sjöbring, 1897) in *Lanius collurio* (Aves) feces.
172. *I. pavlodarica* Nukerbaeva and Svanbaev, 1973 in *Vulpes vulpes* and *Alopex lagopus* (Mammalia) intestine.
173. *I. pavlovskyi* Svanbaev, 1956 in *Mustela eversmanni* (Mammalia) feces.
174. *I. pellerdyi* Dubey and Pande, 1964 (syns., *I. knowlesi* Dubey and Pande, 1963; *I. dubeyi* Patnaik and Ray, 1965) in *Herpestes edwardsi* (Mammalia) feces.
175. *I. peromysci* Davis, 1967 in *Peromyscus californicus, P. maniculatus,* and *P. truei* (Mammalia) small and large intestines.
176. *I. perroncitoi* Carpano, 1937 emend. Pellérdy, 1974 (syn., *I. perronciti* Carpano, 1937) in *Pyrrhula europaea* (Aves) intestine.
177. *I. petrochelidon* Stabler and Kitzmiller, 1972 in *Petrochelidon pyrrhonota* (Aves) intestine.
178. *I. phaeornis* Levine, Van Riper, and Van Riper, 1980 in *Phaeornis obscurus* (Aves) feces.
179. *I. phisalix* Yakimoff and Gousseff, 1934 (syn., *I. phisalixae* Yakimoff and Gousseff, 1934 emend. Pellérdy, 1974; *Eimeria phisalix* Yakimoff and Gousseff, 1933 of Pellérdy [1974] *lapsus calami*) in *Elaphe quatuorlineatus sauromates* (Reptilia) feces.
180. *I. phoenicuri* Schwalbach, 1959 in *Phoenicurus phoenicurus* (Aves) feces.
181. *I. phrynocephali* Ovezmukhammedov, 1971 in *Phrynocephalus helioscopus* (Reptilia) intestinal contents.
182. *I. psittaculae* Chakravarty and Kar, 1947 (syn., *Eimeria psittaculidae* Chakravarty and Kar of Farr [1960] *lapsus calami*) in *Psittacula eupatria nipalensis* (Aves) feces.
183. *I. pycnonoti* Bhatia et al., 1973 emend. Levine, 1982 (syn., *I. pycnonotusi* Bhatia et al., 1973) in *Pycnonotus jocosus* (Aves) feces.
184. *I. pycnonotus* Mandal and Chakravarty, 1964 emend. Levine, 1982 (syn., *I. pycnonotae* Mandal and Chakravarty, 1964) in *Pycnonotus jocosus* and perhaps *Turdoides striatus* (Aves) intestine. (This species might be *Sarcocystis.*)
185. *I. raggianae* Varghese, 1977 emend. Levine, 1982 (syn., *I. raggianai* Varghese, 1977) in *Paradisaea raggiana* (Aves) feces.
186. *I. rajulii* Satyanarayanacharyulu, Subbarao, and Christopher, 1969 in *Acridotheres t. tristis* (Aves) feces.
187. *I. rangiferis* Yakimoff, Matschoulsky, and Spartansky, 1937 in *Rangifer tarandus* (Mammalia) feces.
188. *I. rara* Schneider, 1881 (syn., *I. incerta* Schneider, 1881) (TYPE SPECIES) in *Limax* sp. (Gastropodasida) location uncertain.
189. *I. rastegaiev* Yakimoff and Matikaschwili, 1933 (syn., *I. rastegaievae* Yakimoff and Matikaschwili, 1933 emend. Pellérdy, 1974) in *Erinaceus europaeus* (Mammalia) feces.
190. *I. ratti* Levine and Ivens, 1965 in *Rattus norvegicus* (Mammalia) intestinal contents.
191. *I. rayi* Mandal, 1966 in *Ptyctolaemus* sp. (Reptilia) rectal contents.
192. *I. riccii* Agostinucci, 1957 in *Talpa europaea* (Mammalia) intestinal contents.
193. *I. rivolta* (Grassi, 1879) Wenyon, 1923 (syns., *Coccidium rivolta* Grassi, 1879; *Diplospora bigemina* von Wasielewski, 1904 in part; *I. rivoltae* Dobell, 1919; *I. rivoltai* [Grassi, 1879] *auctores; Lucetina rivolta* [Grassi, 1879] Henry and Leblois, 1926; *I. novocati* Pellérdy, 1974; *Cystoisospora rivolta* [Grassi, 1879] Frenkel, 1977; *C. frenkeli* Arcay, 1981; *Levinea rivolta* [Grassi, 1879] Dubey, 1977) in *Felis catus,* also *F. silvestris* and *F. chaus,* and probably *Leo tigris* and *L. pardus* (Mammalia) small

intestine, rarely cecum and colon; transport hosts (all experimental) *Mus musculus, Rattus norvegicus, Mesocricetus auratus, Bos taurus,* and *Canis familiaris* (Mammalia) and *Gallus gallus* (Aves) spleen, liver, mesentery, and occasionally skeletal muscles.

194. *I. rochalimai* Yakimoff and Gousseff, 1936 emend. Pellérdy, 1974 (syns., *I. rochalimae* Yakimoff and Gousseff, 1936; *Isospora* sp. Nemeséri, 1949 in part) in *Pica pica* (Aves) feces.

195. *I. rodhaini* Yakimoff and Machul'skii, 1938 in *Corvus* sp. (Aves) feces.

196. *I. rustamovi* Ovezmukhammedov, 1977 in *Phrynocephalus reticulatus bannicovi* (Reptilia) feces.

197. *I. samsenis* Svanbaev, 1979 (syn., *I. uralicae* Svanbaev, 1956 of Svanbaev [1960] in *Spermophilus fulvus*) in *Spermophilus fulvus* (Mammalia) feces.

198. *I. schmaltzi* Yakimoff and Gousseff, 1936 in *Erinaceus europaeus* (Mammalia) feces.

199. *I. scholtysecki* Rakhmatullina and Svanbaev, 1971 in *Falco cherrug, F. naumanni,* and *Accipiter nisus* (Aves) feces.

200. *I. schwetzi* Yakimoff and Machul'skii, 1939 in *Corvus c. corone* and *C. corone cornix* (Aves) feces.

201. *I. scorzai* Arcay-de-Peraza, 1967 in *Cacajao rubicundus* (Mammalia) feces.

202. *I. seicerci* Ray et al., 1952 emend. Levine, 1985 (syns., *I. cryptolophae* Ray et al., 1952; *I. seicercusae* Ray et al., 1952) in *Seicercus xantoschistos* (Aves) feces.

203. *I. sengeri* Levine and Ivens, 1964 in *Spilogale putorius ambarvalis* (Mammalia) feces.

204. *I. sittae* Golemansky, 1977 in *Sitta europea* (Aves) small intestine.

205. *I. sofiae* Levine and Ivens, 1979 (syn., *I. talpae* Golemansky, 1978) in *Talpa europaea* (Mammalia) large intestine contents.

206. *I. soricis* Golemanski and Yankova, 1973 in *Sorex araneus* (Mammalia) small intestine.

207. *I. spermophili* Levine, 1984 (syn., *I. laguri* Iwanoff-Gobzem, 1934 of Svanbaev [1962] in *Spermophilus maximus*) in *Spermophilus maximus* (Mammalia) feces.

208. *I. spilogales* Levine and Ivens, 1964 in *Spilogale putorius ambarvalis* (Mammalia) feces.

209. *I. spratti* Grulet, Landau, and Baccam, 1982 in *Passer domesticus* (Aves) feces.

210. *I. stomatici* Chakravarty and Kar, 1944 emend. Levine, 1985 (syn., *I. stomaticae* Chakravarty and Kar, 1944) in *Bufo stomaticus* (Amphibia) intestine.

211. *I. strigis* Yakimoff and Machul'skii, 1936 in *Asio flammeus leucopsis* (Aves) feces.

212. *I. struthionis* Yakimoff, 1940 in *Struthio camelus* (Aves) feces.

213. *I. sturniae* Chakravarty and Kar, 1947 in *Sturnus* (syn., *Sturnia*) *m. malabarica* (Aves) small intestine.

214. *I. suis* Biester, 1934 in *Sus scrofa* (Mammalia) small intestine.

215. *I. sylviae* Schwalbach, 1959 in *Sylvia borin, S. communis,* and *S. atricapilla* (Aves) feces.

216. *I. sylvianthina* Schwalbach, 1959 in *Anthus pratensis* and also possibly *Sylvia borin, S. communis,* and *S. atricapilla* (Aves) feces.

217. *I. talpae* Agostinucci, 1955 in *Talpa europaea* (Mammalia) intestinal contents.

218. *I. tamariscini* Levine, 1984 (syn., *I. laguri* Iwanoff-Gobzem of Svanbaev [1962] in *Meriones tamariscinus* in *Meriones tamariscinus* (Mammalia) feces.

219. *I. temenuchi* Chakravarty and Kar, 1944 emend. Levine, 1982 (syn., *I. temenuchii* Chakravarty and Kar, 1944) in *Temenuchus pagodarum* (Aves) feces.

220. *I. teres* Iwanoff-Gobzem, 1934 (syn., *I. feres* Iwanoff-Gobzem, 1935 of Musaev and Veisov [1965] *lapsus calami*) in *Lagurus lagurus* (Mammalia) intestine.

221. *I. testudae* Davronov, 1985 in *Testudo horsfieldi* (Reptilia) feces.

222. *I. thavari* Else and Colley, 1975 in *Gehyra mutilata* (Reptilia) small intestine.

223. *I. theileri* Yakimoff and Lewkowitsch, 1932 in *Canis aureus* (Mammalia) feces.

224. *I. triffittae* Nukerbaeva and Svanbaev, 1973 emend. Levine, 1985 (syn., *I. triffitti*

Nukerbaeva and Svanbaev, 1973) in *Vulpes vulpes* and *Alopex lagopus* (Mammalia) intestine.

225. *I. tristis* Chakravarty and Kar, 1947 emend. Pellérdy, 1974 (syn., *I. ginginiana* var. *tristis* Chakravarty and Kar, 1947) in *Acridotheres tristis* (Aves) feces.

226. *I. turdi* Schwalbach, 1959 in *Turdus merula* and *T. pilaris* (Aves) feces.

227. *I. turkmenica* Ovezmukhammedov, 1969 in *Eryx miliaris* (Reptilia) feces. (This species might be *Sarcocystis*.)

228. *I. upupae* Chakravarty and Kar, 1947 in *Upupa epops orientalis* (Aves) feces.

229. *I. uralicae* Svanbaev, 1956 in *Apodemus sylvaticus* (Mammalia) feces.

230. *I. vanadica* Musaev and Veisov, 1965 in *Meriones persicus* (Mammalia) feces (?).

231. *I. vanriperorum* Levine, 1982 (syn., *I. cardinalis* Levine, Van Riper, and Van Riper, 1980) in *Cardinalis cardinalis* (Aves) feces.

232. *I. varani* Yakimoff, 1938 in *Varanus griseus* (Reptilia) feces. (This species might be *Sarcocystis*.)

233. *I. vinogradovi* Musaev and Veisov, 1965 in *Meriones vinogradovi* (Mammalia) feces (?).

234. *I. viverrae* Adler, 1924 in *Civettictis civetta* (Mammalia) epithelial cells of small intestine.

235. *I. volki* Boughton, 1937 in *Parotia l. lawesi* and possibly also *Epimachus m. meyeri*, *Lophorina superba minor*, *Manucodia chalybata orientalis*, *Paradisaea a. apoda*, *P. a. salvadori*, *P. guilielmi*, *P. m. minor*, *Paradisornis r. rudolphi*, *Seleucides m. melanoleucus*, and *Uranornis rubra* (Aves) feces.

236. *I. vulpina* Nieschulz and Box, 1933 (syns., *I. vulpina* var. *vulpina* Mantovani, 1965; *I. vulpina* var. *aprutina* Mantovani, 1965; *I. aprutina* Mantovani, 1965 emend. Pellérdy, 1974; *Cystoisospora vulpina* [Nieschulz and Box, 1933] Frenkel, 1977) in *Vulpes vulpes* and also *Canis familiaris* and *Alopex lagopus* (Mammalia) feces; transport host (experimental) *Mus musculus*.

237. *I. vulpis* Galli-Valerio, 1931 in *Vulpes vulpes* (Mammalia) feces.

238. *I. wenyoni* Ray and Das Gupta, 1935 in *Bufo melanostictus* and also presumably *Rana limnocharis* and *R. tigrina* (Amphibia) small intestine.

239. *I. wetzeli* Sugar, 1980 in *Vulpes vulpes* (Mammalia) intestinal contents.

240. *I. wilkiei* Lainson, 1968 in *Crocodylus acutus* (Reptilia) feces.

241. *I. wladimirovi* Yakimoff, 1930 in *Hyla arborea* (Amphibia) intestine.

242. *I. wurmbachi* Schwalbach, 1959 in *Saxicola rubetra, S. torquata,* and possibly *Sylvia atricapilla, S. borin, S. communis, Phylloscopus trochilus, Ficedula hypoleuca,* and *Anthus pratensis* (Aves) feces.

243. *I. xantusiae* Amrein, 1952 in *Xantusia vigilis* and *X. henshawi* (Reptilia) small intestine.

244. *I. xerophila* Barré and Troncy, 1974 in *Quelea quelea, Ploceus capitalis, P. cucullatus, Euplectes oryx, E. afra,* and *Sporopipes frontalis* (Aves) small intestine.

245. *I. yesi* Grulet, Landau, and Baccam, 1982 in *Passer domesticus* (Aves) feces.

246. *I. yukonensis* Hobbs and Samuel, 1974 in *Ochotona collaris* (Mammalia) feces.

247. *I. zorillae* Prasad, 1961 emend. Pellérdy, 1963 (syn., *I. bigemina* var. *zorillae* Prasad, 1961) in *Poecilictis libyca alexandrae* (Mammalia) feces.

248. *I. zosteropis* Chakravarty and Kar, 1947 in *Zosterops palpebrosa* and very dubiously *Megalaima zeylandicus caniceps* (Aves) feces.

Genus *Dorisa* Levine, 1980. Oocysts with variable number of sporocysts, each with eight sporozoites. TYPE SPECIES *D. hoarei* (Yakimoff and Gousseff, 1935) Levine, 1980.

1. *D. aethiopsaris* (Chakravarty and Kar, 1947) Levine, 1980 (syn., *Dorisiella aethiopsaris* Chakravarty and Kar, 1947) in *Acridotheres fuscus* (Aves) intestine.

2. *D. arizonensis* (Levine, Ivens, and Kruidenier, 1955) Levine, 1980 (syn., *Dorisiella arizonensis* Levine, Ivens, and Kruidenier, 1955) in *Neotoma lepida* (Mammalia) feces.
3. *D. bengalensis* (Bandyopadhyay and Ray, 1982) Levine and Ivens, 1987 (syn., *Dorisiella bengalensis* Bandyopadhyay and Ray, 1982) in *Funambulus pennanti* (Mammalia) feces.
4. *D. chakravartyi* (Ray and Sarkar, 1967 emend. Pellérdy, 1974) Levine, 1980 (syn., *Dorisiella chakravartyei* Ray and Sarkar, 1967) in *Lonchura malabarica* and *L. p. punctulata* (Aves) anterior small intestine.
5. *D. hareni* (Chakravarty and Kar, 1944) Levine, 1980 (syn., *Dorisiella hareni* Chakravarty and Kar, 1944) in *Munia m. malacca* and also *M. rubronigra, Amandava amandava, Lonchura malabarica,* and *L. p. punctulata* (Aves) small intestine.
6. *D. harpia* (Sinha and Dasgupta, 1978) Levine, 1981 (syns., *Dorisiella harpia* Sinha and Dasgupta, 1978; Dorisiella sp. Sinha and Dasgupta, 1978) in *Harpiocephalus harpia lasyurus* (Mammalia) small intestine.
7. *D. hoarei* (Yakimoff and Gousseff, 1935) Levine, 1980 (syn., *Dorisiella hoarei* Yakimoff and Gousseff, 1935) (TYPE SPECIES) in *Elaphe quatuorlineata sauromates* (Reptilia) feces.
8. *D. mandali* (Ray and Sarkar, 1967) Levine, 1980 (syn., *Dorisiella mandali* Ray and Sarkar, 1967) in *Zosterops palpebrosa* (Aves) small intestine.
9. *D. passeris* (Ray and Sarkar, 1967) Levine, 1980 (syn., *Dorisiella passeris* Ray and Sarkar, 1967) in *Passer domesticus* (Aves) duodenum and ileum.
10. *D. rayi* (Bray, 1964) Levine, 1980 (syn., *Dorisiella rayi* Bray, 1964) in *Spermestes cucullatus* (Aves) jejunum.
11. *D. vagabundae* (Mandal and Chakravarty, 1964) Levine, 1980 (syn., *Dorisiella vagabundae* Mandal and Chakravarty, 1964) in *Crypsirina vagabunda* (Aves) small intestine.

Genus *Wenyonella* Hoare, 1933. Oocysts with four sporocysts, each with four sporozoites. TYPE SPECIES *W. africana* Hoare, 1933.

1. *W. africana* Hoare, 1933 (TYPE SPECIES) in *Boodon lineatus* (Reptilia) small intestine.
2. *W. anatis* Pande, Bhatia, and Srivastava, 1965 in *Anas platyrhynchos domesticus* (Aves) small intestine.
3. *W. arcayae* Bastardo de San Jose, 1974 in *Tropidurus hispidus* (Reptilia) epithelium of small intestine.
4. *W. baghdadensis* Mirza and Al-Rawas, 1978 (syn., *Wenyonella* sp. Mirza and Al-Rawas, 1978) in *Nesokia indica* (Mammalia) feces.
5. *W. bahli* Misra, 1944 in *Coturnix c. coturnix* (Aves) small intestine.
6. *W. columbae* Haldar and Ray-Choudhury, 1974 in *Columba livia* (Aves) small intestine.
7. *W. gagari* Sarkar and Ray, 1968 in *Anas platyrhynchos domestica* (Aves) intestine.
8. *W. gallinae Ray, 1945 in Gallus gallus* (Aves) intestine.
9. *W. hoarei* Ray and Das Gupta, 1937 in *Sciurus* sp. (Mammalia) small intestine.
10. *W. mackinnonae* Misra, 1947 emend. Levine, 1985 (syn., *W. mackinnoni* Misra, 1947) in *Motacilla alba* (Aves) small intestine.
11. *W. markovi* Grobov and Ven'-Shun', 1963 in *Capreolus capreolus pygargus* (Mammalia) feces.
12. *W. parva* van den Berghe, 1938 in *Paraxerus (Tamiscus) emini* and also *P. anerythrus* (Mammalia) feces.
13. *W. pellerdyi* Bhatia and Pande, 1966 in *Anas querquedula* (Aves) intestine.
14. *W. philiplevinei* Leibovitz, 1968 in *Anas platyrhynchos domesticus* (Aves) intestine.
15. *W. uelensis* van den Berghe, 1938 in *Funisciurus anerythrus* (Mammalia) feces.

Genus *Octosporella* Ray and Raghavachari, 1942. Oocysts with eight sporocysts, each with two sporozoites. TYPE SPECIES *O. mabuiae* Ray and Raghavachari, 1942.

1.   *O. hystrix* Barker, Beveridge, and Munday, 1984 in *Tachyglossus aculeatus* (Mammalia) feces.
2.   *O. mabuiae* Ray and Raghavachari, 1942 (TYPE SPECIES) in *Mabuia* sp. (Reptilia) intestine.
3.   *O. notropis* Li and Desser, 1985 in *Notropis cornutus* (Osteichthys) intestinal epithelium, spleen, and swim bladder.
4.   *O. opeongoensis* Li and Desser, 1985 in *Notemigonus crysoleucas* (Osteichthys) swim bladder.
5.   *O. sanguinolentae* Ovezmukhammedov, 1975 emend. Levine 1985 (syn., *O. sanguinolenti* Ovezmukhammedov, 1975) in *Agama sanguinolenta* (Reptila) feces.
6.   *O. sasajewunensis* Li and Desser, 1985 in *Notemigonus crysoleucas* (Osteichthys) swim bladder.

Genus *Hoarella* Arcay de Peraza, 1963. Oocysts with 16 sporocysts, each with 2 sporozoites. TYPE SPECIES *H. garnhami* Arcay de Peraza, 1963.

1.   *H. garnhami* Arcay de Peraza, 1963 (TYPE SPECIES) in *Cnemidophorus l. lemniscatus* (Reptilia) small intestine.

Genus *Sivatoshella* Ray and Sarkar, 1968. Oocysts with 2 sporocysts, each with 16 sporozoites. TYPE SPECIES *S. lonchurae* Ray and Sarkar, 1968.

1.   *S. lonchurae* Ray and Sarkar, 1968 (TYPE SPECIES) in *Lonchura malabarica* and *L. punctulata* (Aves) small intestine.

Genus *Pythonella* Ray and Das Gupta, 1937. Oocysts with 16 sporocysts, each with 4 sporozoites. TYPE SPECIES *P. bengalensis* Ray and Das Gupta, 1937.

1.   *P. bengalensis* Ray and Das Gupta, 1937 (TYPE SPECIES) in *Python* sp. (Reptilia) intestine.
2.   *P. karakalensis* Glebezdin, 1971 in *Calomyscus bailwardi* (Mammalia) feces.
3.   *P. scelopori* Duszynski, 1969 in *Sceloporus* (Reptilia) feces.

Genus *Gousseffia* Levine and Ivens, 1979. Oocysts with eight sporocysts, each with n sporozoites. TYPE SPECIES *G. erinacei* (Gousseff, 1937) Levine and Ivens, 1979.

1.   *G. erinacei* (Gousseff, 1937) Levine and Ivens, 1979 (syn., *Yakimovella erinacei* Gousseff, 1937) (TYPE SPECIES) in *Erinaceus europaeus* (Mammalia) feces.

Genus *Skrjabinella* Machul'skii, 1949. Oocysts with 16 sporocysts, each with 1 sporozoite. TYPE SPECIES *S. mongolica* Machul'skii, 1949.

1.   *S. mongolica* Machul'skii, 1949 (TYPE SPECIES) in *Allactaga saltator* (Mammalia) feces (?).

Genus *Diaspora* Léger, 1898. Oocysts unknown; sporocysts each with one sporozoite. TYPE SPECIES *D. hydatidea* Léger, 1898.

1.   *D. hydatidea* Léger, 1898 (TYPE SPECIES) in a polydesmid millipede (Diplopodasida) intestinal lumen.

Family BARROUXIIDAE *Léger, 1911*

Homoxenous; oocysts containing different numbers of sporocysts, depending on the genus; sporocysts bivalved, with dehiscence suture.

Genus *Barrouxia* Schneider, 1885. Oocysts with *n* sporocysts, each with one sporozoite; sporocysts with a bivalved wall, with a longitudinal dehiscence suture. TYPE SPECIES *B. ornata* Schneider, 1885.

1.   *B. alpina* Léger, 1898 in *Lithobius forficatus, L. castaneus,* and *L. pilicornis* (Chilopodasida) intestine.
2.   *B. belostomatis* Carini, 1942 in *Belostoma horvathi* and *B. plebejum* (Hemipterorida) intestine.
3.   *B. bulini* Triffitt, Buckley, and McDonald, 1932 in *Isidora tropica* (syn., *Bulinus tropicus*) (Molluska) beneath mantle epithelium.
4.   *B. caudata* Léger, 1898 (syn., *Urobarrouxia caudata* [Léger, 1898] Mesnil, 1903) in *Lithobius martini* (Chilopodasida) midgut.
5.   *B. labbei* (Léger, 1897) Levine, 1981 (syn., *Echinospora labbei* Léger, 1897) in *Lithobius mutabilis* (Chilopodasida) digestive tract.
6.   *B. legeri* Schellack and Reichenow, 1913 (syn., *B. schneideri* Léger, 1897 in part) in *Lithobius impressus* (Chilopodasida) intestine.
7.   *B. ornata* Schneider, 1885 (syn., *Eimeria nepae* Schneider, 1887) (TYPE SPECIES) in *Nepa rubra* (Hemipterorida) intestine.
8.   *B. Schneideri* (Bütschli, 1882) Schellack and Reichenow, 1913 (syns., *Eimeria schneideri* Bütschli, 1882; *B. legeri* Schellack and Reichenow, 1913 in part) in *Lithobius forficatus* (Chilopodasida) midgut and hindgut.
9.   *B. spiralis* Averinzev, 1916 in *Cerebratulus* sp. (Nemertini) intestine.
10.   *B. ventricosa* (Léger, 1898) Levine, 1981 (syn., *Echinospora ventricosa* Léger, 1898) in *Lithobius hexodus* (Chilopodasida) digestive tract.

Genus *Goussia* Labbé, 1896. Oocysts with four sporocysts, each with two sporozoites; dehiscence suture longitudinal. TYPE SPECIES *G. clupearum* (Thélohan, 1894) Labbé, 1896.

1.   *G. alburni* Stankovitch, 1920 (syn., *Eimeria alburni* [Stankovitch, 1920] Yakimoff, 1929; *G. [Goussia] alburni* [Thélohan, 1894] Overstreet, Hawkins, and Fournie, 1984) in *Gobio gobio, Rutilus rutilus,* and *Scardinius erythrophthalmus* (Osteichthys) periintestinal fat tissue.
2.   *G. auxidis* (Dogel', 1948) Dyková and Lom, 1981 (syns., *Eimeria auxidis* Dogel', 1948; *G. [Goussia] auxidis* [Dogiel, 1948] Overstreet, Hawkins, and Fournie, 1984) in *Auxis maru* (Osteichthys) liver.
3.   *G. bigemina* Labbé, 1896 (syns., *Eimeria bigemina* [Labbé, 1896] Yakimoff, 1929; [*non*] *E. bigemina* Labbé, 1896; *G. [Goussia] bigemina* [Labbé, 1896] Overstreet, Hawkins, and Fournie, 1984) in *Ammodytes tobianus* (Osteichthys) intestine.
4.   *G. carpelli* (Léger and Stankovitch, 1921) Lom and Dyková, 1982 (syns., *Eimeria carpelli* Léger and Stankovitch, 1921; *E. cyprini* Plehn, 1924; *G. [Goussia] carpelli* [Léger and Stankovitch, 1921] Overstreet, Hawkins, and Fournie, 1984) in *Carassius carassius* and *Cyprinus carpio* (Osteichthys) intestinal epithelial cells.
5.   *G. caseosa* Lom and Dyková, 1982 (syn., *G. [Plagula] caseosa* [Lom and Dyková,

1982] Overstreet, Hawkins, and Fournie, 1984) in *Macrourus berglax* (Osteichthys) swim bladder, gas gland, and also (oocysts only) gall bladder, intestinal contents, and mesenteric blood vessels.

6.   *G. cichlidarum* Landsberg and Paperna, 1985 in *Sarotherodon galilaeus, Oreochromis aureus, O. aureus-O. niloticus* hybrids, *Oreochromis* sp., and *Tilapia zilli* (Osteichthys) swim bladder epithelium.

7.   *G. clupearum* (Thélohan, 1894) Labbé, 1896 (syns., *Coccidium clupearum* Thélohan, 1894; *Eimeria clupearum* [Thélohan, 1894] Doflein, 1909; *E. wenyoni* Dobell, 1919; *G. [Goussia] clupearum* [Thélohan, 1894] Overstreet, Hawkins, and Fournie, 1984) (TYPE SPECIES) in *Clupea harengus, Sardina* (syn., *Clupea) pilchardus, S. melanosticta, Alosa sardina, Engraulis encrasicholus, Sprattus* (syn., *clupea) sprattus, Etrumeus micropus, Micromesistius poutassou*, and *Scomber scomber* (Osteichthys) liver.

8.   *G. cruciata* (Thélohan, 1892) Labbé, 1896 (syns., *Coccidium cruciatum* Thélohan, 1892; *Eimeria cruciata* [Thélohan, 1892] Yakimoff, 1929; *Eimeria [Goussia]* [Thélohan] of Dogel' [1948]; *G. [Goussia] cruciata* [Thélohan, 1892] Overstreet, Hawkins, and Fournie, 1984) in *Caranx* (syn., *Trachurus) trachurus* (Osteichthys) liver.

9.   *G. degiustii* (Molnar and Fernando, 1974) Dyková and Lom, 1981 (syns., *Eimeria degiustii* Molnar and Fernando, 1974; *G. [Plagula] degiustii* [Molnar and Fernando, 1974] Overstreet, Hawkins, and Fournie, 1984) in *Notropus cornutus, N. heterolepis, Campostoma anomalum, Pimephales notatus, P. promelae*, and *Semotilus atromaculatus* (Osteichthys) spleen, gall bladder, gut, liver, and other viscera.

10.   *G. flaviviridis* (Setna and Bana, 1935) Levine, 1983 (syn., *Eimeria flaviviridis* Setna and Bana, 1935) in *Hemidactylus flaviviridis* (Reptilia) gall bladder.

11.   *G. gadi* (Fiebiger, 1913) Dyková and Lom, 1981 (syns., *Eimeria gadi* Fiebiger, 1913; *G. [Plagula] gadi* [Fiebiger, 1913] Overstreet, Hawkins, and Fournie, 1984) in *Melanogrammus* (syn., *Gadus) aeglefinus, Gadus morrhua, G. virens*, and *Enchelyopus cimbrius* (Osteichthys) swim bladder.

12.   *G. hyalina* (Léger, 1898) Levine, 1983 (syns., *Coccidium hyalinum* Léger, 1898; *Eimeria hyalina* [Léger, 1898] Reichenow, 1921) in unidentified aquatic beetle (Coleoptera) Malpighian tubules.

13.   *G. iroquoina* (Molnar and Fernando, 1974) Paterson and Desser, 1984 (syn., *Eimeria iroquoina* Molnar and Fernando, 1974) in *Nocomis biguttatus, Notropis heterolepis, N. rubellus, N. cornutus, Pimephales notatus, P. promelas, Rhinichthys atratulus*, and *Semotilus atromaculatus* (Osteichthys) anterior intestine.

14.   *G. lacazei* (Labbé, 1895) Levine, 1983 (syns., *Bananella lacazei* Labbé, 1895; *Coccidium schneideri* Schaudinn and Siedlecki, 1897; *Eimeria lacazei* [Labbé, 1895] Moroff, 1908) in *Lithobius forficatus* and *L. martini* (Chilopodasida) intestine.

15.   *G. laureleus* (Molnar and Fernando, 1974) Li and Desser, 1985 (syn., *Eimeria laureleus* Molnar and Fernando, 1974) in *Perca flavescens* and *Lepomis gibbosus* (Osteichthys) small intestine, cecum, gall bladder, and liver.

16.   *G. legeri* Stankovitch, 1920 (syns., *Eimeria stankovitchi* Pinto, 1928 in part; *E. legeri* [Stankovitch, 1920] Pinto, 1928) in *Alburnus lucidus* and *Scardinius erythrophthalmus* (Osteichthys) epithelial cells and subepithelium of intestine.

17.   *G. luciae* Lom and Dyková, 1982 (syn., *G. [Goussia] luciae* [Lom and Dyková, 1982] Overstreet, Hawkins, and Fournie, 1984) in *Mullus barbatus* (Osteichthys) middle part of intestine.

18.   *G. lucida* (Labbé, 1893) Labbé, 1896 (syns., *Coccidium lucidum* Labbé, 1893; *Eimeria lucida* [Labbé, 1893] Reichenow, 1921; *G. [Goussia] lucida* [Labbé, 1893] Overstreet, Hawkins, and Fournie, 1984) in *Mustelus canis, M. vulgaris, Acanthias acanthias, A. vulgaris, Scyllium catulus, S. stellare*, and *Scyliorhynchus canicula* (Chondrichthys) spiral valve and (in *S. canicula* only) posterior intestine.

19. *G. metchnikovi* (Laveran, 1897) Levine, 1983 (syns., *Coccidium metchnikovi* Laveran, 1897; *Eimeria metchnikovi* [Laveran, 1897] Reichenow, 1921; *E. macroresidualis* Schulman and Zaika, 1962; *G. [Goussia] metchnikovi* [Laveran, 1897] Overstreet, Hawkins, and Fournie, 1984) in *Gobio gobio* and *G. albipinnatus* (Osteichthys) spleen, also liver, and less often kidneys and intestinal lumen.
20. *G. minuta* (Thélohan, 1892) Labbé, 1896 (syns., *Coccidium minutum* Thélohan, 1892; *Eimeria minuta* [Thélohan, 1892] Doflein, 1909; *G. [Goussia] minuta* [Thélohan, 1892] Overstreet, Hawkins, and Fournie, 1984) in *Tinca tinca* (Osteichthys) spleen, liver, and kidney.
21. *G. motellae* (Labbé, 1893) Labbé, 1896 (syns., *Coccidium motellae* Labbé, 1893; *Eimeria motellae* [Labbé, 1893] Yakimoff, 1929; *G. [Goussia] motellae* [Labbé, 1893] Overstreet, Hawkins, and Fournie, 1984) in *Motella tricirrata* (Osteichthys) intestine and pyloric ceca.
22. *G. notemigonica* Li and Desser, 1985 in *Notemigonus crysoleucas* (Osteichthys) kidney, spleen, swim bladder, and ureters.
23. *G. notropicum* Li and Desser, 1985 in *Notropus cornutus* (Osteichthys) intestinal wall.
24. *G. polylepidis* Alvarez-Pellitero and Gonzalez-Lanza, 1985 in *Chondrostoma polylepis* (Osteichthys) swim bladder, peritoneum, kidney, and ureter.
25. *G. schaudinniana* (Pinto, 1928) Levine, 1983 (syns., *Coccidium schubergi* Schaudinn, 1900; *Eimeria schubergi* [Schaudinn, 1900] von Wasielewski, 1904; *E. schaudinniana* Pinto, 1928) in *Lithobius forficatus* (Chilopodasida) intestine.
26. *G. siliculiformis* (Schulman and Zaika, 1962) Dyková and Lom, 1981 (syns., *Eimeria siliculiformis* Schulman and Zaika, 1962; *G. [Goussia] siliculiformis* [Schulman and Zaika, 1962] Overstreet, Hawkins, and Fournie, 1984) in *Gobio albipinnatus tenuicorpus* (Osteichthys) air bladder, intestine, and kidneys.
27. *G. stankovitchi* (Pinto, 1928) Levine, 1983 (syns., *G. legeri* Stankovitch, 1920 in part; *Eimeria stankovitchi* Pinto, 1928 in part; *G. [Goussia] stankovitchi* [Pinto, 1928] Overstreet, Hawkins, and Fournie, 1984) in *Alburnus alburnus* (Syn., *A. lucidus), Abramis brama*, and *Scardinius erythrophthalmus* (Osteichthys) intestine.
28. *G. subepithelialis* (Moroff and Fiebiger, 1905) Dyková and Lom, 1981 (syns., *Eimeria subepithelialis* Moroff and Fiebiger, 1905; *G. [Plagula] subepithelialis* [Moroff and Fiebiger, 1905] Overstreet, Hawkins, and Fournie, 1984) in *Cyprinus carpio* (Osteichthys) gut wall.
29. *G. thelohani* Labbé, 1896 (syns., *Coccidium* sp. Thélohan, 1894; *Eimeria thelohani* [Labbé, 1896] Yakimoff, 1929; *G. [Goussia] thelohani* [Labbé, 1896] Overstreet, Hawkins, and Fournie, 1984) in *Labrus* sp. (Osteichthys) liver.
30. *G. (?) truttae* Léger and Hesse, 1919 (syn., *Eimeria truttae* [Léger and Hesse, 1919] Stankovitch, 1924; *G. [Goussia] truttae* [Léger and Hesse, 1919] Overstreet, Hawkins, and Fournie, 1984) in *Salmo fario* and *Salvelinus fontinalis* (Osteichthys) pyloric ceca and anterior intestine.
31. *G. variabilis* (Thélohan, 1893) Labbé, 1896 (syns., *Coccidium variabile* Thélohan, 1893; *Eimeria variabilis* [Thélohan, 1893] Reichenow, 1921; *G. [Goussia] variabilis* [Thélohan, 1893] Overstreet, Hawkins, and Fournie, 1984) in *Cottus bubalis, Crenilabrus melops, Gobius bicolor, G. paganellus, Myoxocephalus bubalis,* and possibly *Anguilla anguilla* and *Lepadogaster gouani* (Osteichthys) pyloric ceca and intestine.

Genus *Defretinella* Henneré, 1966. Early development and gamogony in epidermis; many microgametes formed by microgamonts; oocysts presumably released by rupture of tegument; oocysts with more than 50 bivalved sporocysts, each with several tens of sporozoites; sporocysts with longitudinal dehiscence suture; in polychetes. TYPE SPECIES *D. eulaliae* Henneré, 1966.

1.   *D. eulaliae* Henneré, 1966 (TYPE SPECIES) in *Eulalia viridis* (Polychaetasida)
     epidermis.

Genus *Crystallospora* Labbé, 1896. Oocysts with four sporocysts, each with two sporozoites;
sporocysts resemble crystals, consisting of two pyramids with hexagonal bases, joined base
to base; with an equatorial suture line. TYPE SPECIES *C. cristalloides* (Thélohan, 1893)
Labbé, 1899.

1.   *C. cristalloides* (Thélohan, 1893) Labbé, 1899 (syn., *Coccidium cristalloides* Thé-
     lohan, 1893; *Crystallospora cristalloides* [Thélohan, 1893] Doflein, 1909; *C. thelo-
     hani* Labbé, 1896; *Eimeria [C.] cristalloides* [Thélohan] of Dogel', 1948) (TYPE
     SPECIES) in *Motella* spp. (Osteichthys) intestine and pyloric ceca.

Family ATOXOPLASMATIDAE *Levine, 1982*
     Homoxenous, with merogony and gamogony within the host, sporogony outside; devel-
opment intracellular (merogony in blood and intestinal cells, gamogony in intestinal cells
of the same individual host); transmission by ingestion of sporulated oocysts.

Genus *Atoxoplasma* Garnham, 1950. Sporulated oocysts with two sporocysts, each with
four sporozoites. TYPE SPECIES *A. paddae* (Aragão, 1911) Laird, 1959. (*Remarks.* The
species names and synonymy of this genus will be uncertain until the complete life cycles
of all its members are known. Some species may have been named *Isospora* because their
blood forms and fecal oocysts were not thought to be related.)

1.   *A. adiei* (Aragão, 1911) Baker et al., 1972 (syns., *Haemogregarina adiei* Aragão,
     1911; *Lankesterella adiei* [Aragão, 1911] Lainson, 1959; *L. passeris* Raffaele, 1938;
     L. garnhami Lainson, 1959; *Toxoplasma passeris* Rousselot, 1953) in *Passer domes-
     ticus* (Aves).
2.   *A. amadinae* (Fantham, 1924) Baker et al., 1972 (syns., *Leucocytogregarina amadinae*
     Fantham, 1924; *Hepatozoon amadinae* [Fantham, 1924] Wenyon, 1926) in *Amadina
     erythrocephala* (Aves).
3.   *A. argyae* Garnham, 1950 (syn., *Lankesterella argyae* [Garnham, 1950] Lainson,
     1959) in *Turdoides rubiginosus* (Aves).
4.   *A. avium* (Labbé, 1894) Baker et al., 1972 (syns., *Drepanidium avium* Labbé, 1894;
     Lankesterella avium [Labbé, 1894] Labbé, 1899) in *Lanius excubitor, Pica pica,
     Corvus corax, Buteo buteo, Falco tinnunculus, Asio flammea,* and *Strix aluco* (Aves).
5.   *A. butasturis* (de Mello, 1935) Levine, 1982 (syn., *Toxoplasma butasturis* de Mello,
     1935) in *Butastur teesa* (Aves).
6.   *A. coccothraustis* Corradetti and Scanga, 1963 in *Coccothraustes coccothraustes* (Aves).
7.   *A. corvi* (Baker, Lainson, and Killick-Kendrick, 1959) Levine, 1982 (syn., *Lankes-
     terella corvi* Baker, Lainson, and Killick-Kendrick, 1959) in *Corvus f. frugilegus*
     (Aves).
8.   *A. danilewskii* Zasukhin, Vasina, and Levitanskaya, 1957 in *Carduelis spinus* (Aves).
9.   *A. desseri* Levine, 1982 (syns., *Lankesterella* sp. Khan and Desser, 1971; *Isospora*
     sp. [Khan and Desser, 1971] Desser, 1980) in *Hesperiphona vespertina* and *Pheucticus
     ludovicianus* (Aves).
10. *A. lainsoni* (Dissanaike, 1967) Levine, 1982 (syn., *Lankesterella lainsoni* Dissanaike,
     1967) in *Acridotheres tristis melanosternus* (Aves).
11. *A. liothricis* (Laveran and Marullaz, 1914) Baker et al., 1972 (syn., *Toxoplasma
     liothricis* Laveran and Marullaz, 1914) in *Leiothrix lutea* (Aves).
12. *A. paddae* (Aragão, 1911) Laird, 1959 (syn., *Haemogregarina paddae* Aragão, 1911;

Toxoplasma avium Marullaz, 1913; *A. avium* [Marullaz, 1913] Garnham, 1950; *Lankesterella paddae* [Aragão, 1911] Lainson, 1959; *T. fulicae* de Mello, 1935 [?]) (TYPE SPECIES) in *Padda oryzivora* and possibly *Zosterops lateralis. Z. flavifrons, Z. rennelliana, Woodfordia superciliosa*, and improbably *Gallirallus australis scotti* (Aves).

13. *A. paulasousai* (Correa, 1928) Levine, 1982 (syn., *Haemogregarina paulasousai* Correa, 1928) in *Stephanophorus diadematus* (Aves).

14. *A. pessoai* (Correa, 1928) Levine, 1982 (syn., *Haemogregarina pessoai* Correa, 1928) in *Poospiza thoracica* (Aves).

15. *A. picumni* (Mackerras and Mackerras, 1960) Levine, 1982 (syn., *Lankesterella picumni* Mackerras and Mackerras, 1960) in *Climacteris picumnus* (Aves).

16. *A. serini* (Aragão, 1933) Levine, 1982 (syns., *Haemogregarina serini* Aragão, 1933; *Lankesterella serini* [Aragão, 1933] Lainson, 1959; *Isospora serini* [Aragão, 1933] Box, 1975; *I. lacazei* of Romero Rodriguez [1973]) in *Serinus canarius* (Aves).

17. *A. sicalidis* (Aragão, 1911) Baker et al., 1972 (syns., *Haemogregarina sicalidis* Aragão, 1911; *Hepatozoon sicalidis* [Aragão, 1911] Hoare, 1924) in *Sicalis flaveola* (Aves).

18. *A. spermesti* (Rousselot, 1953) Levine, 1982 (syns., *Hepatozoon spermesti* Rousselot, 1953; *Lankesterella spermesti* [Rousselot, 1953] Bray, 1964; *Toxoplasma* sp. from *Spermestes cucullatus* Wenyon, 1926) in *Lonchura c. cucullatus* (Aves).

19. *A. sporophilae* (Aragão, 1911) Baker et al., 1972 (syns., *Haemogregarina sporophilae* Aragão, 1911; *Hepatozoon sporophilae* [Aragão, 1911] Hoare, 1924) in *Sporophila albogularis* (Aves).

Family LANKESTERELLIDAE *Nöller, 1920*
Heteroxenous or homoxenous, with merogony, gamogony, and sporogony in same vertebrate host; endogenous development intracellular; without metrocytes; oocysts with or without sporocysts, but with eight or more sporozoites; sporozoites in blood cells, transferred without development by an invertebrate (mite, mosquito, or leech) or vertebrate; infection by ingestion of sporozoite-bearing host; microgametes with two flagella so far as is known.

Genus *Lankesterella* Labbé, 1899. Heteroxenous; oocysts produce 32 or more sporozoites; known invertebrate hosts leeches; sporozoites in vertebrate erythrocytes, with about 30 subpellicular microtubules; in amphibia and reptiles. TYPE SPECIES *L. minima* (Chaussat, 1850 (Nöller, 1912.

1. *L. baznosanui* Chiriac and Steopoe, 1977 in *Lacerta vivipara* (Reptilia).

2. *L. bufonis* Mansour and Mohammed, 1962 in *Bufo regularis* and *B. melanostictus* (Amphibia).

3. *L. hylae* Cleland and Johnston, 1910 in *Hyla caerulea* (Amphibia).

4. *L. leptodactyli* Ducceschi, 1914 in *Leptodactylus ocellatus* (Amphibia).

5. *L. millani* Alvarez Calvo, 1975 emend. Levine, 1985 (syn., *Lankesterella* [= *Atoxoplasma] millani* Alvarez Calvo, 1975) in *Lacerta lepida nevadensis* (Reptilia).

6. *L. minima* (Chaussat, 1850) Nöller, 1912 (syns., *Anguillula minima* Chaussat, 1850; *Drepanidium ranarum* Lankester, 1871; *D. monile* Labbé, 1894; *L. monilis* [Labbé, 1894] Labbé, 1899; *D. princeps* Labbé, 1894; *L. ranarum* [Lankester] Labbé, 1899; *Haemogregarina minima* [Laveran] Mathis and Leger, 1911; *H. minima* [Chaussat, 1850] Mathis and Leger, 1911; *H. ranarum* Celli and San Felice, 1890; *L. minima* [Chaussat, 1850] Labbé, 1899 of Pellérdy [1974]) (TYPE SPECIES) in *Rana esculenta, R. catesbeiana, R. clamitans, R. pipiens, R. septentrionalis,* and *R. tigrina* (Amphibia).

7. *L. ranae* R. Ray, 1980 in *Rana cyanophlyctis* (Amphibia).

8. *L. rhacophorae* R. Ray, 1980 in *Rhacophorus maculatus* (Amphibia).

Family DACTYLOSOMATIDAE *Jakowska and Nigrelli, 1955 emend. Levine, 1971*

In erythrocytes of cold-blooded vertebrates; merogony present in erythrocytes, with formation of 4 to 16 merozoites; vectors unknown.

Genus *Dactylosoma* Labbé, 1894. Merogony present; nucleus with endosome; more than four merozoites formed; in amphibia, reptiles, and fish. TYPE SPECIES *D. ranarum* Lankester, 1882) Wenyon, 1926.

1. *D. amaniae* Averinzev, 1914 in *Chamaeleo fischeri* (Reptilia).
2. *D. hannesi* Paperna, 1981 in *Mugil cephalus, Liza richardsoni,* and *L. dumerili* (Osteichthys).
3. *D. lethrinorum* Saunders, 1960 in *Lethrinus nebulosus* and *L. mahsenoides* (Osteichthys).
4. *D. ranarum* (Lankester, 1882) Wenyon, 1926 (syns., *Drepanidium ranarum* Lankester, 1882; *Dactylosoma splendens* Labbé, 1894; *Haemogregarina ranarum* [Lankester, 1871] Celli and Sanfelice, 1891; *H. splendens* [Labbé, 1894] Laveran, 1905; *Laverania ranarum* [Lankester, 1871] Grassi and Feletti, 1892) (TYPE SPECIES) in *Rana esculenta, Rana* spp., *Rappia* spp., and *Bufo* spp. (Amphibia).
5. *D. salvelini* Fantham, Porter, and Richardson, 1942 in *Salvelinus fontinalis* and *S. malma* (Osteichthys).
6. *D. sauriae* Krasil'nikov, 1965 in *Agama caucasica* and *Lacerta armenica* (Reptilia).
7. *D. striata* Sarkar and Haldar, 1979 in *Ophicephalus striatus* (Osteichthys).
8. *D. sylvatica* Fantham, Porter, and Richardson, 1942 in *Rana sylvatica* (Amphibia).
9. *D. taiwanensis* Manwell, 1964 in *Rana limnocharis* (Amphibia).
10. *D. tritonis* (Fantham, 1905) Wenyon, 1926 (syn., *Lankesterella tritonis* Fantham, 1905) in *Triton cristatus* (Amphibia).

Genus *Schellackia* Reichenow, 1919. Heteroxenous or homoxenous; oocyst produces eight sporozoites; in reptiles and amphibia; merogony, gamogony, and sporogony in small intestine, connective tissue, and/or reticuloendothelial system; vectors, if known, mites and Diptera. TYPE SPECIES *S. bolivari* Reichenow, 1919.

1. *S. agamae* (Laveran and Pettit, 1909) Rogier, 1977 (syns., *Haemogregarina agama* Laveran and Pettit, 1909; *Skellackia [sic] agami* Omran, Mandour, and El-Naffar, 1981) in *Agama agama* (syn., *A. colonorum*) and *A. stellio* (Reptilia).
2. *S. balli* Le Bail and Landau, 1974 in *Bufo marinus* (Amphibia).
3. *S. bocagei* Alvarez Calvo, 1975 in *Lacerta hispanica vaucheri* (Reptilia).
4. *S. bolivari* Reichenow, 1919 (TYPE SPECIES) in *Acanthodactylus vulgaris* and *Psammodromus hispanicus* (Reptilia).
5. *S. brygooi* Landau, 1973 (syn., *Schellackia* sp. Brygoo, 1965 in *Chamaeleo brevicornis* [?]) in *Oplurus sebae* and *O. cyclurus* (Reptilia).
6. *S. calotesi* (Ray and Sarkar, 1969) Levine, 1981 (syn., *Gordonella calotesi* Ray and Sarkar, 1969) in *Calotes versicolor* (Reptilia).
7. *S. golvani* Rogier and Landau, 1975 in *Anolis marmoratus* and probably *A. roquet* (Reptilia).
8. *S. iguanae* (Landau, 1973) Levine, 1981 (syn., *Lainsonia iguanae* Landau, 1973) in *Iguana iguana* (Reptilia).
9. *S. landauae* Lainson, Shaw, and Ward, 1976 in *Polychrus marmoratus* (Reptilia).
10. *S. occidentalis* Bonorris and Ball, 1955 in *Sceloporus occidentalis, S. undulatus, S. graciosus,* and *Uta stansburiana* (Reptilia).
11. *S. weinbergi* (Leger and Mouzels, 1917) Levine, 1981 (syns., *Haemogregarina weinbergi* Leger and Mouzels, 1917; *Haemogregarina* sp. Leger and Mouzels, 1917 of

Landau et al. [1974]; *Lainsonia legeri* Landau et al., 1974) in *Tupinambis nigro-punctatus* (Reptilia).

Family CALYPTOSPORIDAE *Overstreet, Hawkins, and Fournie, 1984*
Heteroxenous; oocysts with sporocysts, each containing sporozoites; sporocysts with sporopodia.

Genus *Calyptospora* Overstreet, Hawkins, and Fournie, 1984. Oocysts with four sporocysts, each with two sporozoites; sporocysts with sporopodia. TYPE SPECIES *C. funduli* (Duszynski, Solangi, and Overstreet, 1979) Overstreet, Hawkins, and Fournie, 1984.

1. *C. empristica* Fournie, Hawkins, and Overstreet, 1985 in *Fundulus notti* (Osteichthys) hepatocytes and pancreatic acinan cells; intermediate host probably *Palaeomonetes kadiakensis.*
2. *C. funduli* (Duszynski, Solangi, and Overstreet, 1979) Overstreet, Hawkins, and Fournie, 1984 (syn., *Eimeria funduli* Duszynski, Solangi, and Overstreet, 1979) (TYPE SPECIES) in *Fundulus grandis* (TYPE HOST), *F. pulvereus, F. similis, F. hetero-clitus, F. jenkinsi, Menidia beryllina,* and experimentally *F. parvipinnis* and *F. chry-sotus* (Osteichthys); intermediate hosts *Palaeomonetes pugio, P. kadiakensis, P. paludosus, P. vulgaris,* and *Macrobrachium ohione.*

INDEX

A

Acanthoepimeritus, 90
Acanthospora, 97
Acanthosporidium, 101
Acanthosporinae, 97—102
Acarogregarina, 40
Actinocephalidae, 85—103
  Acanthosporinae, 97—102
  Actinocephalinae, 85—97
  Menosporinae, 102—103
Actinocephalinae, 85—97
Actinocephalus, 85—87
Acuta, 104
Acutidae, 104
Acutispora, 79
Adelea, 115
Adeleidae, 115—117
Adeleinorina, 115—134, see also specific topics
  Adeleidae, 115—117
  Haemogregarinidae, 118—133
    Cyrilia, 133
    Hemogregarina, 118—128
    Hepatozoon, 129—133
    Karyolysus, 128—129
  Klossiellidae, 133—134
  Legerellidae, 117—118
Adeleorina, 115—134
Adelina, 115—116
Adelphocystis, 36
Agamoccoccidiorida, 111
Aggregata, 135—136
Aggregatidae, 135—137
Agramococcidiorida, 111
Agrippina, 94
Aikinetocystidae, 29—30
Aikinetocystis, 29—30
Alaspora, 95
Albertisella, 38—39
Allantocystidae, 41
Allantocystis, 41
Alveocystis, 139
Amphorocephalus, 95
Amphoroides, 93
Ancora, 20—21
Ancyrophora, 97—99
Angeiocystidae, 113
Angeiocystis, 113
Anisolobus, 63
Anthorhynchus, 94
Apigregarina, 104
Apolocystis, 34—35
Arachnocystis, 69
Arborocystis, 39
Ascocephalus, 95
Ascogregarina, 24—25
Aseptatorina, see also specific topics

Aikinetocystidae, 29—30
Allantocystidae, 41
Diplycystidae, 40—41
Enterocystidae, 41—42
Ganymedidae, 42
Lecudinidae, 17—26
Monocystidae, 30—40
  Monocystinae, 30—36
  Oligochaetocystinae, 40—42
  Rhynchocystinae, 37—38
  Stomatophorinae, 38—39
  Zygocystinae, 36—37
Schaudinnellidae, 41
Selenidiidae, 15—17
Urosporidae, 26—29
Asterophora, 88
Astrocystella, 38
Atoxoplasma, 180, 194—195
Atoxoplasmatidae, 194—195

B

Barrouxia, 191
Barrouxiidae, 191—194
Beccaricystis, 39
Bhatiella, 20
Bifilida, 50
Bolivia, 64
Bothriopsides, 92
Brustiospora, 104
Brustiosporidae, 103—104
Bulbocephalus, 82—83

C

Callynthrochlamys, 48—49
Calyptospora, 197
Calyptosporidae, 197
Campanacephalus, 84
Caridohabitans, 47—48
Caryospora, 178—179
Caryotropha, 137
Caryotrophidae, 137—138
Caulleryella, 107
Caulleryellidae, 107
Caulocephalus, 87
Cephalocystis, 35
Cephaloidophora, 45—47
Cephaloidophoridae, 45—48
Cephalolobidae, 48
Cephalolobus, 48
Ceratospora, 29
Chagasella, 117
Chakravartiella, 39
Chilogregarina, 96
Chlamydocystis, 22
Choanocystoides, 39

Cirrigregarina, 64
Clavicephalus, 84
Cnemidospora, 75—76
Cnemidosporidae, 75—76
Coccidia
  Adeleinorina, 115—134, see also specific topics
  Agramococcidiorida, 111
  Eimeriorina, 135—197, see also specific topics
  Ixoheorida, 111
  Protococcidiorida, 111—113
Coccidiasina, 111
Cochleomeritus, 20
Coelotropha, 112
Cognettiella, 65
Coleorhynchus, 93
Colepismatophila, 85
Cometoides, 99
Contospora, 102
Cornimeritus, 87
Coronoepimeritus, 100
Corycella, 97
Craterocystis, 38
Crucocephalus, 96
Cryptosporidiidae, 138
Crystallospora, 194
Cyclospora, 178
Cyrilia, 133
Cystocephaloides, 84
Cystocephalus, 82

D

Dactylophoridae, 78—80
Dactylophorus, 78
Dactylosoma, 196
Dactylosomatidae, 196—197
Defretinella, 193—194
Degiustia, 64
Dendrorhynchus, 80
Deuteromera, 65
Diaspora, 190—191
Didymophyes, 65—67
Didymophyidae, 65—67
Dinematospora, 101
Diplauxis, 22
Diplocystis, 40—41
Diplycystidae, 40—41
Dirhynchocystis, 37—38
Discorhynchus, 91
Ditrypanocystis, 17
Dobellia, 135
Dobellidae, 135
Doliospora, 101
Dorisa, 188—189

E

Echinomera, 78—79
Echinoocysta, 102
Echiurocystis, 40
Eimeria, 139—178

Eimeriidae, 138—191
  Alveocystis, 139
  Caryospora, 178—180
  Cyclospora, 179
  Diaspora, 190
  Dorisa, 188
  Eimeria, 139—178
  Epieimeria, 178
  Gousseffia, 190
  Hoarella, 190
  Isospora, 180—188
  Mantonella, 178
  Octosporella, 190
  Pythonella, 190
  Sivatoshella, 190
  Skrjabinella, 190
  Tyzzeria, 139
  Wenyonella, 189
Eimeriorina, 135—197
  Aggregatidae, 135—137
  Atoxoplasmatidae, 194—195
  Barrouxiidae, 191—194
  Calyptosporidae, 197
  Caryotrophidae, 137—138
  Cryptosporidiidae, 138
  Dactylosomatidae, 196—197
  Dobellidae, 135
  Eimeriidae, 138—191, see also specific topics
  Lankesterellidae, 195
  Pfeifferinellidae, 138
  Selenococcidiidae, 135
  Spirocystidae, 135
Eleutheroschizon, 113
Eleutheroschizonidae, 113
Endomycola, 70
Enterocystidae, 41—42
Enterocystis, 41—42
Epicavus, 96
Epieimeria, 178
Erhardovina, 62
Eucoccidiorida, 115—134, see also Adeleinorina
Euspora, 69
Extremocystis, 26

F

Farinocystis, 109
Faucispora, 64
Filipodium, 21
Fonsecaia, 74
Fusiona, 104
Fusionicae, 104
Fusionidae, 104

G

Gamocystis, 63
Ganapatiella, 117
Ganymedes, 42
Ganymedidae, 42
Garnhamia, 63

Gemmicephalus, 89
Geneiorhynchus, 89—90
Gibbsia, 117
Gigaductidae, 105
Gigaductus, 105
Globulocephalus, 94
Gonospora, 27—28
Gopaliella, 65
Gousseffia, 190
Goussia, 191—193
Grasseella, 137
Grayallia, 38
Grebnickiella, 79
Gregarina, 50—62
Gregarinicae, 45—70
  Cephaloidophoridae, 45—48
  Cephalolobidae, 48
  Didymophyidae, 65—67
  Gregarinidae, 50—64
    Anisolobus, 63
    Bolivia, 64
    Cirrigregarina, 64
    Digiustia, 64
    Erhardovina, 62
    Faucispora, 64
    Gamocystis, 63
    Garnhamia, 63
    Gregarina, 50—62
    Gymnospora, 62
    Molluskocystis, 64
    Spinispora, 64
    Torogregarina, 64
    Triseptata, 63
  Hirmocystidae, 67—70
  Metameridae, 65
  Uradiophoridae, 49—50
Gregarinidae, 50—64
Grellia, 112
Grellidae, 111—112
Grenoblia, 97
Gryllotalpia, 96
Gymnospora, 62—63

H

Haemogregarina, 118—128, 180
Haemogregarinidae, 118—133
  Cyrilia, 133
  Haemogregarina, 118—128
  Hepatozoon, 129—133
  Karyolysus, 128—129
Harendraia, 97
Heliospora, 49
Hentschelia, 21
Hepatozoon, 129—133
Heterospora, 17
Hirmocystidae, 67—70
Hirmocystis, 67—68
Hoarella, 190
Hoplorhynchus, 103
Hyalospora, 68—69

Hyalosporina, 74
Hyperidion, 21

I

Incertae Sedis, 109
Isospora, 180—188
Ithania, 117
Ixoheorida, 111
Ixorheidae, 111
Ixorheis, 111
Ixorheorida, 111

K

Karyolysus, 128—129
Klossia, 116
Klossiella, 133—134
Klossiellidae, 133—134
Kofoidina, 25

L

Lankesterella, 195
Lankesterellidae, 195
Lankesteria, 22—23
Lateroprotomeritus, 26
Lecudina, 17—19
Lecudinidae, 17—26
Lecythion, 21
Legerella, 117—118
Legerellidae, 117—118
Legeria, 90
Leidyana, 74—75
Leidyanidae, 74—75
Lepismatophila, 85
Levinea, 97
Lipocystis, 109
Liposcelisus, 67
Lipotropha, 108
Lipotrophidae, 107—109
Lithocystis, 28—29
Lophocephaloides, 83
Lophocephalus, 83
Lymphotropha, 107

M

Machadoella, 106
Mackinnonia, 112—113
Mantonella, 178
Mastocystis, 35—36
Mattesia, 108—109
Mecistophora, 80
Menospora, 102—103
Menosporinae, 102—103
Menzbieria, 108
Merocystis, 136—137
Metamera, 65
Metameridae, 65
Molluskocystis, 64

Monocystella, 23—24
Monocystidae, 30—42
  Monocystinae, 30—36
  Oligochaetocystinae, 40—42
  Rhynchocystinae, 37—38
  Stomatophorinae, 38—39
  Zygocystinae, 36—37
Monocystinae, 30—36
Monocystis, 30—32
Monoductidae, 76—77
Monoductus, 76
Monoica, 104
Monoicidae, 104
Mukundaella, 102
Myriospora, 112
Myriosporidae, 112—113
Myriosporides, 112

N

Nellocystis, 30
Nematocystis, 33—34
Nematoides, 50
Nematopsis, 43—45
Neogregarinorida
  Caulleryellidae, 107
  Gigaductidae, 105
  Lipotrophidae, 107—109
  Ophryoystidae, 105—106
  Schizocystidae, 106—107
  Syncystidae, 107
Neomonocystis, 40
Neoschneideria, 77—78

O

Octosporella, 190
Odonaticola, 103
Oligochaetocystinae, 40—42
Oligochaetocystis, 40
Oocephalus, 84
Ophioidina, 25
Ophryocystis, 105—106
Ophryoystidae, 105—106
Orbocephalus, 85
Orcheobius, 116—117
Ovivora, 137

P

Pachyporospora, 45
Parachoanocystoides, 39
Paragonospora, 29
Paraophioidina, 25—26
Paraschneideria, 77
Pfeifferinellidae, 138
Phialoides, 90
Phleobum, 76—77
Pileocephalus, 88—89
Pilidiophora, 89
Pintospora, 70

Pleurocystis, 36—37
Polyrhabdina, 19
Pomania, 92
Pontesia, 20
Porospora, 43
Porosporicae, 43—45
Porosporidae, 43—45
Prismatospora, 99
Protococcidiorida, 111—113
Protomagalhaensia, 70
Pseudoklossia, 137
Pterospora, 29
Pythonella, 190
Pyxinia, 90—91
Pyxinioides, 49—50

Q

Quadruhyalodiscus, 67
Quadruspinospora, 101—102

R

Ramicephalus, 99—100
Rasajeyna, 117
Retractocephalus, 70
Rhabdocystis, 34
Rhizionella, 99
Rhopalonia, 79
Rhynchocystinae, 37—38
Rhynchocystis, 37
Rhytidocystidae, 111
Rhytidocystis, 111
Rotundula, 48

S

Sawayella, 109
Schaudinnella, 41
Schaudinnellidae, 41
Schellackia, 196—197
Schizocystidae, 106—107
Schizocystis, 106
Schneideria, 77
Sciadiophora, 94
Selenidiidae, 15—17
Selenidium, 15—16
Selenococcidiidae, 135
Selenococcidium, 135
Selenocystis, 17
Selysina, 137
Septatorina, see also specific topics
  Acutidae, 104
  Brustiosporidae, 103—104
  Fusionicae, 104
  Fusionidae, 104
  Gregarinicae, 45—70
    Cephaloidophoridae, 45—48
    Cephalolobidae, 48
    Didymophyidae, 65—67
    Gregarinidae, 50—64

Hirmocystidae, 67—70
Metameridae, 65
Uradiophoridae, 49—50
Monoicidae, 104
Porosporicae, 43—45
Porosporidae, 43—45
Stenophoricae, 70—104
  Actinocephalidae, 85—103
    Acanthosporinae, 97—102
    Actinocephalinae, 85—97
  Cnemidosporidae, 75—76
  Dactylophoridae, 78—80
  Leidyanidae, 74—75
  Menosporinae, 102—103
  Monoductidae, 76—77
  Sphaerocystidae, 77—78
  Stenophoridae, 70—74
  Stylocephalidae, 80
  Trichorhynchidae, 78
Seticephalus, 80
Sivatoshella, 190
Skrjabinella, 190
Sphaerocystidae, 77—78
Sphaerocystis, 77
Sphaerorhynchus, 84
Spinispora, 64
Spirocystidae, 135
Spirocystis, 135
Steinina, 91—92
Stenoductus, 76
Stenophora, 71—74
Stenophoricae, 70—104
  Acanthosporinae, 97—102
  Actinocephalidae, 85—103
  Actinocephalinae, 85—97
  Cnemidosporidae, 75—76
  Dactylophoridae, 78—80
  Leidyanidae, 74—75
  Menosporinae, 102—103
  Monoductidae, 76—77
  Sphaerocystidae, 77—78
  Stenophoridae, 70—74
  Stylocephalidae, 80
  Trichorhynchidae, 78
Stenophoridae, 70—74
Stictospora, 92—93
Stomatophora, 38
Stomatophorinae, 38—39
Stylocephalidae, 80
Stylocephaloides, 82

Stylocephalus, 80—82
Stylocystis, 93
Sycia, 20
Syncystidae, 107
Syncystis, 107

T

Taeniocystis, 93
Tetractinospora, 102
Tetraedrospora, 99
Tetrameridionospinispora, 102
Tettigonospora, 69
Thalicola, 95—96
Tintinospora, 69
Torogregarina, 64
Trichorhynchidae, 78
Trichorhynchus, 78
Tricystis, 95
Tyzzeria, 139

U

Ulivina, 19—20
Umbracephalus, 88
Uradiophora, 49
Uradiophoridae, 49—50
Urnaepimeritus, 88
Urospora, 26—27
Urosporidae, 26—29

V

Viviera, 20

W

Wenyonella, 189

X

Xiphocephalus, 83

Z

Zeylanocystis, 39
Zygocystinae, 36—37
Zygocystis, 36
Zygosoma, 21

Printed and bound by CPI Group (UK) Ltd, Croydon, CR0 4YY

22/10/2024

01777630-0006